Benchmark Papers
in Genetics

Series Editor: David L. Jameson
University of Houston

PUBLISHED VOLUMES

GENETICS AND SOCIAL STRUCTURE
Paul Ballonoff
GENES AND PROTEINS
Robert P. Wagner
DEMOGRAPHIC GENETICS
Kenneth M. Weiss and Paul Ballonoff
MUTAGENESIS
John W. Drake and Robert E. Koch
EUGENICS: Then and Now
Carl Jay Bajema
CYTOGENETICS
Ronald L. Phillips and Charles R. Burnham
STOCHASTIC MODELS IN POPULATION GENETICS
Wen-Hsiung Li

RELATED TITLES IN OTHER BENCHMARK SERIES

MICROBIAL GENETICS
Morad A. Abou-Sabé
CONCEPTS OF SPECIES
C. N. Slobodchikoff
MULTIVARIATE STATISTICAL METHODS: Among-Groups Covariation
William R. Atchley and Edwin H. Bryant
MULTIVARIATE STATISTICAL METHODS: Within-Groups Covariation
Edwin H. Bryant and William R. Atchley
MOLECULAR BIOLOGY AND PROTEIN SYNTHESIS
Robert A. Niederman
SYNTHESIS OF LIFE
Charles C. Price

Benchmark Papers
in Genetics/6

A BENCHMARK® Books Series

CYTOGENETICS

Edited by

RONALD L. PHILLIPS
CHARLES R. BURNHAM

University of Minnesota

Dowden, Hutchinson & Ross, Inc.

STROUDSBURG, PENNSYLVANIA

LIBRARY OF CONGRESS CATALOGING IN PUBLICATION DATA
Main entry under title:
Cytogenetics.
  (Benchmark papers in genetics ; 6)
  Includes index.
    1. Cytogenetics—Addresses, essays, lectures. I. Phillips, Ronald L.
II. Burnham, Charles Russel. [DNLM: 1. Cytogenetics. W1 BE516 v. 6/
QH605 C997]
QH438.C97   575.2'1   76-55705
ISBN 0-87933-258-1

Exclusive Distributor: **Halsted Press**
A Division of John Wiley & Sons, Inc.
ISBN: 0-470-99046-5

SERIES EDITOR'S FOREWORD

The study of any discipline assumes mastery of the literature of the subject. In many branches of science, even one as new as genetics, the expansion of knowledge has been so rapid that there is little hope of learning of the development of all phases of the subject. The student has difficulty mastering the textbook, the young scholar must tend to the literature near his own research, the young instructor barely finds time to expand his horizons to meet his class preparation requirements, the monographer copes with a wider literature but usually from a specialized viewpoint, and the textbook author is forced to cover much the same materials as previous and competing texts to respond to the user's needs and abilities.

Few publishers have the dedication to scholarship needed to serve the limited market of advanced studies. The opportunity to assist professionals at all stages of their careers has been recognized by the publishers of the Benchmark series and by a distinguished group of Benchmark volume editors knowledgeable in specific aspects of the genetics literature. Some have contributed greatly to the development of that literature, some have studied with the early scholars, and some have developed and are in the process of developing entirely new fields of genetic knowledge. In many cases the judgments of the editors become a historical document that records their opinion of the important steps in the development of the subject. The editors of this volume have selected papers and portions of papers that demonstrate both the development of knowledge and the atmosphere in which that knowledge was developed. There is no substitute for reading great papers. Here you can learn how questions are asked, how they are approached, and how difficult and essential it is to obtain definitive answers and clear writing. My own pleasure in working with this distinguished panel is exceeded only by the considerable pleasure of reading their remarks and their selections. Their dedication and wisdom are impressive.

The editors of this volume have been particularly thoughtful of the reader of the selections. They have provided both a historical perspective and an extensive bibliography of each section. This review serves as an introduction to the important literature that provided the significant background to make possible the research leading to the Benchmark papers.

Drs. Burnham and Phillips have synthesized considerable information and provided special insight to the development of cytogenetics. Clearly they have served their readers well.

DAVID L. JAMESON

PREFACE

The field of cytogenetics emerged from the integration of knowledge concerning the segregation of heritable variations and the behavior of chromosomes in somatic and meiotic cell division. The molecular basis of cytogenetics that is under investigation today will lead to a more complete understanding of genetics. Cytogenetics is not a side branch of genetics but rather a stepping stone to a more complete understanding of the subject.

The volume editors' goal is to guide the reader to an understanding of how the field of classical cytogenetics developed from experiments that utilized both genetics and cytology and how modern cytogenetics couples genetics, cytology, and molecular biology.

Present-day cytogenetics was built from a multitude of benchmark contributions. After four years of working at the interesting task of developing this volume, we have become convinced that the goal of providing a historical perspective of cytogenetics through the reprinting of benchmark papers could not be met solely by means of a necessarily limited number of papers. For this reason we have written a "Historical Perspectives" introduction that precedes each section and refers to other contributions along with references. The editors' intent is that reading this volume will provide cytogenetics students with the historical perspective that will enhance their understanding and enjoyment of a modern course in cytogenetics.

Although we have attempted to survey the entire field of cytogenetics, we are aware that our selection of papers reflects certain biases that result from our own collective knowledge and experience in cytogenetics research, teaching, and personal contacts. Nevertheless, the volume does bring together many important contributions.

We approached the task by listing first the papers that immediately came to mind, then the laboratories that had made major contributions, and finally the various subareas of cytogenetics. By scanning literature lists in textbooks and reviews and by referring to published collections of benchmark papers in genetics and to historical reviews, we obtained a rather lengthy list of papers that were important contributions. After making a tentative selection, we then sought opinions of leading authorities. In general, the papers we finally selected for inclusion present both genetical and cytological evidence. They also are those that had an impact on the field, although not necessarily ones that first advanced a particular idea.

The reader should be aware that the field did not move ahead only through published accounts. Personal contacts and idea exchanges between members of laboratories at meetings and through correspondence were important in stimulating further advances. Early textbooks also were instrumental in disseminating ideas.

This volume is divided into seven sections: Chromosome Individuality, Morphology, Structure, and Number; Chromosome Replication; Chromosome Pairing; Crossing-over; Chromosome Function; Chromosome Behavior; and Genetical and Cytological Mapping. For each section the reprinted papers are preceded by a "Historical Perspectives" narrative and by "Editors' Comments" that emphasize certain important aspects of each paper.

A total of 49 papers is included; 34 in full and 15 from which only excerpts are reprinted because of the lengths of the original papers. Two papers were translated from German and one from French. All three translations were checked by professionals.

The reprinted papers encompass seven decades; 3 papers originally published in the decade of 1901–1910, 4 in 1911–1920, 11 in 1921–1930, 13 in 1931–1940, 5 in 1941–1950, 7 in 1951–1960, and 6 in 1961–1970. The papers represent research conducted in England, France, Germany, Japan, Russia, and the United States. Of the 49 papers, 18 deal with animals, 28 with plants, 1 with both plants and animals, 1 with a prokaryote, and 1 strictly with theory.

We gratefully acknowledge the responses of C. D. Darlington, T. C. Hsu, H. Kihara, E. B. Lewis, B. McClintock, A. Müntzing, M. M. Rhoades, and R. Riley to our request for their opinions on our earlier list of tentative selections. Full responsibility for the contents of this volume, of course, rests on the editors.

We hope the reader enjoys the overall perspective of cytogenetics presented in this volume as much as we enjoyed the collation of the information.

RONALD L. PHILLIPS
CHARLES R. BURNHAM

CONTENTS

Contents

Contents

CONTENTS BY AUTHOR

CYTOGENETICS

Part I

Chromosome Individuality, Morphology, Structure, and Number

HISTORICAL PERSPECTIVES

Although scientists did not recognize the importance of Mendel's discovery published in 1866, until 1900, cytologists made a succession of discoveries during that time that laid the foundation for the chromosome theory of heredity (see Babcock 1949, Sharp, 1934). These discoveries included:

1. In 1875 Strasburger first adequately described the densely staining bodies in the nucleus. The term *chromosome* was not proposed until 1888 when Waldeyer first used it.
2. Schneider (1873), Butschli (1875a, b), and others described the true nature of mitosis, emphasizing that cells derive from previously existing cells through a complicated division process that includes the longitudinal splitting of the chromosomes (Flemming 1879–1881).
3. By 1884 chromosome numbers had been shown to be constant in a species.
4. Fertilization was found to involve the fusion of a nucleus from an egg and a sperm (O. Hertwig 1875–1878 and others).
5. Chromosome continuity in successive cell divisions along with the role of fertilization and cell division in inheritance were recognized (Boveri 1889).
6. Pairs of meiotic chromosomes were understood as comprised of corresponding members of paternal and maternal origin (Roux 1883; Boveri 1889; Weismann 1887).

7. Chromosome number was shown to be halved at meiosis (Henking 1890; Boveri 1893; Rückert 1892; and others). This was a prediction Weismann had made previously.

Hence soon after the rediscovery of Mendel's contributions, Correns (1900) and Strasburger (1901) suggested a connection between Mendelian segregation and the reduction division. Sutton (1903) and Boveri (1904) furnished more complete formulations of the chromosome mechanism of Mendelian inheritance, including the prediction of linkage.

Wilson and his students in America and Boveri, Goldschmidt, and others in Europe added important information about chromosomes (1902–1912). That the chromosomes in a species are qualitatively different was shown in 1902 by Boveri in *Ascaris* and also by Sutton in *Brachystola*. S. Nawaschin (1912) first reported that satellited chromosomes were attached to the nucleolus in *Galtonia candicans*. He also reported heterozygosity for a difference in size of the satellite. Such heterozygosity was known up to then only for sex chromosomes. Carothers, 1913 and again in 1917 (Paper 38), reported heteromorphism for autosomal chromosome pairs in grasshoppers in the wild.

Stebbins (1950) states that Nawaschin and his school

> established firmly the fact that most species of plants and animals possess a definite individuality in their somatic chromosomes, which is evident in their size, shape, position of primary constrictions or centromeres and in such additional features as secondary constrictions and satellites.

In response to a report by Delaunay at a meeting of the Caucasian section of the Russian Botanical Society, S. Nawaschin (1921) stated that for many years in lectures to his students he had named the characteristic composition of the nucleus for a species its "idiogram." This appears to be the paper usually cited as the source of the term, e.g., Sharp (1934). Idiograms representing the chromosome complement have been useful in comparative cytogenetics.

M. Nawaschin (1925), the son of S. Nawaschin, reported comparative studies of the chromosomes of 11 species of *Crepis* and referred to S. Nawaschin's use of the term *idiogram* for the "specific picture of the chromosomes of a species" but cited no reference.

Heitz (1926) also reported comparative studies of several species in three different families of monocots and used the term idiogram without reference to any earlier use. Both Nawaschin and Heitz used schematic representations of chromosome sizes and shapes, with centromeres arranged in a line at the base of the V's,

J's, or rods. Several papers appeared in which line diagrams not designated by any special term were used to represent the chromosomes; for example, Hance (1918) for a trisomic (15-chromosome) plant of *Oenothera scintillans*; Painter (1925) in a comparative study of the chromosomes of man and several other mammals; and Ruttle (1927) for one species of *Nicotiana*. Delaunay reviewed the studies of external chromosome morphology in 1929.

In 1931 two important papers by Lewitsky appeared. One dealt with the karyotype in systematics (1931). The other, from which excerpts are reprinted in this volume, used line diagrams that he termed *idiograms* to represent lengths, centromere positions, and secondary constrictions of the chromosomes in a wide range of species including agronomic crop species. He made all his observations on cells from sectioned material. He measured the somatic metaphase chromosomes by making trigonometric calculations from readings using the micrometer calibrations on the fine-adjustment knob of the microscope. Although the text of the paper was published in both English and Russian, the diagrams illustrating the calculations are only in the Russian text. With the development of smear techniques that flattened the cells, and additional refinements, idiograms have become even more important in chromosome studies of plants and animals. As an outgrowth of the idiogram idea, karyotypes are often displayed pictorially by cutting out mitotic metaphase chromosomes from a photomicrograph and arranging them by length and in pairs and aligned such that the centromeres are at the same level, and the short arms are oriented upward. This method of reporting karyotypes is almost always used for man and other animals and is becoming increasingly popular for plants.

Metaphase chromosomes, as observed in the early days, displayed only limited cytological detail. McClintock's 1930 paper (Paper 2) demonstrated that the pachytene stage of meiosis in maize could be used to distinguish the different chromosomes and also to locate more precisely the breakpoint positions for chromosome translocations (interchanges). She had published (1929a) an idiogram of the maize chromosomes based on their morphology at the first postmeiotic division in the microspores. Among the progeny of a triploid maize plant, she had established trisomics for many of the chromosomes. Since the $n + 1$ microspores from a trisomic plant had one chromosome in duplicate, she was able to identify that chromosome. She also determined its morphology at pachynema in the trisomic plant. She crossed the different trisomic stocks with genetic markers for the different linkage groups. At that time, trisomic plants were identified by chromosome counts

3

made on root tips that had been embedded in paraffin, sectioned with a microtome, and then stained by the crystal violet, $KI + I_2$ procedure. Self-pollinations and reciprocal backcrosses of the trisomic plants resulted in genetic ratios that were easily distinguishable if the genetic marker were carried by the triplicated chromosome, as demonstrated in the paper by McClintock and Hill (1931). Thus each chromosome was soon identified with its linkage group. Pachytene analysis became and still is the standard procedure for obtaining information on a wide range of cytogenetic problems in maize. Although not possible in most species, pachytene analysis has been useful in tomato, *Collinsia, Neurospora*, as well as in man. Wherever pachytene cells can be studied, it is advantageous to do so since the chromosomes are paired and are at the stage in which crossing-over probably occurs.

Cytologists had known for a long time that the sex chromosomes in animals were densely stained in the early stages of meiosis. Gutherz (1907) had described these as being *pycnotic*. Heitz (1928) reported that in the liverwort genus *Pellia* many autosomes also had segments that took on the metaphase appearance in the prophase stages. He stated that chromosomes that are partially or totally heteropycnotic will be designated as *heterochromosomes* and that the word *heterochromatin* will designate that part of a heterochromosome which is heteropycnotic. Hence he used the term *heterochromatin* and also the contrasting term *eu-* (or true) *chromatin* to designate purely morphological features.

Two general classes of heterochromatin, constitutive and facultative, are now recognized (Brown 1966). The former remains condensed throughout all or most of the cell cycle. The latter is pycnotic only under certain conditions or in certain cell lineages. An example of the latter is the X-chromosome, or Barr, body in mammalian females.

In certain species the meiotic chromosomes have heterochromatic regions adjacent to the centromeres which persist in interphase as what have been termed *prochromosomes*, first described by Rosenberg (1909) in *Drosera*. This is found in several of the Solanaceae, e.g., *Physalis* (ground cherry) and *Lycopersicon* (tomato) (Brown 1949). One member of that family, *Datura stramonium*, the subject of extensive cytogenetic investigations by Blakeslee and his coworkers (Avery, Satina, and Rietsema 1959), does not have such regions (Burnham, pers. comm.).

The chromosomes of most species have a single centromere that divides the chromosome into two arms, and the relative lengths of these arms are a diagnostic feature for identifying particular chromosomes. In addition to serving as a spindle fiber attachment region, the centromere also holds the two sister chromatids together at metaphase I of meiosis, behaving as if its division were delayed until division II of meiosis.

The chromosomes of the hemipterous and homopterous insects have what has been described as "a diffuse type of centromere" (Schrader 1935). As Rhoades and Kerr (1949) stated, "In these chromosomes the entire poleward face of the chromosome is involved in fibre formation and a solid sheet of spindle fibres is produced." A polycentric type of centromere is found in the germ line of *Ascaris megalocephala* (Walton 1924). In plants nonlocalized centromeres have been reported in the rush, *Luzula purpurea* (Malheiros, de Castro, and Camara 1947). Brown (1954) found no centromeric-like structures at pachynema using the light microscope, but Braselton's (1971) studies of ultrastructure with the electron microscope have shown that *Luzula* and also *Cyperus* chromosomes have several kinetochores along their length and therefore are actually polycentric. In rye there are strains in which regions other than the true centromere form spindle fibers but only at the two meiotic divisions (Prakken and Müntzing 1942). This has been reported also at meiosis in maize plants that have the abnormal chromosome 10 which has an additional, largely heterochromatic segment at the distal end of the long arm. The knob regions on other chromosomes take on centromeric activity before the beginning of anaphase and move toward the poles (Rhoades and Vilkomerson, Paper 37).

In inversion heterozygotes, Rhoades and Dempsey (1966) showed that in the presence of abnormal 10, chromosome pairing at pachynema is much more intimate than in plants with a normal 10. This could have important applications in plant breeding where recombination is so important.

Chromosome Ultrastructure

The physical nature of the eukaryotic chromosome has been refractory to a variety of analytical techniques. Consensus of opinion has not been reached on the ultrastructure of the chromosome or on the interpretation of genetic behavior based on transmission or scanning electron-microscopic observations. Even before the employment of the electron microscope in chromosome structure studies, cytologists had reported the apparent existence of chromosomal subunits. These observations were not seriously disturbing to cytogeneticists in light of the vast amount of genetic information suggesting that the chromosome behaved as a single unit.

Early electron-microscopic descriptions of chromosomes revealed an extremely complex organizational pattern that seemed almost hopeless to relate to well-defined inheritance patterns. Ris (Paper 5) was first to demonstrate that the basic unit of the chromosome in plant and animal cells is a microfibril about 200Å in diameter. The manner in which these microfibrils organize into a functional chromosome is still elusive. Even the interphase 200Å microfibril is a complex of at least histone and nonhistone proteins and DNA, which itself has a diameter of only 20Å. Recently, different EM preparative techniques have revealed that the basic chromatin fiber actually appears as beads on a string (Olins and Olins 1974). The beads, now termed nucleosomes, are composed of a core of four of the major histone fractions that is associated with about 200 base pairs of DNA that are compressed about five-fold in length. The connecting thread between the beads appears to be DNA. Electron-scattering measurements reveal that the packing ratio (μDNA/μ microfibril) of such interphase 200Å microfibrils may be greater than 50:1. Thus even interphase microfibrils are complicated structures. The chromosome, as observed in mitosis or meiosis, appears to be composed of these 200Å microfibrils folded in a complex manner so as to generate the chromosome typical of that cell type at that stage in the nuclear cycle. Although shifts occur in certain histone and nonhistone chromosomal proteins between interphase and metaphase, the basic unit still appears to be the 200Å microfibril.

Chromosome models, implicating only a single DNA double helix, gained recognition following results of DNase treatments of lampbrush chromosomes performed by Gall (1963). The lampbrush chromosomes were from isolated oocyte nuclei of the newt *Triturus viridescens*. The kinetics of chromosome breakage by DNase indicated that the loop segments are composed of a single DNA double helix while the interchromomeric regions are composed of two such helices; this is consistent with the unineme model of chromosome strandedness.

Using *Drosophila*, Kavenoff and Zimm (1973) have isolated DNA molecules that are long enough to represent the entire DNA from a single chromosome. The longest DNA molecules were obtained from *Drosophila* species with the longest chromosomes. Their research lends strong support to the unineme theory of chromosome structure. Another line of evidence is from DNA renaturation studies. When DNA is isolated, made single stranded, and allowed to renature into double-stranded segments, base sequences that are present in multiple copies (repetitive DNA) will renature much more rapidly than sequences that are present only once (unique DNA)(Britten and Kohne 1968). Unique sequences would be expected with unineme chromosomes. The repetitive portion of the DNA includes highly repetitive as well as moderately repetitive DNA. The highly repetitive DNA may be principally located in pericentromeric regions (areas of constitutive heterochromatin) as in the mouse. There is considerable controversy at the present time over the role of repetitive DNA and particularly the location of the moderately repetitive DNA in relation to structural genes.

The most distinctive ultrastructural feature found along the length of the chromosome is the centromere. Brinkley and Stubblefield (1966) demonstrated that the ultrastructure of the centromeres of somatic chromosomes in the Chinese hamster resembles that of a lampbrush-like filament running for a short distance along the surface of each chromatid. The centromeres of other animals and of various plant species have been described ultrastructurally. Although certain differences in structure exist between plants and animals, in all cases the centromere consists of a distinctive structure quite different from the remainder of the chromosome. Certain cytogenetic phenomena such as the behavior of telocentric chromosomes will be interpreted more easily with an understanding of centromere ultrastructure.

Specialized Chromosomes

The existence of specialized types of chromosomes only in cells of certain tissues or in certain individuals of a species has contributed substantially to cytogenetic theory. Specialized chromosomes of extreme importance to cytogeneticists include polytene, lampbrush, and supernumerary chromosomes.

Polytene Chromosomes: The polytene chromosome most extensively studied is that present in the salivary glands of Diptera. Although the salivary-gland chromosomes in *Drosophila melanogaster* had been described earlier (Balbiani 1881; Kostoff 1930), the papers by Painter (1933, and Paper 3) showed how the cross-banding chromomere patterns in smear preparations could be used to correlate genetical and cytological studies in this species. Prior to this discovery, changes in chromosome structure had been analyzed by studying the somatic chromosomes in nerve ganglion or oogonial cells, utilizing the fact that homologues or homologous segments, although not intimately paired, are aligned more or less parallel to each other. The chromosomes in those cells were several times longer than in meiocytes, but the salivary-gland chromosomes were 100 to 300 times longer, depending on the amount of stretching in smear preparations. The impact of Painter's report on subsequent studies was tremendous. Bridges (1935, 1938) not only mapped the banding patterns in minute detail but also initiated a system of numbers and letters still used to refer to particular segments and the bands within them. Most of the subsequent genetic studies were accompanied by cytological studies to determine if chromosomal aberrations were present. The comparative studies of geographical races, not only of *D. melanogaster* but also of other species of *Drosophila*, *Sciara*, and *Chironomus*, were another outcome of this discovery. One disadvantage of studying these chromosomes in *Drosophila* is that there are very few bands representing the Y-chromosome and the other heterochromatic segments that are relatively long in nonpolytene cells (see Part II). Another disadvantage is that these and the centromeric regions form a single chromocenter with the arms of the various chromosomes radiating out from that body. Hence if chromosome breakpoints are in heterochromatic regions, polytene analysis is difficult or impossible. A chromocenter is not present in the salivary-gland nuclei of *Chironomus*.

Supernumerary Chromosomes: Another type of chromosome, the supernumerary or B-chromosome, shows no apparent homology with the normal set of chromosomes. They occur in certain individuals or races in a wide variety of plant and animal species. Individuals possessing B-chromosomes have nearly normal vigor and fertility. Some of the original flint, dent, and sugary commercial varieties of corn had one or more of these chromosomes (Randolph 1928). The presence and behavior of supernumerary chromosomes confused early attempts to determine the chromosome number of certain species, as evidenced in early papers dealing with maize. These chromosomes may be present in varying numbers, from zero to more than the number of normal (termed A) chromosomes. Pachytene analysis (McClintock, Paper 16) showed B-chromosomes to be small, largely heterochromatic with an almost terminal centromere. In maize these chromosomes provide a technique for placing genes to chromosome (Roman, Paper 40). The method involves the use of translocations between standard and B-chromosomes and exploits their propensity for nondisjunction which occurs with high frequency at the second postmeiotic division of the microspore, i.e., the division that produces the two sperm nuclei.

The supernumerary chromosomes found in certain varieties of rye have been the subject of extensive investigations by Müntzing (1954). In rye these chromosomes undergo nondisjunction in post-meiotic divisions in both micro- and megasporogenesis.

Lampbrush Chromosomes: In certain animals an interesting type of specialized chromosome occurs in oocyte nuclei during prophase of meiosis. At this particular stage, loops extend laterally along each chromosome's central axis. These are called lampbrush chromosomes. They occur in amphibia, birds, fish, and reptiles; but most studies have been on the first group. The chromosomes of *Triturus viridescens*, for example, are extremely large, about 3 times the total length of dipteran salivary-gland chromosomes and more than 10 times the length of maize pachytene chromosomes. The chromosomes exist in a large nucleus that can be isolated and disrupted with jeweler's tweezers, and thus they can be isolated without special equipment. Another advantage in studying the lampbrush chromosomes in amphibia is that the chromosomes are in the diplotene stage of meiosis and may remain in the lampbrush state for a year or more. The lampbrush loops represent lateral projections of the linear continuity of the chromosomes; loops can be straightened out by pulling on the ends of the chromosome.

Lampbrush chromosomes were identified very early, and their name was assigned by Rückert in 1892; the descriptive cytology that accented the usefulness of these chromosomes as tools for cytogenetic study was provided by Gall (1952). Since then the lampbrush chromosome has been utilized as a model system to provide basic information on the structure and activity of chromosomes.

Chromomeres

The thought had often been expressed before that each chromomere (crossband) of the salivary-gland chromosome of *Drosophila* represented a single gene. In fact, Bridges had counted more than 5000 bands in *D. melanogaster*, and that number had served as the basis for one estimate of the number of genes in *Drosophila*. Work in *Chironomus* suggested that puffs arise from single bands and may be involved in the synthesis of a specific RNA. A significant advance was made in understanding the functional aspects of single bands when Judd, Shen, and Kaufman (1972) determined the cytological location of 16 complementation groups plus *zeste* and *white*, all localized in the 3A–3C region of the X-chromosome. Genetic and cytological placement of the various complementation groups by recombination and deletion mapping indicated a one chromomere-one complementation group relationship. Since the average-sized chromomere in *D. melanogaster* contains enough DNA to code for 30 genes (each 1000 nucleotides in length), finding only one complementation group per chromomere raises numerous questions about the function of the remainder of the DNA. Is it of a regulatory nature influencing a single structural gene? Does it represent multiple copies of the structural gene? Is it present solely for structural reasons to maintain the integrity of that chromosomal region? As often occurs in science, an elegant experiment concerning certain questions raises many more. In any event, these findings pave the way for a better understanding of the structure and function of chromosomal units.

Changes in Chromosomes: Mutations, Aberrations, and Polyploidy

Many different kinds of chromosome changes have occurred naturally in related species or in different races of the same species. Many have been produced in experimental cultures by treatments with X-rays and other radiations, chemicals, and other agents. These chromosome changes have been important in elucidating cytogenetic principles concerning chromosome behavior. The changes include increases or decreases in chromosome number, interchanges between nonhomologous chromosomes, transpositions within the chromosome, inversions, duplications, and deficiencies. Many different agents were used over the years in attempts to produce mutations, but not until the late 1920s did Muller (1927), using X-rays and radium, produce them in *Drosophila*; and Stadler (1928), using X-rays, produce them in maize and barley. Muller reported chromosome changes as well as gene changes. Stadler's (1931) later report included photographs of maize ears showing aleurone color deficiencies and a ring-of-four chromosomes at diakinesis first observed by Randolph.

Stadler (1941) also conducted breeding tests with the mutations produced by X-ray and by ultraviolet radiations. Of the plants with changes from *A* to *a* (one of the complementary factors for aleurone color) produced by X-rays, only 2 percent had normal pollen. When Stadler tested those with normal pollen, they showed reduced transmission. Ultraviolet rays produced changes of *A* to *a* which had normal pollen and normal transmission through pollen and ovules. X-irradiation apparently caused more chromosomal damage than ultraviolet radiations. During World War II the mutagenic effect of the mustard gases was discovered, but it was not reported until after the war (Auerbach and Robson 1946).

One of the races of *Datura* (B-white) being used in cytogenetic studies was found to be homozygous for an interchange between end segments of two nonhomologous chromosomes (Belling and Blakeslee, Paper 11). The interchanges in maize reported by Burnham (Paper 13) also arose spontaneously.

Another change in structure that occurs in certain species as they evolve is the duplication of certain segments. Bridges (1935) noted in *Drosophila*, particularly in one arm of chromosome II, that when the salivary-gland chromosomes were smeared and stretched, there were arc-like connecting strands between certain regions that had identical banding patterns (termed *ectopic pairing*). He concluded that these were duplications. Sturtevant (1925) had previously studied the genetic behavior of the Bar eye gene and its various "alleles." From a series of cleverly devised genetic experiments using a closely linked genetic marker on each side of Bar, he demonstrated that the mutations were accompanied by crossing-over between those markers. He also showed how unequal crossing-over following oblique synapsis between duplicated segments in the members of the pair could explain the mutational behavior at the Bar locus including different orders of the segments involved. In 1936 three different reports showed that the original Bar mutation was itself a duplication (Muller, et al.; Dubinin and Volotov; and Bridges, Paper 4). Bridges referred to the extra segment as an *insertion*. Muller, et al., suggested that the original Bar mutant duplication had arisen as the result of an exchange between the members of the homologous pair of chromosomes but not exactly at the same point. One of the resulting chromosomes would have a duplication and the other a deficiency. Their proposal actually described an unequal cross-over event.

Sturtevant (1926) was the first to show by comparison of the linear orders of several genes in *D. melanogaster* that the order in a "cross-over–reducer" stock was the reverse of the normal order. The order in chromosome III of another species, *D. simulans*, was also the reverse of that in *D. melanogaster*. McClintock's (1931) studies in maize were the first to show that in an inversion heterozygote, reverse pairing occurred at pachynema to bring homologous parts together. Painter (Paper 3) showed similar configurations in the polytene salivary-gland chromosomes of inversion heterozygotes in *Drosophila*. The statement often made was that the

11

inversions eliminated crossing-over. Ingenious genetic experiments in *Drosophila* (Sturtevant and Beadle 1936, and Sidorow, Sokolow, and Trofimow 1936) designed to recover deficient chromosomes, demonstrated that these cross-overs occurred in many inversion heterozygotes at relatively high frequencies, often at frequencies comparable to normal values. Hence the reduced genetic crossing-over was the result of inviability of zygotes receiving the deficiency-plus-duplication chromatids produced by crossing-over within the inversion (if pericentric), or the result of orientation of cross-over chromatids into the polar bodies, away from the functional terminal egg cell (if the inversion is paracentric).

Analyses of polytene chromosomes in *Drosophila* also showed that inversions, mostly the paracentric type, had become established in wild populations and were useful in determining phylogenetic sequences (Sturtevant and Dobzhansky, 1936).

Following the discovery of the constancy of the chromosome number for a particular species (1882–1884), Guignard (1891) and also Strasburger found that different genera of Liliaceae had different numbers, for example, $n = 12$ in *Lilium* and *Fritillaria*, and $n = 8$ in *Allium*. In the numerous investigations of chromosome number that followed, it was obvious also that species of the same genus might differ in number by a ratio of 1:2 or other multiples. Apogamous species often had the higher number. Tischler (1915) published a list of chromosome numbers that had been reported in plants, but in many cases there were different numbers for the same species. With the cytological techniques then in use, the numbers in many species, especially those with high numbers, were difficult to determine. The first extensive investigation of many species in the same genus was for *Chrysanthemum* with gametic numbers of 9, 18, 27, 36, and 45, all multiples of 9 (Tahara 1915). Winge (1917) reported results of his own studies of chromosome numbers and also summarized others that he considered reliable.

Phylogenetic reduction in chromosome number was studied by Tobgy (1943) in the hybrid between *Crepis neglecta* ($n = 4$) and *C. fuliginosa* ($n = 3$). Two chromosomes of *C. neglecta* were represented by only one in *C. fuliginosa*, the result of a very unequal interchange followed by the loss of one of the new largely heterochromatic chromosomes.

Phylogenetic increases in chromosome number have come about in several ways. In certain groups of related species, differences in number were the result of differences in the number of telocentric and metacentric chromosomes, but the total number of arms was similar. The explanation Robertson (1916) offered from his studies of several groups of species of grasshoppers was that a metacentric chromosome had arisen by fusion of two telocentric ones. These have come to be known as *Robertsonian fusions*. Other examples are found in certain *Drosophila* species.

In other species the change to $4n$ was the result of doubling the diploid $2n$ number to produce an autopolyploid. Probable examples are alfalfa and the potato. Occasionally this doubling occurs naturally. For example, Lutz (Paper 6) found a stocky, large form of *Oenothera*, called *gigas*, to have twice the diploid number, i.e., $4n$. In *Datura* an occasional stocky plant appeared that bred true but could not be crossed with normal plants and hence was referred to as a "new species" (Blakeslee 1917). In 1920 Belling found that these were tetraploids. Tetraploid sectors or branches also occurred in diploids in *Datura* (Avery, Satina, and Rietsema 1959). In maize triploids occur occasionally and are recognized by their sterility (McClintock 1929b). In general, tetraploids are rare among the progeny of diploids.

Polyploids occur in animals but more rarely than in plants. Fankhauser (Paper 7), the pioneer in finding such individuals, reported that n, $2n$, $4n$, and $5n$ larvae occurred naturally in the newt (*Triturus viridescens*), an amphibian. Triploids could be induced regularly by temperature shock treatments immediately after fertilization. This induction of triploids occurs because the second maturation division in amphibian eggs is not completed until after fertilization. In the chicken occasional individuals are triploid (Abdel-Hameed and Shoffner 1971). Since the first one reported was an intersex with a distinctly different phenotype, others with such phenotypes have been identified as triploids.

Another method by which individuals with multiples of lower chromosome numbers might have arisen phylogenetically was proposed by Winge (1917). Doubling the chromosome number of an F_1-hybrid between two species with $n = 9$ would produce an individual with $n = 18$ (i.e., $9 + 9$), an allopolyploid. If the two sets were so different the chromosomes would not pair, he believed the F_1 embryo could not perpetuate itself. We now know that this is a favorable situation that leads to fewer irregularities at meiosis in the allopolyploid. Examples in which this doubling did occur in untreated experimental material are: (1) tetraploid *Primula kewensis* which arose as a fertile branch on an otherwise sterile hybrid between two different species (chromosome pairs were formed in the hybrid, but it was sterile, Digby 1912); (2) in a sterile hybrid between the cabbage and radish from which the fertile *Raphanobras-*

sica arose (Karpechenko 1927); and (3) in a sterile hybrid between two species of insects (Federley 1913).

Winkler (1916) had produced polyploids in grafts by cutting off the shoot at the point of union between the scion from one species of *Solanum* and the stock from a different species. Among the shoots that formed, many were chimeras, others tetraploids, but some may have resulted from the fusion of cells from both species. This is of interest especially in relation to the current surge of attempts to fuse protoplasts from different species. The Marchals (1907–1911) accomplished chromosome doubling in mosses by inducing a 2*n* protonema to develop by wounding the diploid sporophyte. In certain species this 2*n* protonema produced a 4*n* sporophyte. This also demonstrated that a gametophyte was not necessarily a haploid.

The report that chromosome doubling results from treatments with colchicine (Blakeslee 1937 and Blakeslee and Avery, Paper 8) initiated a wave of studies that established polyploids in a wide range of species. Certain polyploids have become important commercially.

Another change in chromosome number that occurs naturally in plants is the haploid that appears occasionally among the progeny of diploids. The first haploid was reported in *Datura*, a diploid species, by Blakeslee, et al. (1922). Haploids have been of considerable interest over the years as a possible tool in breeding. As East (1930) first pointed out, doubling the chromosome number of a haploid would be a method of rapidly producing homozygous lines. One source of haploids in certain species is twin seedlings, one member of the pair sometimes being a haploid or a triploid. Haploids have been used in maize breeding, using genetic markers to screen for them (Chase 1969); and haploids are used currently in barley from crosses between *Hordeum vulgare* and *H. bulbosum* (Kasha and Kao 1970). In addition, haploids have been produced and used in alfalfa (Bingham 1971) and in the potato (Hougas and Peloquin 1958). The first report of a haploid in an allopolyploid species was in hexaploid wheat, *Triticum vulgare*, (Gaines and Aase 1926). More recently haploids in many species are arising from anthers placed on a culture medium (Nitsch and Nitsch 1969).

Smear Techniques

The development of smear techniques for studying chromosomes at meiosis in plants was a very important advance since it avoided the time-consuming steps (i.e., embedding in paraffin, sectioning, and staining) between fixation and the final step at which cytological observations could be made. Since cell walls are absent in animals, smear techniques had been used much earlier in studies of meiosis, e.g., in *Drosophila* (Stevens 1905). Taylor (1924) reported a technique in which a fresh anther was smeared on the slide followed by fixation and staining. Belling (1921) discovered that when iron was added to the drop of acetocarmine stain in which anthers had been smeared, the chromosomes in the pollen mother cells took up the stain immediately. McClintock (1929c) discovered later that heating the prepared slide greatly improved the contrast between cytoplasm and chromosomes; she also developed a method of making the acetocarmine smear slides permanent. Other smear techniques were developed for studying chromosomes in somatic cell division in root tips and other meristematic tissues for a wide range of plant species. Treatment of these tissues with colchicine, 8-hydroxyquinoline, or cold shortens the chromosomes and also makes the centromere constrictions more prominent, enabling the researcher to more easily obtain accurate counts of chromosome numbers.

Human Chromosome Methodology

Advancements in human chromosome methodology have been remarkable during the last 20–25 years. The chromosome number of man ($2n = 46$) was uncertain until 1956 (Tjio and Levan).

Cell swelling and spindle apparatus inhibition at metaphase were accomplished when mammalian tissue culture materials were immersed for 20–30 minutes in a warmed hypotonic salt solution (Hsu and Pomerat 1953). In tissue culture there are usually single cell layers, and therefore the method is much more successful than when tissues not in culture are used. Tjio and Levan (1956) reported modifications of the Hsu and Pomerat (1953) technique. The modifications included hypotonic pretreatment for only 1 or 2 minutes, following a colchicine dose to the culture medium 12–20 hours before fixation. Moorehead, et al. (1960), described a technique for mammalian chromosome studies using leucocyte cultures from peripheral blood. Phytohemagglutinin was used to agglutinate the red blood cells, thus separating the leucocytes before culturing; and also to stimulate leucocyte cell division.

Even with the technical ability to prepare excellent metaphase spreads of human chromosomes facilitating accurate counts, the problem remained of identifying individual chromosomes of similar size within different size groups. Certain chromosomes similar in size could be identified based on differing times of DNA replication, but the autoradiographic technique used for this purpose was time consuming and expensive. There was a need for a rapid, inexpensive, and accurate method of identifying all homologous pairs of chromosomes. This need was met in 1971 by several laboratories (see Patil, Merrick, and Lubs 1971) at about the same time through the development of modified Giemsa staining, using Giemsa at pH 9.0 instead of the usual 6.8. The technique has been improved by coupling the Giemsa staining procedure with a trypsin treatment (Seabright 1971).

Caspersson, et al. (1970), reported that human metaphase chromosomes from blood cultures treated with quinacrine mustard showed a banded fluorescence pattern that was constant and reproducible, at least for the most strongly fluorescent regions.

Editors' Comments
on Papers 1 Through 8

The first portion of Paper 1 reviews the studies by S. Nawaschin and his students (1910–1928) on the individuality of chromosomes and their gross morphology. For comparisons of the gross morphology of the chromosomes within and between species, they initiated the use of schematic representations which they termed *idiograms*. These became widely used and are similar to those used today for karyotype analysis. The second portion of Paper 1 is Lewitsky's analysis of one crop species, rye (*Secale cereale*), in which he uses line diagrams to represent the chromosomes.

Paper 2 is the first published demonstration that an analysis of meiosis at pachynema in an interchange heterozygote furnishes important information about the structure of paired normal chromosomes as well as changes in that structure. This study supplemented the cytological information from studies of other stages in the chromosome cycle and added another dimension to correlative studies of cytology and genetics.

Paper 3 by Painter describes the salivary-gland chromosomes of *Drosophila* and gives examples of the chromosome configurations in heterozygotes for inversions, deletions, and translocations. The paper also shows how the positions of genes in the chromosomes could be determined. Painter's paper did for *Drosophila* what McClintock's paper on analysis at pachynema did for maize; but Painter's was even more striking because of the distinctive banding patterns that characterize the different chromosomes in the salivary glands. Painter's paper has had and continues to have a tremendous impact on studies of *Drosophila* and of other species that have similar chromosomes, e.g., *Sciara*.

Paper 4 is one of three published in that year that furnished cytological confirmation of the fact that the original Bar mutant was a duplication, a conclusion Sturtevant had reached many years earlier from his genetic experiments in which markers on each side of the Bar gene served to identify cross-over events. This paper was selected for inclusion in this volume because of its detailed description and illustration of the banding pattern in the salivary-gland X-chromosome. The study is a good example of the usefulness of cytological analysis of salivary-gland chromosomes.

With its limited resolution, light-microscopic examination of chromosomes furnished little information on their substructure. This fact was recognized early in the study of chromosomes, but no suitable alternatives existed until the advent of the electron microscope with its high resolving power. In this important paper published in 1956 (Paper 5), Ris showed how to prepare plant and animal cells for chromosome ultrastructure studies with the electron microscope and showed how to interpret the observations and relate them to those made with the light microscope. He showed that the basic structure of chromosomes consists of microfibrils about 200Å in diameter. Ris's paper initiated a new era of chromosomology.

Paper 6 by Lutz was the first report that a naturally occurring mutant form had double the normal chromosome number, i.e., it was tetraploid. Paper 7 by Fankhauser was the first report on naturally occurring polyploids in an animal species. Paper 8 was the first report on the highly successful technique of using colchicine to obtain plants with doubled chromosome number in many different genera. O. J. Eigsti had previously observed polyploid cells in roots treated with colchicine and had suggested the use of this chemical for chromosome doubling (Blakeslee 1937). Colchicine acts to block anaphase movement of chromosomes. A surge of interest followed for many years in the use of colchicine to produce polyploids, both auto- and amphipolyploids. The technique has led to studies that have furnished information on the structure of the spindle and on the mechanics of chromosome movements.

REFERENCES

Abdel-Hameed, F., and R. N. Shoffner. 1971. Intersexes and sex determination in chickens. *Science* **172**:962–964.

Auerbach, C., and J. M. Robson. 1946. Chemical production of mutations. *Nature* **157**:302.

Avery, A. G., S. Satina, and J. Rietsema. 1959. *Blakeslee: The genus Datura*. The Ronald Press Co., New York. 289 pp.

Babcock, E. B. 1949. The development of fundamental concepts in the science of genetics. *Port. Acta Biol.* Ser. A., R. B. Goldschmidt Vol.:1–50.

Balbiani, E. G. 1881. Sur la structure du noyau des cellules salivaraires chez les larves de *Chironomus*. *Zool. Anz.* **4**:637–641.

Belling, J. 1921. Counting chromosomes in pollen-mother cells. *Am. Nat.* **55**:573–574.

Bingham, E. T. 1971. Isolation of haploids of tetraploid alfalfa. *Crop Sci.* **11**:433–435.

Blakeslee, A. F. 1917. The inheritance of germinal peculiarities. Flowering plants. *Carnegie Inst. Washington Yearb.* **16**:125–127.

_____. 1937. Dédoublement du nombre de chromosomes chez les plantes par traitement chimique. *C. R. Seances Acad. Sci., Paris.* **205**:476–479.

_____, J. Belling, M. E. Farnham, and A. D. Bergner. 1922. A haploid mutant in the Jimsonweed, "*Datura stramonium.*" *Science* **55**:646–647.

Boveri, Th. 1889. Die Vorgänge der Befruchtung und Zellteilung in ihrer Beziehung zur Vererbungsfrage. *Beitr. Anthropol. Bayerns* (München Anthropol. Ges. Verh.) **8**:27–39.

_____. 1893. Über die Entstehung des Gegensatzes zwischen den Geschlechtszellen und den somatischen Zellen bie *Ascaris*. *Sitzungsber, Ges. Morph. Phys. München.* **8**:114–125.

_____. 1904. *Ergebnisse über die Konstitution der chromatischen Substanz des Zellkerns.* Jena.

Braselton, J. P. 1971. The ultrastructure of the non-localized kinetochores of *Luzula* and *Cyperus*. *Chromosoma* **36**:89–99.

Bridges, C. B. 1935. Salivary chromosome maps. *J. Hered.* **26**:60–64.

_____. 1938. A revised map of the salivary gland X-chromosome of *Drosophila melanogaster*. *J. Hered.* **29**:11–13.

Brinkley, B. R., and E. Stubblefield. 1966. The fine structure of the kinetochore of a mammalian cell in vitro. *Chromosoma* **19**:28–43.

Britten, R. J., and D. E. Kohne. 1968. Repeated sequences in DNA. *Science* **161**:529–540.

Brown, S. W. 1949. The structure and meiotic behavior of the differentiated chromosomes of tomato. *Genetics* **34**:437–461.

_____. 1954. Mitosis and meiosis in *Luzula campestris* DC. *Univ. Calif. Publ. Bot.* **27**:231–277.

_____. 1966. Heterochromatin. *Science* **151**:417–425.

Bütschli, O. 1875a. Vorläufige Mittheilung über Untersuchungen betreffende die ersten Entwicklungsvorgänge in befruchteten Ei von Nematoden und Schnecken. *Z. Wiss. Zool.* **25**:201–213.

_____. 1875b. Vorläufige Mitteheilung einiger Resultate von Studien über die Conjugation der Infusorien und die Zelltheilung. *Z. Wiss. Zool.* **25**:426–441.

Carothers, E. E. 1913. The Mendelian ratio in relation to certain Orthopteran chromosomes. *J. Morph.* **24**:487–511.

Caspersson, T., L. Zech, C. Johansson, and E. J. Modest. 1970. Identification of human chromosomes by DNA-binding fluorescent agents. *Chromosoma* **30**:215–227.

Chase, S. S. 1969. Monoploids and monoploid-derivatives of maize (*Zea mays* L.). *Bot. Rev.* **35**:117–167.

Cleland, R. E. 1956. Analysis of trends in biological literature—Plant Sciences. *Biol. Abstr.* **30**:2459–2462.

Correns, C. 1900. G. Mendel's Regel über das Verhalten der Nachkommenschaft der Rassenbastarde. *Ber. Dtsch. Bot. Ges.* **18**:158–168.

Delaunay, L. 1929. Kern und Art. Typische Chromosomenformen. *Planta* **7**:100–112.

Digby, L. 1912. The cytology of *Primula kewensis* and of other related *Primula* hybrids. *Ann. Bot.* **26**:357–388.

Dubinin, N. P., and E. N. Volotov. 1936. Mutations arising at the Bar locus in *Drosophila melanogaster*. *Nature* **137**:869.

East, E. M. 1930. The origin of plants of maternal type which occur in connection with interspecific hybridizations. *Proc. Nat. Acad. Sci. U.S.A.* **16**:377–380.

Federley, H. 1913. Das Verhalten der Chromosomen bei der Spermatogenese der Schmetterlinge *Pygaera anachorata*, *curtula* und *pigra* sowie einiger ihrer Bastarde. *Z. Indukt. Abstamm.-Vererbungsl.* **9**:1–110.

Flemming, W. 1879. Beiträge zur Kenntnis der Zelle und ihrer Lebenserscheinungen I. *Arch. Mikrosk. Anat.* **16**:302–346.

_____. 1880. Beiträge zur Kenntnis der Zelle und ihrer Lebenserscheinungen II. *Arch. Mikrosk. Anat.* **18**:151–259.

_____. 1881. Beiträge zur Kenntnis der Zelle und ihrer Lebenserscheinungen III. *Arch. Mikrosk. Anat.* **20**:1–86.

Gaines, E. F., and H. C. Aase. 1926. A haploid wheat plant. *Am. J. Bot.* **13**:373–385.

Gall, J. G. 1952. The lampbrush chromosomes of *Triturus viridescens*. Proc. Symp. Chemistry and Physiology of the Nucleus. *Exp. Cell. Res.*, Sup. **2**:95–102.

_____. 1963. Kinetics of deoxyribonuclease action on chromosomes. *Nature* **198**:36–38.

Guignard, L. 1891. Nouvelles études sur la fécondation. *Ann. Sci. Nat. Bot.* **14**:163–296.

Gutherz, S. 1907. Zur Kenntnis der Heterchromosomen. *Arch. Mikrosk. Anat.* **69**:491–514.

Hance, R. T. 1918. Variations in the number of somatic chromosomes in *Oenothera scintillans* de Vries. *Genetics* **3**:225–275.

Heitz, E. 1926. Der Nachweis der Chromosomen. Vergleichende Studien über ihre Zahl, Grösse und Form im Pflanzenreich I. *Z. Bot.* **18**:625–681.

_____. 1928. Das heterochromatin der Moose. I. *Jahrb. Wiss. Bot.* **69**:762–818.

Henking, H. 1890. Über Reduktionsteilung der Chromosomen in den Samenzellen von Insekten. *Int. Monatsschr. Anat. Phys.* **7**:243–248.

Hertwig, O. 1875. Beiträge zur Kenntnis der Bildung, Befruchtung und Teilung des tierischen Eies I. *Morphol. Jahrb.* **1**:347–434.

_____. 1877. Beiträge zur Kenntnis der Bildung, Befruchtung und Teilung des tierischen Eies II. *Morphol. Jahrb.* **3**:1–86.

_____. 1878. Bieträge zur Kenntnis der Bildung, Befruchtung und Teilung des tierischen Eies III. *Morphol. Jahrb.* **4**:176–213.

Hougas, R. W., and S. J. Peloquin. 1958. The potential of potato haploids in breeding and genetic research. *Am. Potato J.* **35**:701–707.

Hsu, T. C., and C. M. Pomerat. 1953. Mammalian chromosomes in vitro II. A method for spreading the chromosomes of cells in tissue culture. *J. Hered.* **44**:23–29.

Judd, B. H., M. W. Shen, and T. C. Kaufman. 1972. The anatomy and function of a segment of the X chromosome of *Drosophila melanogaster*. *Genetics* **71**:139–156.

Karpechenko, G. K. 1927. Polyploid hybrid of *Raphanus sativus* L. X *Brassica oleracea*. *Z. Indukt. Abstamm.-Vererbungsl.* **48**:1–85.

Kasha, K. J., and K. N. Kao. 1970. High frequency haploid production in barley (*Hordeum vulgare* L.) *Nature* **225**:874–876.

Kavenoff, R., and B. H. Zimm, 1973. Chromosome-sized DNA molecules from *Drosophila*. *Chromosoma* **41**:1–27.

Kostoff, D. 1930. Discoid structure of the spireme. *J. Hered.* **21**:323–324.

Lewitsky, G. A. 1931. The "karyotype" in systematics. *Bull. Appl. Bot., Genet. Plant Breeding* **27**(1):187–240. (Eng. 220–240).

Malheiros, N., D. de Castro, and A. Camara. 1947. Chromosomas sem centrómero localizado. O. casa da *Luzula purpurea* Link. *Agron. Lusit.* **9**:51–79.

Marchal, El. and Em. 1911. Aposporie et sexualité chez les mousses. III. *Bull. Acad. R. Belg.* **1911**:750–778.

McClintock, B. 1929a. Chromosome morphology in *Zea mays*. *Science* **69**:629.

_____. 1929b. A cytological and genetical study of triploid maize. *Genetics* **14**:180–222.

_____. 1929c. A method for making aceto-carmin smears permanent. *Stain Technol.* **4**:53–56.

_____. 1931. Cytological observations of deficiencies involving known genes, translocations and an inversion in *Zea mays*. *Missouri Agr. Exp. Stn. Res. Bull,* **163**:1–30.

_____ and H. E. Hill. 1931. The cytological identification of the chromosome associated with the *R–G* linkage group in *Zea mays*. *Genetics* **16**:175–190.

Moorehead, P. S., P. C. Nowell, W. J. Mellman, D. M. Battips, and D. A. Hungerford. 1960. Chromosome preparations of leukocytes cultured from human peripheral blood. *Exp. Cell Res.* **20**:613–616.

Muller, H. J. 1927. Artificial transmutation of the gene. *Science* **66**:84–87.

_____, A. A. Prokofjeva-Belgovskaja, and K. V. Kossikov. 1936. Unequal crossing-over in the Bar mutant as a result of duplication of a minute chromosome section. *C. R. (Doklady) Acad. Sci. URSS.* **10**:87–88.

Müntzing, A. 1954. Cyto-genetics of accessory chromosomes (B-chromosomes). *Proc. 9th Int. Congr. Genet. (Bellagio) Caryologia* **6**, Sup. I: 282–301.

Nawaschin, M. 1925. Morphologische Kernstudien der *Crepis*-Arten in Bezug auf die Artbildung. *Z. Zellf. Mikrosk. Anat.* **2**:98–111.

Nawaschin, S. 1912. Sur le dimorphisme nucléaire des cellules somatiques de *Galtonia candicans*. Russian, only the title in French. *Bull. Acad. Imp. Sci. St. Petersburg* Ser. 5, **6**:373–385.

_____. 1921. "Resumé of the rejoinder to the lecture of L. N. Delaunay" (In Russian) *J. Bot. Soc.* **6**:171–172. From Supplements to the *Proceedings of the Caucasian Section of the Russian Botanical Society.*

Nitsch, J. P., and C. Nitsch. 1969. Haploid plants from pollen grains. *Science* **163**:85–87.

Olins, A. L., and D. E. Olins. 1974. Spheroid chromatin units (v bodies). *Science* **183**:330–332.

Painter, T. S. 1925. A comparative study of the chromosomes of mammals. *Am. Nat.* **59**:385–409.

_____. 1933. A new method for the study of chromosome rearrangements and the plotting of chromosome maps. *Science* **78**:585–586.

Patil, S. R., S. Merrick, and H. A. Lubs. 1971. Identification of each human chromosome with a modified Giemsa stain. *Science* **173**:821–822.

Prakken, R. and A. Müntzing. 1942. A meiotic peculiarity in rye, simulating a terminal centromere. *Hereditas* **28**:441–482.

Randolph, L. F. 1928. Chromosome numbers in *Zea mays* L. *Cornell Univ. Agric. Exp. Stn. Mem. 117.*

Rhoades, M. M., and E. Dempsey. 1966. The effect of abnormal chromosome 10 on preferential segregation and crossing over in maize. *Genetics* **53**:989–1020.

Rhoades, M. M., and W. E. Kerr. 1949. A note on centromere organization. *Proc. Nat. Acad. Sci. U.S.A.* **35**:129–132.

Robertson, W. R. B. 1916. Chromosome studies I. Taxonomic relationships shown in the chromosomes of Tettigidae and Acrididae: V-shaped chromosomes and their significance in Acrididae, Locustidae, and Gryllidae: Chromosomes and Variation. *J. Morph.* **27**: 179–331.

Rosenberg, O. 1909. Cytologische und morphologische Studien an *Drosera longifolia* X *rotundifolia*. *K. Sven. Ventenskapsakad. Handl.* **43**(11):1–65.

Roux, W. 1883. *Über die Bedeutung der Kernteilungs figuren*. 19 pp. Leipzig. W. Engelmann.

Rückert, J. 1892. Über die Verdoppelung der Chromosomen in Keimbläschen des Selachiereies. *Anat. Anz.* **8**:44–52.

Ruttle, M. L. 1927. Chromosome numbers and morphology in *Nicotiana* I. The somatic chromosomes and nondisjunction in *N. alata* var. *grandiflora, Univ. Calif. Publ. Bot.* **11**:159–176.

Schneider, A. 1873. Untersuchungen über Platelminthen. *Oberhess. Gesell. Natur-Heilk. Giessen, Ber. Oberhess.* **14**:69–140.

Schrader, F. 1935. Notes on the mitotic behavior of long chromosomes. *Cytologia* **6**:422–430.

Seabright, M. 1971. A rapid banding technique for human chromosomes. *Lancet* **2**:971–972.

Sharp, L. W. 1934. *Introduction to Cytology*. 3rd ed. McGraw-Hill Book Co., N.Y.

Sidorow, B. N., N. N. Sokolow, and I. E. Trofimow. 1936. Crossing-over in heterozygoten Inversionen. *Genetica* **18**:291–312.

Stadler, L. J. 1928. Mutations in barley induced by X-rays and radium. *Science* **68**:186–187.

_____. 1931. The experimental modification of heredity in crop plants. I. Induced chromosomal irregularities. *Sci. Agric.* **11**:557–572.

_____. 1941. The comparison of ultra-violet and X-ray effects on mutation. *Cold Spring Harb. Symp. Quant. Biol.* **9**:168–178.

Stebbins, G. L., Jr. 1950. *Variation and Evolution in Plants*. Columbia Univ. Press, New York. p. 443.

Stevens, N. M. 1905. Studies in spermatogenesis with special reference to the "accessory chromosome." *Carnegie Inst. Washington Publ.* **36**:1–75.

Strasburger, E. 1901. Über Befruchtung. *Bot. Zeitung* **59**:352–367.

Sturtevant, A. H. 1925. The effects of unequal crossing-over at the Bar locus in *Drosophila. Genetics* **10**:117–147.

_____. 1926. A crossover reducer in *Drosophila melanogaster* due to inversion of a section of the third chromosome. *Biol. Zentralbl.* **46**:697–702.

_____ and G. W. Beadle. 1936. The relations of inversions in the X chromosome of *Drosophila melanogaster* to crossing over and disjunction. *Genetics* **21**:554–604.

Sturtevant, A. H., and T. Dobzhansky. 1936. Inversions in the third chromosome of wild races of *Drosophila pseudoöbscura* and their use in the study of the history of the species. *Proc. Nat. Acad. Sci. U.S.A.* **22**:448–450.

Sutton, W. S. 1903. The chromosomes in heredity. *Biol. Bull.* **4**:231–251.

Tahara, M. 1915. Cytological studies on *Chrysanthemum. Bot. Mag., Tokyo* **29**:48–50.

Taylor, W. R. 1924. The smear method for plant cytology. *Bot. Gaz.* **78**:236–238.

Tischler, G. 1915. Chromosomenzahl, -Form und -Individualität im Pflanzenreich. *Progr. Rei Bot.* **5**:164–284.

Tjio, J. H., and A. Levan. 1956. The chromosome number of man. *Hereditas* **42**:1–6.

Tobgy, H. A. 1943. A cytological study of *Crepis fuliginosa, C. neglecta,* and their F$_1$ hybrid, and its bearing on the mechanism of phylogenetic reduction in chromosome number. *J. Genet.* **45**:67–111.

Walton, A. C. 1924. Studies on nematode gametogenesis. *Z. Zell.-Gewebelehre* **1**:167–239. Vol. 2 became *Z. Zellforsch, Mikrosk. Anat.*

Weismann, A. 1887. Über die Zahl der Richtungskörper und über ihre Bedeutung für die Vererbung. Jena.

Winge, Ø. 1917. The chromosomes. Their numbers and general importance. *C. R. Trav. Lab. Carlsberg* **13**:131–275.

Winkler, H. 1916. Über die experimentelle Erzeugung von Pflanzen mit abweichenden Chromosomenzahlen. *Z. Bot.* **8**:417–531.

1

Reprinted from pp. 103–104, 105, 106–107, 108–109, 111, 113–114, 121–122 of *Tr. Prikl. Bot., Genet., Sel.* **27**(1): 103–174 (1931)

The Morphology of the chromosomes.

History. Methods. Facts. Theory.

G. A. Lewitsky.

I. Introduction.

As early as 1882, Strasburger noticed that the chromosomes of one of the plants investigated by him, namely in *Funkia Sieboldiana*, were very different in size. Analogous relations were afterwards discovered in a whole series of plants and animals [1]. In the work of Cl. Müller (1912) we have already an investigation devoted entirely to this question. The above mentioned differences were not mere variation, but were characterizing definite types of larger and smaller chromosomes strictly repeated in nuclear plates of the respective species. Each type was represented in the somatic cells by a pair of equal elements, evidently of „paternal" and „maternal" origin.

Data as such were, with the constancy of chromosome number together, a telling instance of the „individuality" of the chromosomes, to be understood in the sense of definite peculiarities, characteristic of each pair. Further investigations have shown, however, that such individuality is still richer in contents, as being expressed in special m o r p h o l o g i c a l d i f f e r e n t i a t i o n o f t h e c h r o m o s o m e s t h e m s e l v e s.

For an exact statement of this fact — i. e. that the chromosomes in different organisms, as well as in the same cell, besides their absolute and relative size, may show special constant and characteristic differences in structure of their body — science is indebted to the works of S. G. Navashin (1910, 1911, 1912, 1914), Agar (1913) and Sakamura (1915, 1920).

The discovery of a definite morphological differentiation of the chromosomes has opened to us a new domain of questions of a general biological character (comp. espec. the works of Delaunay, M. Navashin, I. Sveshnikova, Darlington). Up to date, however, the investigations on these lines are carried on chiefly with choice objects — „classical" in cytological respect, i. e. possessing chromosomes, morphology of which is exceptionally well expressed, the chromosomes being of large dimensions and showing marked individual differences. The while, a whole series of most important genetical objects to which most widespread cultivated plants belong, as well as such plants as *Oenothera* and

---

[1] With regard to the latter the first data belong to Montgomery (1901) and Sutton (1902).

Datura genetics of which is closely connected with their karyology, remain o u t s i d e t h e c i r c l e o f m o r p h o- k a r y o l o g i c a l s t u d y. Up to now their chromosomes represent the usual traditional picture of straight or bent rods and loops, of purely accidental dimensions depending on the position of real chromosomes in the preparation.

Since my scientific work has become closely connected to cultivated plants, I have made it my purpose to make these „bad objects" suitable for a minute cytological investigation as well.

The investigation of such a s p e c i a l problem is simultaneously a beginning of the investigation of a question of a g e n e r a l character, namely, whether the differentiation of chromosomes described for a f e w objects is not a g e n e r a l l a w o f chromosome structure, and what should be the meaning of its different manifestations from cytological, genetical and evolutional point of view. This question has never been subjected to detailed special investigation.

As regards cultivated plants in particular, the problem of chromosome morphology has been put forth by life itself. Up to now cytological characterictics in cultivated plants were exclusively reduced to chromosome n u m b e r s. These data have proved rather valuable in classification, as well as in every kind of hybridization work. But as whole groups of species — to say nothing of smaller subdivisions — have proved to possess equal numbers of chromosomes, possibilities for further cytological investigation have ceased at this. Only a study of the peculiarities of the chromosomes themselves would ensure further progress in cytological investigation — this with purely systematical purposes and in analysis of segregation products. As far as the chromosomes are bearers of hereditary factors, such special investigations inevitably lead to questions of a general character — heredity and evolution. As regards our investigation in particular, its purposes concretely fall into two parts: 1) a m a n i f e s t a - t i o n o f t h e c h r o m o s o m e m o r p h o l o g y and 2) i t s m o r e e x a c t q u a n t i t a t i v e c h a r a c t e r i s t i c s. The first task was to devise me- thodics for fixation, the second — those for measuring chromosomes.

[*Editors' Note:* Material has been omitted at this point.]

2. The investigations of S. G. Navashin (1910—1914).

[*Editors' Note:* Material has been omitted at this point.]

In his address reported at the „Société des Naturalistes" à Kiev, on 3/ɪᴠ, 1910, S. N a v a s h i n states that according to his observations „the chromosomes of *Fritillaria tenella* are distinctly double, i. e. „gemini" (H ä c k e r) and distinctly divided in the place where the bundles of spindle fibers are attached".

On February I-st, 1912, S. Navashin presents his famous report (to the former Imperial Academy of Sciences) „On the dimorphism of the nuclei in somatic cells of *Galtonia candicans*". This work is the first to state for this already repeatedly investigated object the presence of special, minute, but perfectly constant appendages attached by means of a „thread" to two (out of our) „medium" chromosomes characteristic of this plant.

These appendages were called by Navashin „satellites". During the division of the nucleus, these small bodies split — together with the remaining body of the chromosome whose normal, morphological constituent part they are. Here the possibility of distinguishing the chromosomes by the peculiarities of their structure was shown for the first time. In one race of *Galtonia* two „medium" chromosomes had each an entirely alike satellite, while in another pair the satellite was perceptible only in one chromosome, the other showing only a thread; later on however, a satellite was detected on this thread though of considerably smaller dimensions.

The above stated means the discovery of „nuclear heterozygousness" known up to then only as to the „sexual chromosomes" of animals. The formation of „heteromorphous pairs" by ordinary chromosomes, was established for the latter only in the following year by Carothers (1913).

In the anthers of the „asymmetric" individuals of *Galtonia*, there are formed two kinds of pollen grains — with big and with small satellites (Navashin 1913), which is in a way a cytological illustration of Mendelian segregation.

In his address reported on September 19-th 1912, Navashin mentions another detail of structure observed in the chromosomes of *Galtonia*, namely, the presence in each of them — on the end turned towards the interior of the nucleus — of special appendages which he calls here „palps", coming to the conclusion that the chromosomes are not merely „segments" of chromatin substance, but „obviously organized bodies".

In 1914[1], in a collective work dedicated to K. A. Timiryasev on occasion of his jubilee, there appeared a work of S. G. Navashin representing a summary and theoretical interpretation of the data obtained by him with regard to „external signs of the internal organization of the chromosomes" (1916). The paper is illustrated by two plates showing a

---

[1] To this year the separate prints must be referred, the book itself appeared in 1916.

fineness of representation of details quite inordinary for cytological litera ture of that time. The objects of investigation are once more *Fritillaria* and *Galtonia*. With regard to the first the author proves that transversal division of the chromosomes in the place where the bundles of spindle fibers are attached „neither by its origin, nor by its position on the chromosome, may be regarded as accidental, depending on a tendency of the chromosome to segmentation, especially under the influence of fixation, as was supposed by the authors who noticed such structural peculiarities in their preparations without attributing any special importance to them". „Transversal slits" in the chromosome could be also readily detected in the prophases of division, i. e. when the nuclear wall had not yet disappeared and, consequently, the spindle fibers could not penetrate into the nucleus (fig. 1), and at this in the same, quite definite places as in the metaphase. Thus the places of articulation of the two chromosome branches arise independently of the spindle threads which are in somewhat way attracted by them and quite accurately attached to these „constrictions" in form of bundles. As has been mentioned above, N a v a s h i n distinguished already in 1911 t h r e e t y p e s of chromosomes i n *Fritillaria tenella* (see our fig. 1 on the page 24), differing by the position of the constriction: with a medial one — i. e. in the middle of an V - shaped, equal-armed or almost equal-armed chromosome (fig. 1A); with a subterminal one, i. e. shifted to one of the ends of a markedly unequal-armed chromosome (fig. 1B), and finally with a „terminal", i. e. end attachment—to the „rod-shaped" chromosome (fig. 1C).

„Properly speaking" the author remarks, „such mode of attachment d o e s n o t e x i s t a t a l l, as I have been able to establish on irreproachable preparations: here a minute part of the body of the chromosome i s s e p a r a t e d f r o m i t b y a t r a n s v e r s e s l i t, and the spindle threads are attached b e y o n d i t, i. e. once more in the place of the achromatic interruption" (page 10).

An excellent confirmation of the statement just mentioned were the minute structural relations found by the author in *Galtonia*. The rod-shaped chromosomes of the latter seemed a classical example of strictly „terminal" attachment of the spindle threads. „With a careful study of the chromosome extremities of this plant", as the author writes, „we find in any chromosome a special small body, as if a head-piece, separated from the body of the chromosome by a deep constriction. Owing to this fact all chromosomes, with a favourable arrangement in the late prophase and especially in the equatorial plate, seem to be provided with small round heads of sharp regular outlines (fig. 12, ibid. page 13).

The spindle threads in the same manner as in *Fritillaria* are attached not to the head itself „but to the constriction between it and the body of the chromosome" (ibid., page 14).

The satellite is another extreme particle of the chromosome and is attached by a thread to its head (ibid., fig. 12).

[*Editors' Note:* Material has been omitted at this point.]

Next year, 1913, S. G. Navashin presents an address on the chromosomes of *Galtonia* at the meeting of „German Naturalists and Physicians" in Vienna, where he demonstrates two plates represented on pl. I D. That to the right has been fixed with a mixture (introduced into cytological practice by S. G. Navashin) of chromic and acetic acids with formalin („chrom-acetic formalin"). The chromosomes are of somewhat swollen aspect, but in distinction to a publication of 1912 they show distinct heads on their „proximal" (i. e. turned towards the centre) ends. Both X-chromosomes are with satellites of equal size. The left plate is fixed with „medium Flemming", i. e. a mixture of chromic, acetic and osmic acids. The chromosomes are fine, showing distinct splitting. The heads are distinctly disjoined from the body of the chromosome, sometimes even on threads and clearly show splitting (if only one half does not cover the other). In X-chromosomes to these heads are attached the „Satellites": a larger satellite with distinct splitting but with indistinct thread and a small one with intensively stained thread. It is the same „asymmetric" race as in fig. C, but with a minute satellite at the end of the thread. Below, the attachment of the spindle threads is shown (b e t w e e n head and body of the chromosome). The number of chromosome types is here already five: 3 large chromosomes (A, B, C), 1 — somewhat shorter one (D), 1 of medium length (e), 1 slightly shorter (X), 2 small ones (*m*). Exact dimensions in microns are given for each type. All these data exposed by S. G. Navashin as early as 1913, could be published in a form accessible to foreign scientists only after 17 years had elapsed, namely in 1924 in Ber. der Deutsch. Bot. Gesellschaft. The same data, even more detailed, though with less perfectly reproduced illustrations, appeared in 1916 in Navashin's paper. „On some characters of the internal organization of chromosomes" (in Russian), published in a collective work dedicated to K. A. Timiryasev. This work being a summary of the data obtained by S. G. Navashin and of his theoretical conclusions, as to morphology of chromosomes, has remained entirely unknown abroad, due to the language in which it was published, to the special character of the publication, as well as to subsequent political perturbations. Thus, the facts stated by Navashin were gradually discovered during these years by foreign scientists—so to say, for a second time. With regard to *Galtonia*, in particular, we find data in two works published in 1924 by Newton and by Taylor (see pl. I E and F). Both authors are acquainted with only the first data of Navashin 1912. Both discover the „heads" of the chromosomes and the attachment of spindle threads to the constriction. Both represent satellites but regard them as the s a m e h e a d s, only somewhat more separated from the body of the chromosome.

As follows from comparison of their drawings with those of Navashin, this may be explained by the fact that they did not succeed in obtaining such exact and perfect differentiation of the chromosomes as was attained by the Russian author.

29

[*Editors' Note:* Material has been omitted at this point.]

5. The school of Navashin (1915—1928).

Having extended his investigations over a series of new objects, S. N a-v a s h i n discovered there analogous data. Satellites were detected in *Muscari Albuca* (in the preparations of C l. M ü l l e r), *Najas.* In former two, the satellites proved to be attached to an other end than in *Galtonia*, namely, to the outer or „distal" one. In *Najas* the satellites attached as usual by threads on one of the pairs of chromosomes are so large that the preceding investigator (C l. M ü l l e r 1912) considered them to be separate chromosomes [2].

A detailed investigation of the first and of the last plant was entrusted to the pupils of N a v a s h i n, D e l a u n a y and T s c h e r n o y a r o w which have further developed the study of chromosome structure.

After a considerable interruption due to war and to political circumstances, the problems concerned with morphology of chromosomes once more make their appearance in form of a whole series of works—again chiefly by Russian cytologists. Now the chief attention is directed to the use of an exact knowledge of the chromosome shape for investigation of systematical questions on one side, and of general problems of evolution and phylogenetics on the other. The former „law of the constancy in chromosome number“ becomes now a more general „law in constancy of composition (morphological) of the nucleus“ (D e l a u n a y 1922) or of the constancy of its „idiogram“, as a „graphic representation of a species“ by its characteristic nuclear plate (N a v a s h i n 1921).

The rich content of the morphological characteristic of the nucleus gives to it a great importance when discussing questions of systematical affinity and differences, as D e l a u n a y has tried to show for the genera *Muscari* and *Bellevalia* (1922)[1].

On the other hand, c h a n g e s of the „idiogram“ in a series of successive species, especially in connection with their external morphology, affords material for drawing up particular phylogenetic lines and processes (D e l a u n a y — for *Muscari* and *Ornithogalum* 1925), as well as for establishing general regularities of evolution (D e l a u n a y 1926).

In some cases the analysis of such changes leads to a sharp conflict with the current theories of heredity (D e l a u n a y 1926, N a - v a s h i n 1925, p. 108). With all contestableness of many of the propositions set forth by the above mentioned authors, being sometimes based on a too small number of facts, the „karyo-phyletical“ and „karyo-evolutional“ tendency in science is a serious fact and has great prospects before it.

A series of investigations on these lines, more or less in connection to the present work, has been also carried out in our laboratory.

Among them an extensive monograph of A v d u l o v (1930) embraces the whole family of *Gramineae*. He alters to an extent their generally accepted classification, throwing quite a new light on the evolution of the grasses.

The same, based on detailed morphology of chromosomes, is done by S e n j a n i n o v a - K o r c z a g i n a for the genus *Aegilops* (1931), as well as with regard to the origin of *Vicia Faba* (1931).

Our work on the „Karyotype in systematics—on the basis of the investigation of the subfamily *Helleboreae*“ (1931), as well as the work „Zur Frage der karyotypischen Evolution der Gattung *Muscari*“ (L e w i t s k y und T r o n 1930), are devoted to an elucidation of the principles of using „karyotypical“ characters in systematics.

Of special importance in the domain we are interested in, is the genus *Crepis* — due to investigations of M. N a v a s h i n (junior). The small number of chromosomes and their distinct characteristics make the species of this genus an exceptionally expedient object. Different current hypotheses starting from the number of chromosomes and tending to explain the differences in this regard in akin species („repetition“, „segmentation“, „doubling“ etc. of the chromosomes), find their final appreciation on the basis of an exact study of chromosome morphology (N a -

---

[1] Comp. another work of the same author on the genus *Iris* (1928), as well as S w e s h n i k o v a on *Vicia* (1927).

v a s c h i n 1925). A special investigation of the different types of variations of the nuclear relations from point of view of their importance for species formation was carried out by M. N a v a s c h i n (1926). Besides the multiplying of the whole chromosome set (tri- and tetraploids), and its separate chromosomes (tri- and polysomics), there were found within the limits of the species (*C. tectorum*) still deeper changes in form of transformation of one of the chromosomes to two new, shorter ones, and, even of an e n t i r e l y n e w f o r m a t i o n of a particular type of chromosomes not peculiar to the species studied. In another species there were established three races differing by dimensions of satellites in one of the chromosome types. Such small and smallest, continually accumulating differences appear to the author as a real basis to species formation (N a v a s c h i n 1926, the same in D e l a u n a y 1926, and 1926 a, russian).

To such small differences in shape and sometimes in number of chromosomes, observed within a species, in connection with external peculiarities of corresponding races, S w e s c h n i k o v a devotes her communication on *Vicia* at the International Congress of Genetics (1928).

Of especial importance is the exact study of the chromosome shapes in hybrids between species differing in this regard.

In this case it is possible to trace in the offspring every chromosome type of the one or the other chromosome type, as has been shown by M. N a v a s h i n for the hybrids in the genus *Crepis* (1927). The possibility of distinguishing the individuality of chromosomes has helped to discover a very important fact, namely the a l t e r a t i o n o f t h e s h a p e in some chromosomes under the influence of hybridisation, which in the author's opinion realizes thus its truly creative part in process of evolution (M. N a v a s c h i n 1927, 1928).

A new era in the investigation of the chromosome morphology has been opened by works, whose purpose is the study of the changes in chromosomes, induced by the influence of X-rays. The application of the latter to plant objects, among them *Crepis*, carried out in our laboratory, has revealed the same sharp changes in the morphology of separate chromosomes, as has been shown recently for *Drosophila* by P a i n t e r and M ü l l e r (1929) and D o b z h a n s k y (1929), („Fragmentation“, „Translocation“, etc.) (L e w i t s k y and A r a r a t y a n 1931). Analogous changes have been revealed by us in *Crepis* and other plants even without any experimental influences, which points to the evolutional importance of such transformations of the chromosomes.

[*Editors' Note:* Material has been omitted at this point.]

B. Special part.

Secale cereale L.

Though the number of chromosomes typical of rye (14) is known since 1918 (Sakamura) we find the first representation of its somatic plates but in 1924, in a work of Gotoh (fig. 10). Here as in the subsequent work of Stolze (1925, Fig. 14) the chromosomes of rye are represented in form of more or less twisted rods of equal or irregularly varying thickness. In the work of Emme (1927), the author mentions vaguely that „some chromosomes are articulate", „the constrictions are not everywhere perceptible", „in two cases satellites have been found", „constrictions in long chromosomes are of frequent occurrence" and so on. The majority of plates and chromosomes are nevertheless of the same character as in the works of the two preceding authors.

For my investigations I have chosen the variety „Vyatka" No 4448 (from the collection of the Institute of Applied Botany). The best results were obtained in fixing with a mixture of 10% formalin and 1% chromic acid in equal proportions („5—5").

The plate investigated by us, with measurings, is represented in pl. II fig. A and B. The purpose of the drawing in shades is to give an idea of the degree of inclination of the separate chromosomes or of their parts, towards the horizon. The horizontal areas are the lightest, the steeper the inclination, the darker the shade—correspondingly to greater thickness of the stained layer in the body of the chromosome, through which the light penetrates while examining the preparation. All chromosomes show extremely marked constrictions. The daughter halves of the chromosomes, which have arisen in result of longitudinal splitting, are also perceptible — with all their mutual curvatures.

Alongside with the drawing in shades, we give the same drawing of the same plate in black, the separate chromosomes being marked with the numbers of their pairs established in result of measurings.

The results of the measuring of chomosomes are represented in fig. C (Pl. II).

The chromosomes here are arranged in the order of sequence of their decreasing length; the chromosomes most akin to one another in their general dimensions and in the ratio of their arms, are placed along-side and are evidently to recognize as „homologons", or „allelomorphs".

The separate „pairs" thus obtained are marked with Roman numerals.

The Arabic numerals placed above denote the whole length of the chromosomes in μ, those near the arms—the length of the latter. The transverse line corresponds to the place of the constriction.

One pair of chromosomes may be exactly recognized by the presence in them of special appendages separated from the remaining part of the arm by a non-stainable „achromatic area" or „zone". This shall be discussed further on with more detail. In both chromosomes (II' and II'') of the given plate, the arms with appendages show a very steep inclination; the latter are therefore not at once discerned and are represented with less clearness . In view of the sharply oblique position of the appendages, the dimensions of the latter, as well, as the size of the achromatic areas could not be sufficiently clearly established and have been marked out in the scheme (by a transverse line) only approximately. The remaining six chromosomes may be characterized but by their general length and by relative dimensions of the arms. Summarily they may be divided into three groups, each of them containing two chromosomes: two (I and III) larger (10 and 9.6 μ), distinctly unequally armed ones (showing approximately the arm ratio 1.5); two pairs (IV—V) of medium length (9.6 and 8.9) and almost equi-armed, and two pairs (VI and VII) small (8.9 and 8.1), sharply unequally armed ones (ratio—1.7 and 2.1).

[*Editors' Note:* Material has been omitted at this point. Plates I and III through V are not reproduced here.]

EXPLANATION OF PLATES.

Plate II.

Fig. 8.—[1]) A. S e c a l e   c e r e a l e L. Measured nuclear plate. The drawing with shading which indicate the degree of inclination of chromosomes to the horisontal plane. Enl. 5000

B. S e c a l e   c e r e a l e L. Measured nuclear plate with numerated pairs of chromosomes.

C. — S e c a l e   c e r e a l e L. The quantitative scheme of the measured plate. Enarg. 10000.

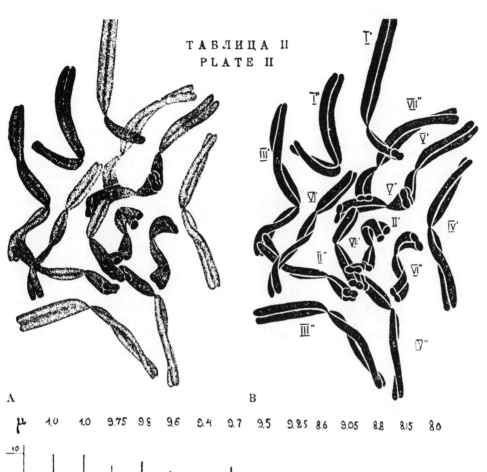

ТАБЛИЦА II
PLATE II

A
B

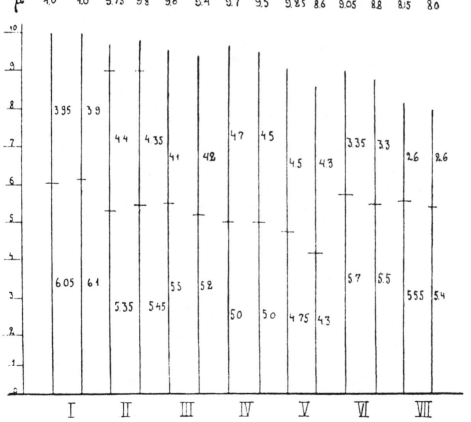

C

REFERENCES

Agar, W. E. 1913 (1912). Transverse segmentation and internal differentiation of chromosomes. *Q. J. Microsc. Sci.* **58**:285–298.

Avdulov, N. 1930 (1931). Karyosystematische Untersuchung der Familie Gramineae. *Sup. Bull. Appl. Bot., Genet. Plant Breeding.*

Carothers, E. 1913. The Mendelian ratio in relation to certain Orthopteran chromosomes. *J. Morph.* **24**:487–511.

Delaunay, L. 1926. Phylogenetische Chromosomenverkurzung. *Z. Zellforsch. Mikrosk. Anat.* **4**:338–364.

_____. 1926a. Reference not given by author.

Dobzhansky, T. 1929. Genetical and cytological proof of translocations involving the third and the fourth chromosomes of *Drosophila melanogaster. Biol. Zentralbl.* **49**:408–419.

Emme, H. 1927. Zur Zytologie der Gattung Secale. *Bull. of Appl. Bot., Genet. Plant Breeding.* **17**, No. 3 (Summary).

Gotoh, K. 1924. Über die Chromosomenzahl von *Secale cereale* L. *Bot. Mag.* **38**:135–151.

Kuhn, E. 1928. Zur Zytologie von *Thalictrum. Jahrb. Wiss. Bot.* **68**:382–430.

Lewitsky, G. A. 1931. The "karyotype" in systematics. *Bull. Appl. Bot., Genet. Plant Breeding* **27**:187–240 (Eng. 220–240).

_____, and A. Araratian. 1931. Transformations of chromosomes under the influence of X-rays. *Bull. Appl. Bot., Genet. Plant Breeding* **27**:265–303 (Eng. 289–303).

Lewitsky, G. A. and E. J. Tron. 1930. Zur Frage der karyotypischen Evolution der Gattung *Muscari* Mill. *Planta* **9**:760–775.

Montgomery, T. H. 1901. A study of the chromosomes of the germ cells of Metazoa. *Trans. Am. Phil. Soc.* **20**:154–236.

Müller, H. A. C. 1912. Kernstudien an Pflanzen. *Arch. Zellforsch.* **9**:1–51.

Nawaschin, M. 1925. Morphologische Kernstudien der Crepis-arten in Bezug auf die Artbildung. *Z. Zellforsch. Mikr. Anat.* **2**:98–111.

_____. 1926. Variabilität der Zellkernes bei Crepis-arten in Bezug auf die Artbildung. *Z. Zellforsch. Mikrosk. Anat.* **4**:171–215.

_____. 1927. Über die Veranderung von Zahl und Form der Chromosomen infolge der Hybridization. *Z. Zellforsch. Mikrosk. Anat.* **6**: 195–233.

_____. 1928. "Amphiplastie"—eine neue karyologische Erscheinung. *Verhand. V. Int. Kongr. Vererbungswiss.* **2**:1148–1152.

Nawaschin, S. G. 1913. Reference not supplied by author.

_____. 1921. Resumé of the rejoinder to the lecture of L. N. Delaunay (in Russian). *J. Bot. Soc.* **6**:171–172. From *Supplements to the Proceedings* of the Caucasian Section of the Russian Botanical Society.

Newton, W. C. F. 1924. Studies on somatic chromosomes. I. Pairing and segmentation in *Galtonia. Ann. Bot.* **38**:197–206.

Painter, T. S., and H. J. Muller. 1929 . Parallel cytology and genetics of induced translocations and deletions in *Drosophila. J. Hered.* **20**:287–298.

Sakamura, T. 1915. Über die Einschnürung der Chromosomen bei *Vicia Faba* L. *Bot. Mag.* **29**:287–300.

_____. 1920. Experimentelle Studien über die Zell-und Kernteilung mit besonderer Rücksicht auf Form Grosse und Zahl der Chromosomen. *J. Coll. Sci. Imp. Univ. Tokyo.* **39**:1–221.

Senjaninova-Karczagina, M. V. 1931 (1932). Karyological investigation of *Vicia Faba* and related species. *Bull. Appl. Bot. Genet. Plant Breeding* **28**:91–118.

Strasburger, E. 1882. Über den Theilungsvorgang der Zellkerne und das Verhältniss der Kerntheilung zur Zelltheilung. *Arch. Mikrosk. Anat.* **31**:476–590.

Sutton, W. S. 1902. On the morphology of the chromosome group in *Brachystola magna*. *Biol. Bull.* **4**:24–39.

Sweschnikova, I. 1928 (1927). Die Genese des Kerns im Genus *Vicia*. *Verhand. V. Int. Kongr. Vererbungswiss.* **2**:1415–1421.

Taylor, W. R. 1924. Cytological studies on *Gasteria* L. Chromosome shape and individuality. *Am. J. Bot.* **11**:51–59.

Additional References

[*Editors' Note:* The following references were not supplied in the original article.]

Delaunay, L. 1922. Vergleichende karyologische Untersuchungen einiger *Muscari* Mill.-und *Bellevalia* Lapeyr.-Arten. *Moniteur du jardin botan. de Tiflis*, ser. II. 1 (Russisch).

_____. 1925. The S-chromosomes in *Ornithogallum* L. *Science* **62**:15–16.

_____. 1927. Die Anwendung der karyologischen Methode zum Auflösen der Problem der speziellen Systematik. *Nawaschin's Festschrifft, Moskau* (Russisch).

Navaschin, S. G. 1910. *Proc. Conf. Kiev Soc. Nat. Hist.* 6–7, 12–15, 25–27, 27–31.

_____. 1911. About individual and specific characteristics (variations) of chromosomes. *Proc. Conf. Kiev Soc. Nat. Hist.*, 1911.

_____. 1912. Sur le dimorphisme nuclèaire des cellules somatiques de *Galtonia candicans* (title only in French, text in Russian). *Bull. Acad. Imp. Sci. (St. Petersburg)* Ser. 5, **6**:373–385.

_____. 1914 (1916). On some characters of the internal organization of chromosomes (in Russian). Collection of scientific articles dedicated to K. A. Timiryasev, pp. 185–214.

_____. 1927. Zellkerndimorphismus bei *Galtonia candicans* Des. und einigen verwandten Monokotylen. *Ber. Dtsch. Bot. Ges.* **45**:415–428.

Sakamura, T. 1918. Kurze Mitteilung über die Chromosomenzahlen und die Verwandtschaftverhältnisse der *Triticum* Arten. *Bot. Mag. (Tokyo)* **32**:150–153.

Senjaninova-Korczagina, M. V. 1932. Karyo-systematical investigations of the genus *Aegilops*. *Bull. Appl. Bot. Gen. Plant Breeding*, ser. 2.

Stolze, K. V. 1925. Die Chromosomenzahlen der hauptsächlichsten Getreidearten. *Bibliotheca Genet.* **8**:1–168.

2

Reprinted from *Natl. Acad. Sci. (U.S.A.) Proc.* **16**(12):791–796 (1930)

A CYTOLOGICAL DEMONSTRATION OF THE LOCATION OF AN INTERCHANGE BETWEEN TWO NON-HOMOLOGOUS CHROMOSOMES OF ZEA MAYS

By Barbara McClintock

Department of Botany, New York State College of Agriculture

Communicated November 6, 1930

It has been suggested (Brink,[1] Brink and Burnham[2]) that semisterility in maize is associated with some form of chromosomal change involving non-homologous chromosomes. Burnham[3] reported the presence of a ring of four chromosomes in diakinesis in such semisterile plants which could be explained by assuming either translocation or segmental interchange. Plants showing a ring of four chromosomes in diakinesis and 50% sterility in pollen and eggs gave, when crossed with normal plants, an F_1 generation, one-half of which were normal and one-half of which were 50% sterile. A semisterile plant when selfed gave, again, one-half semisterile plants and one-half non-sterile plants, but one-half of the non-sterile plants were homozygous for the translocation or interchange. When the latter plants were crossed to normals all the F_1 individuals were 50% sterile and showed a ring of four chromosomes at diakinesis.

The present investigation of semisterile-2 (Burnham[3]) was undertaken to determine which two chromosomes of the haploid set of ten were involved, and whether a simple translocation or a reciprocal one (segmental interchange) had occurred.

A comparison of the sizes of the four chromosomes constituting the ring with those of the remaining chromosomes of the complement indicated that the chromosomes involved in semisterile-2 were two of the four smallest chromosomes. To determine which two chromosomes were involved plants homozygous for the translocation were crossed with individuals trisomic $(2n + 1)$ for (a) the smallest and (b) the fourth smallest chromosome. Examination of meiosis in F_1 $2n + 1$ individuals showed, in both cases, a ring of four chromosomes and also a trivalent, indicating that the two chromosomes of the ring were independent of the smallest and the fourth smallest chromosomes. The chromosomes involved in what proved to be a reciprocal translocation, or segmental interchange, were, therefore, the second and third smallest chromosomes.

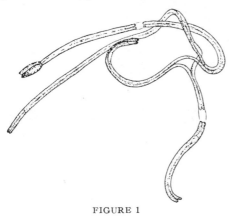

FIGURE 1

Interchange complex in mid-prophase, before opening out of four parasynapsed members to form ring; outline drawing made with the aid of a camera lucida. Magnification, 1875×. The clear portions represent the achromatic spindle fiber attachment regions. No attempt has been made to show the chromonemata in detail. For further explanation, see figure 2.

Open rings in late meiotic prophase do not show the nature or extent of synaptic association present in the earlier prophase period. In consequence, early prophases were sought in which the chromosomes, as long threads, were synapsed throughout their entire length. The microsporocyte membrane in *Zea* is very delicate in the early prophase stages. Aceto-carmine smears were made, the cover glass being placed over the sporocytes after removing all excess tissue, and the slides gently heated. With this method the sporocyte flattens, the nuclear membrane disappears and the long thread-like parasynapsed chromosomes are mostly spread out in a horizontal plane. It is frequently easy, therefore, to observe the full length of a parasynapsed bivalent, or an interchange complex which in an unflattened condition would be exceedingly difficult to trace.

Fortunately, the second smallest chromosome possesses, in certain

39

strains of maize, a very conspicuous accumulation of stainable substance at the end of the short arm; this is more prominent in early and mid-prophases than in later stages.* It is a constant feature of the chromosome, being regularly passed on from one cell generation to another. In the material used this was the only chromosome which possessed such a terminal knob. Consequently, this chromosome, which was involved in the interchange, could be distinguished readily from all the other chromosomes

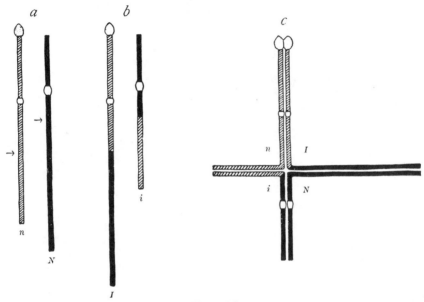

FIGURE 2

a.—Diagram of the two normal chromosomes which were involved in the segmental interchange. The clear portions in the chromosomes represent the spindle fiber attachment regions. The smaller chromosome terminated in an enlarged, deeply staining knob. The arrows indicate the places in the chromosomes at which the interchange occurred to produce the situation shown in *b*. *b.*—The two chromosomes produced as the result of the segmental interchange. *c.*—The type of synaptic complex in mid-prophase of meiosis obtained by combining a normal chromosome complement with an interchange complement through crossing. *N*, larger normal chromosome; *n*, smaller normal chromosome; *I*, larger interchange chromosome; *i*, smaller interchange chromosome.

of the prophase group. Similarly, the interchange complex in a semi-sterile plant could readily be distinguished from other chromosomes of the group.

In mid-prophase the two parental chromosomes in a normal bivalent lie side-by-side throughout their entire length. The most conspicuous structural feature of each chromosome is the spindle fiber attachment region, or so-called constriction. In general, it is a long, relatively clear region, frequently appearing slightly swollen with the methods used.   In some cases a more deeply staining spot is visible at each margin in the mid-region. This spindle fiber attachment region is achromatic; the stainable chromonemata do not pass through it.   Furthermore, the relative size of this region is a constant feature of the morphology of the chromosome.   In maize, the second smallest chromosome possesses a rather short spindle fiber attachment region.   In the third smallest chromosome this region is nearly twice as long.   When the two chromosomes are found together in an interchange complex, the contrast is evident.

If a segmental interchange had occurred one would expect, during early meiotic prophase in plants heterozygous for the interchange, a cross-shaped synaptic complex made up of two normal and two interchanged chromosomes (figure 2, *c*).   The interchange point in each chromosome would be at the center of the cross.   The relative length of the four arms would depend upon the location of the interchange points in the two chromosomes involved.   A number of such complexes were observed (figure 1).   In some of these the cross was so perfect that it could be photographed readily.

A morphological comparison of the knobbed chromosome in normal plants and in plants homozygous for the interchange showed the length of the longer arm of the knobbed chromosome to be much greater in the latter plants.   This marked difference allowed the interchange chromosome (*I*, figure 2, *c*) to be distinguished from the normal chromosome (*n*) in the prophase synaptic complex in plants heterozygous for the interchange. Thus, with the aid of the knob at the end of the short arm of the second smallest chromosome (*n*) and the obvious spindle fiber attachment regions, each of the four chromosomes in the cross-shaped synaptic complex was interpretable.   By means of a camera lucida, outline drawings of a number of clear figures were made and the length of each arm of each cross-shaped synaptic complex measured.   A close agreement, with regard to the relative lengths of the arms, was found to exist among the figures.   The diagrams in figure 2 were constructed after averaging these measurements.

It is clear that an unequal reciprocal translocation has taken place between the long arms of the two chromosomes, and that the interchanged pieces maintain the same orientation with respect to the spindle fiber attachment regions as they did in their previous, normal arrangement.

In later prophase an opening out of the members of the synaptic complex occurs, destroying the cross-like structure and forming the characteristic ring of diakinesis and metaphase *I*.

The distribution of the individual chromosomes in the ring at meiosis could not be observed directly but could be inferred from an analysis of the chromosome complements in the microspores.   As a general rule, the chromosomes in the ring are distributed two-by-two in anaphase *I*.   Consequently, each spore contains ten chromosomes.   Genetic analysis indicates that any 10-chromosome carrying spore possessing one interchange chromosome is sterile, since its nucleus lacks some part of the haploid complement.

The unequal interchange produced chromosomes of two new morphological types.   The presence of the conspicuous end knob on the second smallest chromosome (*n* in figure 2, *a*) and the long interchange chromosome (*I* in figure 2, *b*) made recognition of these two chromosomes simple and sure.

If homologous spindle fiber attachment points go always to opposite poles, only four types of spores with regard to chromosome complement would be expected, two fertile and two sterile.   Of the two fertile types, one would contain the normal chromosome complement (*N*,*n*) and one the interchange complement (*I*,*i*).   Of the sterile complements, one would possess the long interchange chromosome with the end knob (*I*) plus a normal third smallest chromosome (*N*); the other would possess the normal chromosome with the end knob (*n*) and the small interchange chromosome (*i*).   Each spore should contain, then, only one knobbed chromosome.   On the contrary, many 10-chromosome-carrying spores were seen which contained the two-knobbed chromosomes (*I*,*n*) and no normal third smallest chromosome.   Likewise, chromosome complements with no knobbed chromosome were observed.   It is obvious, therefore, that in the distribution of the four members of the ring, chromosomes possessing homologous spindle fiber attachments can go to the same pole.   Hence, there should be four types of sterile spores (*I*,*n*; *n*,*i*; *i*,*N*; *N*,*I*) besides the two fertile ones (*I*,*i*; *N*,*n*).   Since the sterility is 50%, it is assumed that in half of the sporocytes any two adjacent chromosomes in the ring go to the same pole, forming sterile combinations, and in the other half of the sporocytes the adjacent members go to opposite poles, forming fertile combinations.

No numerical relationship has been established between observed and expected microspore types because of the difficulty of analyzing all types equally well.   Two types of sterile combinations, those with the two knobbed chromosomes (*I*,*n*) and those without any (*i*,*N*) are easy to detect under the microscope, whereas, the other types (*n*,*i*; *N*,*I*) are more difficult and require better figures to be properly interpreted.   The two readily identifiable sterile types occur frequently enough to support the interpretation of anaphase *I* distribution given above.

Summary.—1.   A case of semisterility in *Zea mays* was found to be associated with a reciprocal translocation (segmental interchange) between the second and third smallest chromosomes.

2. Through observations of chromosome synapsis in early meiotic prophases of plants heterozygous for the interchange it has been possible to locate approximately the point of interchange in both chromosomes. The interchange was found to be unequal.

3. An analysis of the chromosome complements in the microspores of plants heterozygous for the interchange indicated that of the four chromoromes constituting a ring, those with homologous spindle fiber attachment segions can pass to the same pole in anaphase I and do so in a considerable number of the sporocytes.

The author is indebted to Dr. C. R. Burnham for furnishing the plants for this investigation, to Dr. L. W. Sharp for aid in the revision of the manuscript, and to Miss H. B. Creighton for assistance in the preparation of the material.

* Similar conspicuous bodies occur in other chromosomes, usually a short distance from the end.

[1] Brink, R. A., *J. Hered.*, **18**, 266–70 (1927).

[2] Brink, R. A., and C. R. Burnham, *Am. Nat.*, **63**, 301–16 (1929).

[3] Burnham, C. R., *Proc. Nat. Acad. Sci.*, **16**, 269–77 (1930).

3

Reprinted from *J. Hered.* **25**(12):465–476 (1934)

SALIVARY CHROMOSOMES AND THE ATTACK ON THE GENE

Theophilus S. Painter
University of Texas

EVER since the formulation of the chromosome theory of heredity cytologists and geneticists alike have dreamed of the day when some one would find somewhere an organism in which the chromosomes were so large that it would be possible to see qualitative differences along their lengths corresponding to the different genes which we know must reside there. With the methods and material in common use up to two years ago, the realization of this dream seemed very remote. With the best of optical equipment, very refined techniques and detailed studies by the most competent observers over the world, little was to be seen at the time of nuclear division except deeply staining rods, *v*'s or *j*'s; if earlier prophase stages were examined, knotted strands or threads. Patient work had revealed in both of these stages abundant evidence for the individuality of the chromosomes, but for any single element the landmarks were few and really of little practical value. My own work on chromosome rearrangements and the plotting of chromosome maps of *D. melanogaster*, as well as the splendid work and maps of Dobzhansky, had convinced me that if any real progress was to be made in this field we would have to use new material and possibly devise new methods from those which had heretofore been in common use. Where was this new material to be sought?

All cytologists have known for a long time that in the salivary glands and other tissues of the larvae of Diptera in general there occurs what has been called a "permanent spireme." In the nuclei of such tissues structures called chromosomes are very large and show very conspicuous bands or disks of deep-ly staining material alternating with clearer areas. (In the following paper I shall call these structures bands without prejudice as to their real nature). No one knew, however, whether these banded elements corresponded to the elements seen in the metaphase plate or were nuclear differentiations related, in some way, to the functional activity of the cells. In looking for new material for a study of chromosomes it was quite natural that I began by investigating the possibilities of the large structures found in the salivary gland of *D. melanogaster*.

Work was begun in the early fall of 1932. The first problem was to develop a technique for separating the elongated threads intertwined within the nuclear wall, so that they could be studied. This did not prove a very difficult task, indeed, we are frequently able to separate a whole element from the nucleus. (See especially Figure 1 and the photomicrograph of the right arm of the third chromosome). The chromosome consists of an elongated more or less cylindrical rod made up of lightly staining material, while running apparently across each may be seen a great variety of "bands," some broad and deeply staining, others narrow or made up of a series of dots. A study was begun at once to determine if the patterns of bands and lines, which are so conspicuous, were constant morphological characteristics of a given element. It turned out that the landmarks are constant to a most extraordinary degree so that we were able to recognize the same element in the nuclei of different individuals and ultimately to follow any characteristic bit of a chromosome as it is shifted here and there to other chro-

[*Editors' Note:* The photomicrograph of the right arm of the third chromosome and Figure 4 are not reproduced here.]

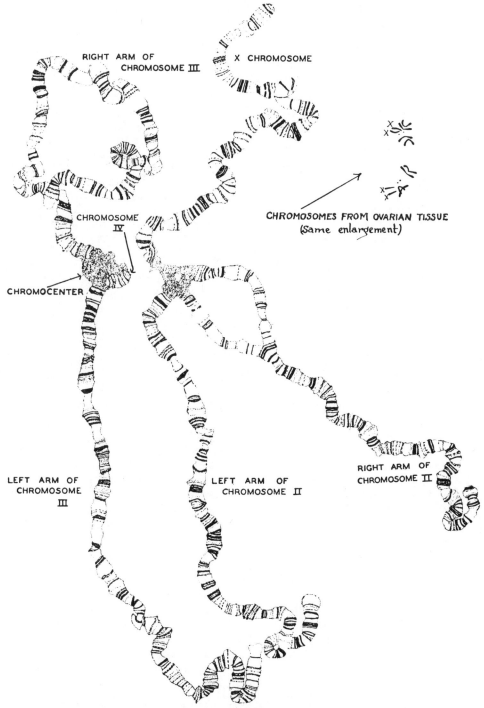

RIGHT ARM OF CHROMOSOME III

X CHROMOSOME

CHROMOSOMES FROM OVARIAN TISSUE
(Same enlargement)

CHROMOSOME IV

CHROMOCENTER

LEFT ARM OF CHROMOSOME III

LEFT ARM OF CHROMOSOME II

RIGHT ARM OF CHROMOSOME II

GIANT CHROMOSOMES COMPARED WITH "NORMAL"
Figure 1

The main figure is a camera lucida drawing of the elements found in a salivary gland nucleus of a female larva, showing the chromosomes just as they lay on the slide. The chromocenter material, to which all elements or arms are apparently anchored, is composed of inert chromatin. All homologous elements or parts have undergone somatic synapsis except

(Continued on next page)

mosomes through the agency of irradiation.

After the individual chromosomes had been studied until each element could be recognized, we were confronted by an enigma. The fruit fly has eight chromosomes consisting of four pairs of like elements. In salivary gland nuclei there are six chromosomes which are all attached to a common substrate. These six elements are individually quite distinct, five being very long and one quite short (see Figure 1). The short element, we guessed, represented the dot-like fourth chromosome of metaphase plates, but why was it single? After all, did these nuclear structures really represent the chromosomes as we know them at metaphase?

Obviously the next step was to study some chromosome rearrangement which would allow us to identify an individual element. The one selected is known as the "forked-Bar" break, discovered several years ago by my colleague, Mr. Wilson Stone. My own earlier cytological study of this break[7] had shown that the metaphase X chromosome was broken into two approximately equal parts. Female larvae homozygous for the aberration were examined and in the salivary glands one of the six elements was broken and thus we could identify the broken element as representing the X chromosome. But there were two puzzling features in this case: first, there was only one X element in the cell and, second, the break was not in the middle of the X but well towards one end. The final elucidation of this case led to two separate discoveries, both essential for an understanding of the condition existing within the salivary gland. One

is the occurrence of somatic synapsis and the other is a realization that, in salivary glands, it is the relatively active genetic part of the chromosome which shows the banded form.

The discovery of somatic synapsis was forced on me by the following considerations. The larva studied was a female and hence must carry two sex or X chromosomes. Therefore the element identified as the X must be bivalent in nature. It had been observed when studying small salivary nuclei, at the beginning of this work, that often an element showed, as I then thought, the prophase split. It was now clear that what had been seen was a step in a fusion process. Just why the forked-Bar break came so near to one end in the salivary gland did not become clear until later.

Once the X's and, by inference, the fourth chromosomes were known, by means of chromosomal aberrations, involving the other elements, the remaining chromosomes were identified. It is to be noted that in the salivary gland the two arms of the second and third chromosomes, respectively, appear as separate elements.

With these four discoveries before us (i. e., constant and distinctive patterns, somatic synapsis, the behavior of active and inactive regions and the separation of the arms of the large autosomes) it was clear that we had within our grasp the material of which every one had been dreaming. We found ourselves out of the woods and upon a plainly marked highway with by-paths stretching in every direction. It was clear that the highway led to the lair of the gene.

Legend to Figure 1 (continued)

the right arm of the second chromosome near the point where this is attached to the chromocenter. By checking one can see that the patterns of the unsynapsed parts are similar. The synapsed fourth chromosomes are somewhat bent and more or less covered by chromocenter material.

In order that the reader may compare the salivary gland chromosomes with the chromosomes of *D. melanogaster* that have been studied heretofore, in the upper right hand corner of this plate are two camera lucida drawings of these chromosomes as they appear in the ovary. When we take into consideration the inert chromatin, the salivary gland elements are over 100 times as large as the oogonial elements. Actual measurements give a ratio of 1:110

Upon consultation with my colleagues, it was decided to withhold publication until we had at least looked at the house in which the gene lives, and explored some of the by-paths. We all well knew that at the first hint of these discoveries Drosophila workers over the world would rush into this field and exploit it, and we desired that the credit for both the discovery and the orderly presentation of the main facts should come to the University of Texas. My colleagues, Dr. J. T. Patterson, Mr. Wilson Stone, and Dr. Patterson's technical assistants Miss Bedichek and Miss Suche, isolated and genetically analyzed hundreds of translocations which were placed in my hands so that really worthwhile cytological maps of Drosophila chromosomes could be presented. I wish to stress that, aside from the initial discovery, what has been accomplished here is largely the result of scientific cooperative effort in its very best sense. To my colleagues named I hereby express my obligations.

To date five papers or reports have been published by the writer and one by a student working under him. I have three articles now in press. At the invitation of the Editor of the JOURNAL OF HEREDITY it is proposed to give a summary of what has been accomplished here at the University of Texas, up to the date of this paper. If the information given is not covered by my published works I shall indicate it as in press or unpublished data.

How the Position of a Gene Is Located

The first step in the location of genes and the plotting of cytological maps is an accurate knowledge of the morphological characteristics of each chromosome or arm. This is gained by a careful study of normal chromosomes. In many regions of the chromosomes the patterns are extremely complex and often the lines are extremely fine. This being the case, the reader must realize that the figures I have given of the various elements may contain minor errors. In some cases extremely fine details have been omitted. We are dealing with labile material, and while as a whole the patterns are very constant and definite, different degrees of staining and doubtless different preservations and stains may somewhat alter the image. Certain regions of certain definite chromosomes seem somewhat variable. In my several papers I have called attention to these cases.

Once the normal morphology is known, the fact that homologous chromosomes or parts undergo an intimate union, during which they pair up line for line, band for band, is of the very greatest service in determining the position of gene loci. This process I have termed "somatic synapsis" and it must not be confused with the tendency of homologous chromosomes of all Diptera to lie together in metaphase plates, though in the end, both processes may be the expression of the same force. As a result of somatic synapsis two homologous chromosomes unite into one apparently single structure.

There are three general ways of determining the position of gene loci. Simple mutual translocations or inversions in which we know genetically between what genes the break or breaks have occurred; short deletions where we know what genes are missing; and a study of a series of breaks all falling between the same two gene loci. These will be taken up in turn.

When we study a larva heterozygous for a normal, and a broken chromosome, we find that, beginning usually at the spindle fiber ends, the normal and aberrant elements synapse line for line, band for band, up to or nearly to the point where the latter is broken, where the aberrant element (if we are dealing with a mutual) diverges. By comparing the adjacent patterns of the two elements it is easy to determine where the break has occurred. Knowing which two genes are separated by the break, we can

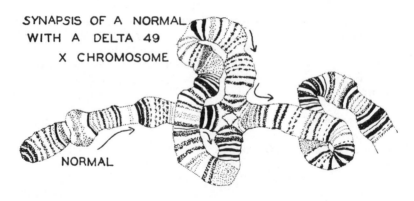

SYNAPSIS OF A NORMAL WITH A DELTA 49 X CHROMOSOME

NORMAL

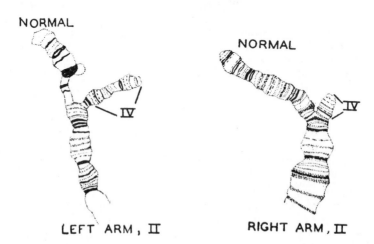

NORMAL

NORMAL

IV

IV

LEFT ARM, II

RIGHT ARM, II

SYNAPSIS OF NORMAL AND INVERTED CHROMOSOMES
Figure 2

The upper figure is a camera lucida drawing, taken from a female larva, which was heterozygous for a normal and a "delta 49" X chromosome. Synapsis has been complete except at the points where the delta 49 X is inverted. The extreme right hand ends of the paired X's are not shown.

The two lower figures are camera lucida drawings to illustrate the way the point of breakage of a chromosome is determined. In each example, the normal element lies on the left and the aberrant one to the right. The aberrant element synapses with the normal right up to the point of breakage. These two cases involve a mutual exchange with a fourth chromosome. The latter is indicated in the drawings.

place these to either side of the point of breakage. Figure 2 illustrates cases of this sort, in addition a figure of an inversion (delta 49) in the X chromosome is shown. This method is relatively crude; nevertheless, by using it I was able to circumscribe a number of genes along the X chromosomes within the confines of an area covered by two or three bands. It was realized very early that short deletions would be more serv-

iceable for the exact location of genes, and fifteen months ago, Mr. Mackensen, working under my direction, began a study of short deletions. The method was clearly indicated in my abstract given to the American Society of Zoologists and appeared early in December, 1933. At the meetings of the Genetics Society of America, 1933, Mr. Mackensen gave in his abstract both the method and his results up to that time.

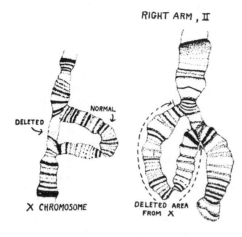

RIGHT ARM, II

NORMAL

DELETED

X CHROMOSOME

DELETED AREA FROM X

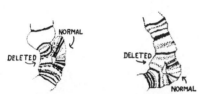

NORMAL

DELETED

DELETED

NORMAL

SYNAPSIS OF DELETED AND NORMAL CHROMOSOMES
Figure 3

These are camera lucida drawings showing the appearance of deleted and normal X chromosomes after they have undergone somatic synapsis. The upper left hand figure shows the normal and deleted X, while the figure to the right shows the deleted area inserted into the right arm of the second chromosome.

The lower figure shows short sections of normal and deleted X's after somatic synapsis.

When a normal and a deleted chromosome synapse they usually pair throughout their lengths except in the region of the deletion, where the normal element usually buckles. Figure 3 illustrates cases of this type, drawn by Mr. Mackensen and used by me with his consent. It is very fortunate that as a result of irradiation, as my colleague Dr. Patterson has shown,[6] short deletions are extremely common; knowing then what genes are missing and what bands are lacking we can say that the missing genes lie somewhere in the deleted area. We had in our stocks a number of dele-

tions isolated by Dr. Patterson. These were given to Mr. Mackensen for study and since then Mr. Mackensen has isolated many cases of his own.

The third method is to study a number of breaks which have occurred between genetically adjacent gene loci. It has been pointed out elsewhere by my colleagues[17] and in my study of the third chromosome (in press) that translocations between the fourth chromosome and the large autosomes are often bunched within narrow limits both morphological and genetic. Mr. Mackensen has isolated two deletions which removed the locus of vermilion, these overlapped a little so that Mackensen could restrict the locus of vermilion to the left part of a broad compound band. (See Figure 4, X chromosome). Mr. Mackensen's study is now in press.

By the simple methods outlined, cytological maps of all the chromosomes of *D. melanogaster* have been produced, and in Figure 4 the maps, as they stand at the moment of writing, are reproduced. Especial attention is called to the map of the X chromosome. Above this figure lines show breaks which have allowed me to restrict the positions of loci. Below the figure I have indicated further localization of specific genes by Mr. Mackensen. Full credit is due Mr. Mackensen for this work, which is now in press. It will be seen that there are large uncharted areas along all of the maps of the autosomes, but this laboratory proposes to concentrate its efforts for the next year or two in filling in the gaps. Other investigators will, it is hoped, pursue the same line of work so that ultimately we shall know the locus of every known gene.

Where Are the Genes?

The methods have been described by which it has been possible to restrict the morphological position of gene loci to the area covered by one or even a part of one of the deeply staining bands on the X chromosome. Where then are the genes? Are they represented by the deeply staining material or by some other part of that region of the chromosome? To answer these questions we are led to a consideration of one of the oldest problems in cytology, namely, the ultimate structure of these cross bands and of the chromosomes. At this point I wish to state that while I have used the term "band" in all of my papers, I was very careful to state in my article in *Science*[9] that these cross striations "appear to run around an achromatic matrix." As to the ultimate nature of these "bands," I have not been, nor am I now, willing to commit myself finally.

Earlier Work

So far in this account little mention has been made of the work of other investigators because we have been utilizing and interpreting morphological structures in terms of modern genetics. The application is new, but much of the morphological background has been known for a long time. Several investigators have examined the chromosomes of *D. melanogaster*. In this JOURNAL, Kostoff published a photograph of these discoidal chromosomes in 1930, and in 1931 Kaufmann described these elements. Both of these studies were mainly morphological and did little more than show that in *D. melanogaster* the conditions were much the same as had been described for other dipteran chromosomes.

As I was in the midst of my first year's work, an article appeared by Heitz and Baur[2] dealing with the salivary chromosomes of *Bibio hortulanus*. There are 10 chromosomes in this fly but in the salivary glands there are only five. Each of these five is made up of two elements which, as the figures show, are twisted together. At times the union appears more intimate. Heitz and Baur did not regard this union as of any particular importance, apparently. However, they did make it clear that partner strands were morphologically alike in the arrangement of the bands, and they state that each of the five groups of

chromosomes can be recognized from cell to cell on the basis of length, association with a plasmosome, the shape of the ends and the pattern of the bands. No genetic application of these facts was made by them.

Again in a paper[3] which appeared later in 1933 Heitz figured parts of the chromosomes as they appear in the salivary gland of *D. melanogaster,* but this work need not concern us here.

Balbiani in 1881 was the first to describe the banded chromosomes in the salivary gland nuclei of the larvae of the Chironomus fly. Within these nuclei he found a long thick thread which showed throughout its length cross striations or bands. Balbiani interpreted these cross striations as disks of deeply staining material and he conceived of the chromosomes as being made up of disks of chromatin and what he called a "zwischensubstanz" or in modern terms, matrix or plastin.

At approximately the same period Leydig (1883) also worked on larvae of Chironomus; he saw the cross striations on the chromosomes, but he interpreted their structure and that of the chromosomes in quite a different way. According to Leydig "the cross striations are restricted to the periphery of the thread cylinder without cutting through it or having connections through it." (The writer's translation of a passage cited by Alverdes[1] p. 173). He describes the cross striations as showing shallow indentations "put together out of small sticks comparable to the elements of muscle disk." "The finer dividing lines of the dark striations formed by the little sticks are stretched out through the light between zone (zwischenzone) so that we get the impression of faint longitudinal lines being present." (The writer's translation of passage cited by Alverdes[1]).

In 1884 Carnoy described the presence of cross striations in a number of Arthropods. He conceived of the chromosome as being made up of a hollow cylinder of plastin on the inner wall of which the chromatin was deposited. When the chromatin is localized in definite areas, the cross striation effect is produced. Carnoy saw the chromatin of the dark areas divided into short segments and granules. In some forms Carnoy found that the chromatin lying on the inner wall was arranged in a corkscrew fashion, so that a spiral thread of chromatin ran from one end of the cylinder to the other.

In 1884 Korschelt published upon the structure of the chromosomes in Chironomus larvae and he visualized the condition observed in this way: The chromosome is not made up of alternate layers of material but the cross striations are due to an optical effect. There are ridges and valleys running across the chromosome and with transmitted light these give the impression of light and dark bands.

After this early period, comparatively little was added to our knowledge of the banded chromosomes until 1910 when van Herwerden took up Chironomus again. She thought the cross striations were due to the presence of a spiral band or rod of chromatin running around an inner axis.

Shortly after this Erhard (1911) studying the same type of material concluded that the chromosome was made up of alternate disks of light and deeply staining material.

The next study which requires our attention is that of Alverdes[1] in 1912. (References to work cited above are given by Alverdes). As we have seen there were two general explanations for the crossbands on Chironomus chromosomes, one was that they were due to disks or rings of chromatin separated by relatively clear material, and the other was that there was a spirally wound chromatic thread running in some way from one end of the chromosome to the other, and that the banded form was the result of viewing the imperfectly preserved spiral from the side. Alverdes proposed to settle this controversy by studying the way in which the crossbanded chromosomes arose in the ontogeny of the individual. He found in the earliest stages of the salivary gland that the chromosomes showed the usual chromomeric structure, that is, were made up of little knots of chromatin lying along a thin thread. As the nucleus enlarged, the achromatic matrix increased in thickness and adjacent chromomeres sent out processes and ultimately fused so that in the end a typical spiral chromonema thread was produced. Subsequently, these chromomeres, or at least parts of the chromatic spiral, separated forming heavy disks which by subdivision later formed narrower disks which frequently appeared as a series of more or less connected dots running around the chromosome. The numerous detailed drawings which Alverdes gives of this period show a large number of lines or striae running through the clearer areas between the more darkly staining landmarks. Thus it will be seen that Alverdes would explain the various interpretations given for the structure of Chironomus chromosomes as being due largely to the age of the larvae studied.

Kaufmann[4] has discussed the probable relation between the discoidal bands on the salivary chromosomes of D. melanogaster and the chromonema theory of chromosome structure and he concludes "that the appearance of the discoid or discontinuous structure is the expression of an inadequate technique, the spiral bands presenting a more accurate picture of the normal arrangement of the chromatic material." (p. 556).

The Structure of the Salivary Chromosomes

This was the status of the question of the nature of the cross-bands, principally in Chironomus, when I began work. The picture before me, in D. melanogaster was so like what the earlier workers had described that there was little doubt in my mind as to their ontogenetic and structural similarity. Although I have used the word "bands" to describe the deeply staining land-marks, the most superficial examination of a lightly stained chromosome shows the bands to be made up of smaller parts. Any of my drawings, except the schematic ones, will show much evidence of this. Even the heaviest bands have crenelated edges just as Leydig described, and the lighter ones appear to be made up of from five to seven smaller areas running dash-like across the uppermost surface of the chromosome, making a total of more than double these numbers. This appearance is best seen on the fourth chromosome. In my paper which appeared in May, 1934, I called attention to the latter fact, and stated that in a later article I would take up the question of the nature of the chromosome, in detail.

As a result of my observations on somatic synapsis I knew that the apparent single chromosomes were composite in nature and represented the union of at least four similar strands very closely apposed to each other. (The two homologues which unite in somatic synapsis often show in turn prophase splits). It therefore seemed to me that little more was to be gained by a direct study of these cross structures in old larvae. Who can say whether we are dealing with disks or with rings, because four closely apposed rings viewed from the side would look like disks running through the composite chromosome. The safer avenue of approach appeared to be the ontogenetic. When we know how the constituent parts of the simple unsplit and unsynapsed elements arise, then we can more certainly interpret what we see in the cross bands of old larvae. This is what I had in mind when I wrote[12] in the summer of 1933: "Foremost stand questions about the intimate structures of chromosomes, whether the chromatic bands lie on or within the matrix? What part carries the genes? What is the relation between the genes and the chromatic substance?

and a host of similar questions. From the first, such questions have dominated our interest in these studies; but as tempting as it may be to speculate upon them, at present it seems unprofitable to consider these points until we know more about the genesis and ontogenetic history of these peculiar chromosomes."

In the fall of 1933 I set one of our graduate students, Mr. Sigmund Hayes, to work on the early history of these peculiar chromosomes. Our preliminary study was brought to a close this past summer. As far as we have been able to determine, we do not have in *D. melanogaster* the series of complicated changes in the chromosomes described by Alverdes for Chironomus. The earliest stages show small chromomeres lying along a relatively achromatic thread. As the nuclei increase in diameter, the threads hypertrophy and we can see more and more details. Whether these details become visible just because of their size, or whether the chromomeric thread is undergoing changes is not known. I rather expected to find that the chromomeres of the earlier stage would prove to be "compound chromomeres," in the sense that Belling used this term, and that these would spread out, in some way, along the matrix giving the final pattern, and that what Alverdes had seen were steps in this process. A great deal needs to be done and the nature of the problem requires the best optics and a very intensive study of minute details.

Mention has been made in the popular press recently of the work of Dr. Calvin B. Bridges in this field based on a lecture given at Woods Hole, Mass. Since some of the short accounts are misleading, out of fairness to Dr. Bridges, and to myself the following statement is made with the knowledge of Dr. Bridges and based on a personal letter from him. Following my lead, Dr. Bridges took up a study of salivary chromosomes in December, 1933. He confirms my work on the constancy of pattern down to the finest detail and has directed his efforts among other things along three lines: (1) By studying chromosome aberrations, he has been able to localize three genes in the right arm of the second chromosome to an area covered by four bands. (2) He has made a detailed census of the lines on the X and fourth chromosomes. (3) From a detailed study of the cross bands he concludes that they are made up of unitary components apparently definite and usually sixteen in number. The dots and dashes visible as light lines are arranged longitudinally in spiral rows so that we have the effect of a twisted cable made up of 16 strands.

It would be out of place to discuss the work of Dr. Bridges in advance of his paper (in press in U. S. S. R. and a Carnegie report), which it is hoped will not be long delayed in publication.

As the final draft of this paper was being typed a short article by Koltzoff has come to hand (*Science,* Oct. 5, 1934), giving his views on the structure of the salivary chromosomes of *D. melanogaster*. Like Bridges, Koltzoff thinks that these chromosomes are made up of 16 strands running spirally the length of the chromosome. Without going into the question at this time, it may be remarked that the morphological observations on which both men, apparently, base their ideas of chromosome structure are much like those reported by Leydig somewhat more than half a century ago.

To return now to the question of where the genes are, let me emphasize that so far we have been able to show that they reside in that area of the chromosome covered by one or a part of one band. We have no definite proof that they are a part of any of the structures which have been seen in these bands: they might lie in the matrix. But since the bands are obviously particulate and show a complex structure, it seems more probable that the genes are to be sought here. But after all, as the Editor of *Science News Service* aptly expresses it: "Whether these tiny units are the genes themselves, or only the 'genophores' or gene bearers, is a matter of relative unimportance. The important thing is that they have been unveiled, so that the searching finger of science may probe a little further into the secrets of life." (Sept. 29, 1934, p. 196.)

By-Paths

We have stressed, in the foregoing pages, how the salivary chromosomes have allowed us to trace the genes to definite regions of the chromosomes, and this is, perhaps, from a theoretical point of view, the outstanding feature of the work to date. There are, however, a number of both related and unrelated problems which can be attacked by this new method. Some of these will be briefly discussed.

Because the new cytological maps are really pictures of the several chromosomes with labels showing the approximate or exact location of many genes they can be used very advantageously for the study of all types of chromosome rearrangements. These maps really take the guess-work out of a genetic study of chromosome aberrations and give us a very much more accurate picture of what has occurred than could be obtained by any but the most exhaustive breeding tests. This is true of simple mutual translocations and especially of complex rearrangements which are proving much more numerous than was first suspected. Once a translocation or deletion is isolated, a cytological examination of heterozygous larvae allows us to read right off just where the break is with reference to known gene loci, or what parts (and genes) have been deleted. Ordinarily, this can be done within an hour after the larvae is killed, often in much less time. The practical value of the new maps for this type of work can scarcely be overestimated.

Again, as a by-product of the X-ray work, is the realization that in many cases the formation of new species is preceded by some type of chromosome rearrangement which sets up a physiological isolation within the parent species. By crossing *D. melanogaster* with *D. simulans,* for example, and examining the salivary chromosomes of the hybrid larvae, we can tell at once just what rearrangements have occurred. In this way, we are now preparing a full set of maps of *D. simulans* here at the University of Texas. The possibilities of this type of work seem unlimited. Perhaps equally valuable, from the evolutionary standpoint, is the fact that with the salivary chromosome method we can make a careful study of the effects of various aberrations on the individual, e. g., the duplication of short pieces.

Spacing of the Genes

Another field upon which this new method has thrown light is crossing over. Thanks to somatic synapsis we can actually see how the chromosomes and genes pair in diploid and triploid cells and we now enter an era of testing out some of the speculations with which this general field is overrun. One example will suffice to illustrate what can be done. Long ago, Dr. H. J. Muller called attention to the bunching of the genes along the chromosomes. In one case at least, this seems to have a simple explanation. In the X chromosome (see Fig. 4) from crossveinless on the right, the crossover map and the cytological map agree rather closely. But to the left of crossveinless the genes tend to be much farther apart physically than they are on the crossover map. The explanation would seem to be simple. As we approach the left hand (free) end of the X there is actually less crossing over per physical unit than in the middle (due possibly to the fact that the strands slip out to the end rather than break) and so the bunching of genes is not real but only the effect of reduced crossing over. The same method of approach has been used to explain other cross-

over phenomena in various papers now in press by my colleagues and by myself.

A study of the second and third chromosomes has led to the discovery that both of these elements carry relatively inert material in the region of the spindle fiber. We have been able to confirm and prove, by genetic and cytological methods, the brilliant work of Heitz in this field. A good deal of the material which we see in the metaphase chromosomes does not carry any known active genes. (For full discussion of this topic, see my study on the third chromosome, now in press.)

Finally, mention should be made of the work now in press (Painter and Stone) dealing with the fusion of the X and fourth chromosomes. It has been found that about half of the translocations between the X and fourth chromosomes are really fusions at or very close to the point of spindle fiber attachment. Similarly, many translocations between the second (or third) and fourth chromosomes are simple exchanges by which one whole arm is replaced by a fourth, and the displaced arm persists as an independent element. The theoretical aspects of these cases (see paper in press) have a broad bearing not only on the more general problem of chromosomes and speciation but on the nature of the mitotic spindle itself.

In this review it has been possible to touch only the high spots of what has been accomplished since this new lead was taken up and the salivary chromosomes made available for study. New vistas open before us on every hand. But after all is said and done we must realize that the utilization of these giant chromosomes for solving many vexing problems of cytology and genetics is comparable to the forging of a new tool and the important thing is to *use* this tool. Like the X-ray, it is a new method of attack and, while we can scarcely hope that the results which may follow will have as profound an influence on biology as Dr. Muller's discovery, it is nevertheless clear that cytology, as a science, enters a new era in its development.

Literature Cited

1. ALVERDES, F. Die Kerne in den Speicheldrüsen der Chironomus-Larve. *Arch. f. Zellforsch.* 9: 168-204, 1912.

2. HEITZ, E., and BAUR, H. Beweise für die Chromosomennatur der Kernschleifen in die Knauelkerner von *Bibio hortulanus. Zeitsch. f. Zellforsch. u. mikr. Anat.* ₁17: 67-82, 1933.

3. HEITZ, E. Die somatische Heteropyknose bei *Drosophila melanogaster* und ihre genetische Bedeutung. *Zeitsch. f. Zellforsch. u. mikr. Anat.* 20: 237-287, 1933.

4. KAUFMANN, B. P. Chromosome Structure in Drosophila. *Amer. Nat.* 65: 555-558, 1931.

5. KOSTOFF, D. Discoid Structure of the Spireme. *Jour. Hered.* 21: 323-324, 1930.

6. MACKENSEN, O. A Cytological Study of Short Deficiencies in the X Chromosome of *Drosophila melanogaster. Amer. Nat.* 67: 76, 1934.

7. MULLER, H. J., and PAINTER, T. S. The Differentiation of the Sex Chromosomes of Drosophila into Genetically Active and Inert Regions. *Zeitsch. f. ind. Abs. u. Verer.* 62: 316-365, 1932.

8. PAINTER, T. S. A method for the Qualitative Analysis of the Chromosomes of *Drosophila melanogaster. Anat. Rec. Sup.* No. 4, 57:90, 1933.

9. PAINTER, T. S. New Method for the Study of Chromosome Rearrangements and the Plotting of Chromosome Maps. *Science* 78: 585-586, 1933.

10. ———— A New Type of Cytological Map of the X Chromosome in *Drosophila melanogaster. Amer. Nat.* 68: 75-76, 1934.

11. ———— A New Method for the Study of Chromosome Aberrations and the Plotting of Chromosome Maps in *Drosophila melanogaster. Genetics* 19: 175-188, 1934.

12. ———— The Morphology of the X chromosome in Salivary Glands of *Drosophila melanogaster* and a New Type of Chromosome Map for this Element. *Genetics* 19: 448-469, 1934.

13. ———— The Morphology of the Third Chromosome in the Salivary Gland of *Drosophila melanogaster* and a New Cytological Map of this Element. In press.

14. PAINTER, T. S., and WILSON STONE. Chromosome Fusion and Speciation in Drosophila. In press.

15. PAINTER, T. S. The Morphology of the Second Chromosome in the Salivary Gland of *Drosophila melanogaster* and a New Cytological Map of This Element. Ready for press.

16. PATTERSON, J. T. Lethal mutations and Deficiencies Produced in the X Chromosome of *Drosophila melanogaster* by X-radiation. *Amer. Nat.* 64: 193-206, 1932.

17. PATTERSON, J. T., et al. The Production of Translocations in Drosophila. *Amer. Nat.* 68: 359-369.

4

Reprinted from *Science* **83**(2148):210–211 (1936)

THE BAR "GENE" A DUPLICATION

Calvin B. Bridges

California Institute of Technology

THE nature of the Bar gene has been the subject of extensive investigation and speculation since February, 1913, when Tice[1] found this reduced-eye mutant as a single male in the progeny of normal-eyed parents. The eye-reduction behaves as a sex-linked dominant, with a locus at 57.0, and has been one of the most important of all the sex-linked characters of *D. melanogaster*. A remarkable peculiarity of the mutant is that occasionally the homozygous stock gives rise to a fly indistinguishable in appearance and genetic behavior from wild-type.[2] More rarely the stock gives rise to an even more extreme reduction in eye-size, a type which was called Ultra-Bar by Zeleny,[3] who found it.

Sturtevant and Morgan[4] and Sturtevant[5] found that these two-way changes were the result of a novel type of "'unequal" crossing-over, by which the two genes originally present in the two parental chromosomes both emerged in the same chromosome (Bar-double) while the other resultant chromosome was without Bar (Bar-reverted). The change from Bar to Bar-double was considered to be a single gene duplication, while the converse change, from Bar to Bar-reverted, corresponds to a one-gene deficiency. Since the Bar-reverted type proved to be indistinguishable from the normal unmutated wild-type, the gene present in Bar and lost in Bar-reverted must have itself correspond to a new addition or one-gene duplication.[6]

Sturtevant[5] found the unexpected relation that two Bar genes in the same chromosome (BB/B⁺) gave a greater reduction in the size of the Bar eye than did two Bar genes in opposite chromosomes (B/B), an intensification of action which he formulated as a "position effect." Dobzhansky[7] interpreted his allelic Baroid mutant as a position effect due to the substitution of material at or near the Bar-locus (in the normal X) by material translocated from the right limb of chromosome 2, and the reduction in the Bar eye to the interaction between a gene in the X chromosome and the duplication.

A chance to clear up some of the puzzles as to the origin and behavior of Bar was offered by the salivary chromosomes. Study of the banding in a stock of Bar (forked Bar) showed that an extra, short section of bands is present in excess of the normal complement, forming a duplication. The insertion point of this duplication is in the bulbous "turnip" segment, not far from the basal end of the X.[8]

The exact point of the insertion is ambiguous, for a reason which will appear below. The normal X in this region (see revised map in Fig. 1) shows in sub-sec-

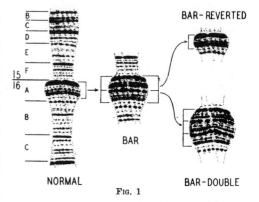

FIG. 1

tion 16A a heavy band, which in well-stretched chromosomes, or with certain fixations, is a clear doublet, usually with the halves united in a capsule, but occasionally completely separate. This is followed by a very faint dotted line, which can be seen only in the most favorable conditions. Next follows a fairly weak line which is distinctly "dotted" in texture, with the separate dots loosely connected across the width of the chromosome. Next follows closely a still fainter, diffuse, continuous-textured doublet, with the doubleness generally appearing as mere broadening. The last line of sub-section 16A is again a very faint dotted singlet. Sub-section 16B starts with a sharply discontinuous line of fairly heavy dots or vesicles and is a line very easy to recognize. The greatest width of the bulbous segment 16A is at the two fairly weak bands, while a very sharp change in size occurs at the transition from 15F to 16A.

In the Bar chromosome the condition may be described observationally as the repetition of section 16A, with the exception of the final very faint dotted line. But the whole region of this bulb has undergone changes in the Bar chromosome as follows: the "puff" of the bulbous segment is more pronounced and its size is increased; the banding is more discontinuous by being broken into blocks and vesicles, and the regularity of synapsis is disturbed by oblique junctions. Thus, in Bar the heavy doublet following the last faint dotted line of 15F is more segmented than normal and more rarely shows its doubleness clearly. This tendency is more pronounced in the heavy broken line of

[1] S. C. Tice, *Biol. Bull.*, 26: 221–51, 1914.
[2] H. G. May, *Biol. Bull.*, 33: 361–95, 1917.
[3] C. Zeleny, *Jour. Exp. Zool.*, 30: 293–324, 1920.
[4] A. H. Sturtevant and T. H. Morgan, SCIENCE, 57: 746–7, 1923.
[5] A. H. Sturtevant, *Genetics*, 10: 117–47, 1925.

[6] S. Wright, *Amer. Nat.*, 63: 479–80, 1929.
[7] Th. Dobzhansky. *Genetics*, 17: 369–92, 1932.
[8] C. B. Bridges, *Jour. Hered.*, 26: 60–4, 1935.

the repeat seriation to the right. All the lines of the repeat seriation to the right differ from the corresponding lines of the initial seriation by being somewhat less intense, more broken, more diffuse and more confused in their synapsis relations.

In a forked non-Bar stock recently derived from the above forked Bar stock by breeding from the rare Bar-reversions, the banding was found to be precisely identical with that of unrelated normals as far as could be observed in excellent permanent preparations of well-stretched chromosomes.

In a forked Bar-double stock, similarly derived from the same f B stock by breeding from the very rare "Ultra-Bar" type of eye, it was found that the extra section observed in Bar was present still again, giving a thrice-repeated seriation in direct sequence. The changes differentiating Bar from normal were carried further in Bar-double, as follows: The size and puffiness of the bulbous regions was still greater, as well as the blockiness of the banding and irregularity and obliqueness of the synapses. These disturbances were greatest in the middle one of the three seriations.

These findings enable the Bar "gene" to be reinterpreted as a section of inserted genes—a duplication. The production of Bar-double and of Bar-reverted is seen to be the insertion of this extra section twice, or conversely, its total loss—both presumably by a process of unequal crossing-over. That the section of bands should behave as a unit in this process is perhaps accounted for by the observation of oblique synapsis, especially frequent in Bar-double, where presumably one entire sequence synapses with another of a different position in the series of three. The oblique synapses were even more frequent in BB/B⁺, where one series in B⁺ has a choice of three series in BB, apparently usually synapsing with one or the other end series.

According to this interpretation the source of the duplication is the material directly adjacent to the repeat. But whether the point of insertion preceded the heavy doublet of 16A1 or the very faint final singlet of 16A5, can not be determined. If Bar is itself a repeat, a reason is thereby provided for its unique behavior of giving rise to Bar-double and Bar-reverted by oblique synapsis. Perhaps half of the Bar-reversions carry the original series and the other half the subsequent repeat restored to its original position.

On this interpretation, the "position effect"—the reinforcement of the action of one Bar gene by another in direct sequence next to it—has a visible cytological accompaniment in the increased size and puffiness, and the change in the character of the banding of both series in Bar as compared with normal and of all three series in Bar-double as compared to Bar itself. Part of this is presumably due to the "rounding-up" tendency of the synaptic attraction *along* the chromosome in addition to the oblique attractions and the straight-across attractions.

The Bar-eye reduction is thus seen to be interpretable as the effect of increasing the action of certain genes by doubling or triplicating their number—a genic balance effect. But "position effects" are never excluded when duplications or other rearrangements are present, either in the wedging further apart of genes normally closer, or by the interaction with new neighbors. The respective shares attributable in the total effect to the genic-balance change and to the position-effect change seems to be at present a matter of taste.

Study of the Baroid translocation apparently shows that the break in X comes between the two halves of the heavy doublet of 16A1. The break in 2R follows directly after the heavy capsular doublet of 48C1. Thus a demonstrable basis is laid for Dobzhansky's interpretation of the Baroid eye-reduction as a position effect.[7]

The previously reported finding[8] of the presence of "repeats" as a normal part of the chromosomes of *D. melanogaster*, and the suggestion that unequal crossing-over is probably the mechanism of production of some short repeats, thus have received ample verification by these direct observations on these processes in the case of Bar and its derivatives.

5

Reprinted from *J. Biophys. Biochem. Cytol.* **2**(4), Pt. 2:385–390, 392 (1956)

A STUDY OF CHROMOSOMES WITH THE ELECTRON MICROSCOPE*

By HANS RIS, Ph.D.

(From the Department of Zoology, University of Wisconsin, Madison)

PLATES 130 AND 131

"The central problem of biology is the physical nature of living substance. It is this that gives drive and zest to the study of the gene, for the investigation of the behavior of genic substance seems at present our most direct approach to this problem" (Stadler (14)). This genic substance is localized in the chromosomes. The morphological, chemical, and physiological investigation of chromosomes is therefore one of the main approaches to the nature of the genic substance. Lately the chemical analysis of chromosomes has provided interesting information (1). Most significant is perhaps the behavior of the deoxyribonucleic acid (DNA) and the non-histone protein fraction. The DNA is present in constant amounts per chromosome set of a species, it is remarkably stable biochemically, and has a very interesting molecular structure which has suggested a useful model for speculations on the process of replication of genes and chromosomes (15). The non-histone protein and ribonucleic acid (RNA) on the other hand vary greatly in amount and this variation is closely related to the physiological activity of the nucleus. Ultimately we want to describe the behavior of chromosomes and genes in terms of these molecules. However, there is still a large gap between the level of the DNA molecule and that of the chromosome as we know it, a fact too often neglected by those who have tried to explain chromosome behavior with the Watson-Crick model of DNA. To help fill this gap we need information on chromosome organization that may be obtained with the electron microscope.

During mitosis, when they can be clearly recognized, chromosomes are generally too thick for electron microscopy. Suspensions of interphase chromosomes mechanically isolated from vertebrate tissues are not suitable, since it is difficult in the electron microscope to distinguish chromosome material from contaminations and it is not known how the isolation procedures affect the structure of chromosomes. Thin sections on the other hand are very difficult to interpret by themselves. We found that lampbrush chromosomes dissected from oocyte nuclei of amphibians and pachytene chromosomes of insects and certain plants are favorable material. They can be followed from the light

* Supported in part by the Research Committee of the University of Wisconsin Graduate School from funds supplied by the Wisconsin Alumni Research Foundation.

microscope level into the electron microscope and pictures of whole chromosomes or at least short regions of them can be obtained (8–10). These prophase chromosomes consist of bundles of microfibrils. The microfibrils are helically coiled (in lampbrush chromosomes) or twisted less regularly, and the bundle in turn forms a helix which corresponds to the chromonema visible in the light microscope. The microfibrils are from 500 to 600 A thick in all prophase chromosomes studied and are split into two finer fibrils which are about 200 A thick. Fragments of interphase nuclei were also obtained in Feulgen-squash preparations and were seen to contain twisted fibrils about 200 A thick. For a detailed study and higher resolutions such preparations are not suitable however, and one has to resort to thin sections. In the following I should like to give a brief report on work now in progress on the organization of chromosomes in interphase nuclei and at various stages of mitosis and meiosis.

Material and Methods

Anthers of *Lilium longiflorum* were chosen for a study of meiotic chromosomes. The development of sporogenous cells from premeiotic stages to mature microspores is synchronous in all anthers of a bud (4). One anther was analyzed in the light microscope and the others fixed for 1 hour in 1 per cent OsO_4 buffered with acetate-veronal to pH 7.5, embedded in methacrylate, and sectioned with a Porter-Blum microtome. Anthers of *Tradescantia paludosa*, onion root tips, oocytes of *Triturus viridescens*, and testis of the rat were fixed and sectioned in the same way. Small pieces of onion root were fixed also in 10 per cent neutral formalin and treated with deoxyribonuclease (1 mg./cc. in 0.003 M $MgSO_4$ at 40°C. for 2½ hours) (7). A sample was stained with the Feulgen reagent and the nuclei found to give a negative reaction. The material was then fixed in buffered OsO_4, embedded in methacrylate, and sectioned.—In some of the sections the methacrylate was removed with amylacetate and the preparation shadowed with uranium.

The micrographs were taken with an RCA electron microscope type EMU-2 on Ilford process plates N.40 and developed in D-19.[1]

OBSERVATIONS

Intact Leptotene Chromosome.—Part of a leptotene chromosome from a formalin-fixed Feulgen-squash is shown in Fig. 1. The chromosome is made of fibrils, about 500 A thick. In a few places the double nature of these fibrils is visible (arrow). The thicker regions labelled *ch* correspond to the chromomeres of the light microscopist. They apparently contain the same kind of fibrils as the interchromomeric regions, but seem more closely coiled and packed together. The exact number of fibrils per chromosome is not known, but appears to be around 8 in some interchromomeric regions. While it cannot be proven that these fibrils are continuous for the length of the chromosome, no other continuous structures have been found. Since genetic data and the longitudinal

[1] The shadow casting and electron microscopy were done in the Electron Microscope Laboratory of the University of Wisconsin. I wish to thank Dr. P. Kaesberg for his advice and practical help.

division of chromosomes require some continuous unit through the length of the chromosome it is assumed at present that these fibrils are the continuous subunits of chromosomes and that their number is a multiple of two.

Sections of Meiotic Prophase Chromosomes.—Knowing what the intact free chromosome looks like in the electron microscope, it becomes possible to interpret thin sections through entire nuclei. We expect to find random sections through bundles of coiled microfibrils. They should be about 200 A thick and in prophase chromosomes two of them should be closely associated into 500 A fibrils.

Fig. 6 shows a section through a leptotene nucleus of lily. Instead of the expected 200 A fibers we see much thinner lines, about 60 A thick, but they occur generally in pairs of parallel lines. The width of the double unit is 200 A. In addition to the double lines we find circles with a diameter of 200 A (Text-fig. 1 *A*). Similar structures are visible also in sections through metaphase chromo-

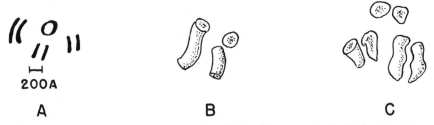

A **B** **C**

Text-Fig. 1. Diagrammatic representation of the chromosomal microfibrils in sections. *A*, methacrylate not removed. *B*, methacrylate removed, shadowed with uranium, somatic interphase. *C*, methacrylate removed, shadowed, prophase.

somes (Fig. 7). In order to decide whether we are dealing with double fibrils or cylinders with the osmium deposited on the outside only, the methacrylate was removed and the sections shadowed with uranium. In Fig. 4 we see that after shadowing the 200 A fibrils have a solid contour and there is no evidence of any split. Where the fibrils are cut across, the sections look like doughnuts, the center being less electron-dense than the outer part of the fibrils. Since Fig. 4 represents a section through a late prophase chromosome we find always two of the elementary fibrils running parallel and closely associated (Fig. 4, brackets, Text-fig. 1 *C*). These pairs correspond to the 500 A fibrils seen in the intact leptotene chromosome (Fig. 1). While this association in pairs is clearly visible after shadowing, it is less obvious in micrographs of unshadowed preparations with their confusing array of black lines.

Somatic Interphase Nuclei.—Sections through somatic interphase nuclei from lily anthers show similar double lines and circles (Text-fig. 1 *A*). After shadowing they appear as random sections through coiled fibrils 200 A in thickness. These fibrils again are less electron-dense in the center with a denser outer region (Figs. 2 and 3, arrows, Text-fig. 1 *B*). In the interphase nucleus the microfibrils are spaced farther apart than in mitotic chromosomes where they appear to be as closely packed as possible (compare Figs. 2 and 4). The spacing between the microfibrils is affected by fixation. Fixatives which produce a coarse nuclear structure cause the microfibrils to stick together in clumps. After fixation with osmium, which preserves the life-like optically homogeneous appearance of nuclei, the microfibrils are separated by a space about equal to their width or larger. The shrinking and swelling of chromosomes (12) appear to be due largely to the closer or wider spacing of the microfibrils. Between the chromosomal areas there are irregular spaces which contain scattered dense granules and rodlets. Their significance is not known.

Interphase Nuclei of Other Tissues.—The pattern described for nuclei of the lily is found also in other tissues. Interphase nuclei from onion roots, from somatic cells in anthers of *Tradescantia* (Fig. 5), from thecal cells in the ovary of *Triturus*, lampbrush chromosomes in oocyte nuclei of *Triturus* (Fig. 8), and spermatid nuclei of the rat (Fig. 9) show the double lines and circles in sections with the methacrylate left in and segments of microfibrils 200 A thick after shadowing. These microfibrils seem to be the universal and basic component of chromosomes in both animal and plant cells.

Appearance of Microfibrils after Treatment with Deoxyribonuclease.—Sections through onion root nuclei treated with deoxyribonuclease show pieces of microfibrils that are not different from those of control sections. The removal of DNA from the microfibrils therefore does not visibly alter their structure.

DISCUSSION

In previous publications (8–10) we have shown that amphibian lampbrush chromosomes and pachytene chromosomes of various insects and plants are made of microfibrils of similar thickness in all species. There was some evidence that these units were double. Interphase nuclei were found to contain microfibrils half as thick. It was suggested that the double units in prophase chromosomes were the result of duplication of the 200 A fibrils. This duplication appeared to occur before visible prophase.

Thin sections through prophase and interphase nuclei have confirmed these findings. Interphase nuclei contain masses of microfibrils 200 A thick. Similar fibrils are present in chromosomes during mitosis, but they occur here as pairs which correspond to the 500 A fibrils described earlier. In mitotic chromosomes the microfibrils are tightly packed together while during interphase they are more widely spaced.

It is now generally agreed that chromosomes are multiple structures. Even the light microscope has revealed two or even four units (chromonemata) in anaphase chromosomes. With the electron microscope these chromonemata are found to be still further subdivided. In the 200 A fibrils we seem to have reached a definite structural unit, since it is not further split into thinner fibers, but appears to have a more complex organization. While the exact number of these units per chromosome has not been determined yet, there is evidence that it may not be the same in all species (9, 10).

For some time now the nature of the longitudinal differentiation of chromosomes into chromomeres, interchromomeric fibers, hetero- , and euchromatin has been debated. Some ten years ago I suggested that these chromosomal regions did not differ in kind, but in the spatial arrangement or degree of coiling of the same unit, the chromonema (11). The electron microscopic study of chromosomes has confirmed this point of view. The morphological unit of chromosomes is a coiled fibril. Chromomeres and interchromomeric regions contain the same fibrils, but they are differently arranged in space. This does, of course, not exclude chemical differences along their length. No doubt such differences are very important, one striking example being the loops and chromomeres in lampbrush chromosomes (8). Both structures consist of similar fibrils, but in chromomeres they contain DNA, while loops are Feulgen-negative and are made essentially of protein and RNA.

The organization of the elementary microfibril in terms of its chemical constituents is of special interest. Speculation on the nature of the gene, gene action, and gene duplication, or chromosome physiology and reproduction must take into account the organization of these chromosomal units. What is the arrangement of DNA, histone, and non-histone protein in the fibril? While little factual evidence is yet available, the appearance of the fibrils as described in this paper suggests an interesting model. Superficially at least there is a striking resemblance to the structure of the tobacco mosaic virus as described recently (5, 13). Apparently the virus particle contains a core of RNA surrounded by a sheath of protein. Perhaps the chromosomal fibril contains a core of DNA surrounded by protein. The data so far are in agreement with the model: The DNA can be removed without changing the outer appearance of the microfibril; osmium is present in the outer layer, but not in the core of the fibril, which is expected if OsO_4 combines with protein but not with nucleic acid (2); x-ray diffraction studies suggest that the protein is arranged around the DNA double helix (6). Such an organization of the chromosomal unit could bridge the gap between the duplication of the DNA macromolecule and that of the chromosomal unit. The DNA double helix could be duplicated in the core of the fibrils according to one or the other of the schemes proposed (see for instance reference 3), followed by a reorganization of the protein shell resulting in a doubling of the microfibril. It should be noticed that since most chromosomes contain more than one elementary fibril, separation of daughter chromosomes at anaphase does not separate the "old" and "new" microfibril, but the two halves (chromatids) of the bundle of fibrils which were already separate units in previous anaphase. Daughter chromosomes therefore must have equal numbers of "old" and "new" microfibrils.

SUMMARY

Amphibian lampbrush chromosomes and meiotic prophase chromosomes of various insects and plants consist of a bundle of microfibrils about 500 A thick. These fibrils are double, being made of two closely associated fibrils 200 A thick. Fragments of interphase nuclei contain a mass of fibrils 200 A thick. Ultrathin sections through nuclei in prophase or interphase show sections of these double or single fibrils cut at various angles. A comparison of sections with the methacrylate left in and sections that were shadowed after removing the methacrylate suggests that the OsO_4 reacts only with the outer part of the fibrils either because it does not penetrate, or as a result of a chemical difference of the inner core and the outside of the fibril. It is suggested that in analogy to the structure of the tobacco mosaic virus the chromosomal microfibril may have an inner core of DNA surrounded by a shell of protein.

BIBLIOGRAPHY

1. Allfrey, V. G., Mirsky, A. E., and Stern, H., *Advances Enzymol.*, 1955, **16,** 411.
2. Bahr, G. F., *Exp Cell Research*, 1954, **7,** 457.
3. Bloch, D., *Proc. Nat. Acad. Sc.*, 1955, **41,** 1058.
4. Erickson, R. O., *Am. J. Bot.*, 1948, **35,** 729.
5. Hart, R. G., *Proc. Nat. Acad. Sc.*, 1955, **41,** 261.
6. Feughelman, M., Langridge, R., Seeds, W. E., Stokes, A. R., Wilson, H. R., Hooper, C. W., Wilkins, M. H. F., Barclay, R. K., and Hamilton, L. D., *Nature*, 1955, **179,** 834.
7. Kaufmann, B. P., McDonald, M. R., and Gay, H., *J. Cell. and Comp. Physiol.*, 1951, **38,** suppl. 1, 71.
8. Lafontaine, J., and Ris, H., *Genetics*, 1955, **40,** 579.
9. Ris, H., *Genetics*, 1952, **37,** 619.
10. Ris, H., Svmposium on the Fine Structure of Cells, Groningen, Noordhoff Ltd., 1956, 121.
11. Ris, H., *Biol. Bull.*, 1945, **89,** 242.
12. Ris, H., and Mirsky, A. E., *J. Gen. Physiol.*, 1949, **32,** 489.
13. Schramm, G., Schumacher. G., and Zillig, W., *Z. Naturforsch.*, 1955, **10b,** 481.
14. Stadler, L. J., *Science*, 1954, **120,** 811.
15. Watson, J. D., and Crick, F. H. C., *Nature*, 1953, **171,** 964.

EXPLANATION OF PLATES

PLATE 130

FIG. 1. Part of a leptotene chromosome of *Lilium*. Feulgen-squash. The chromosome consists of microfibrils, 500 A thick. In some regions they are visibly double (arrow). Regions in which the bundle of fibrils is doubled up and more closely packed correspond to the chromomeres of the light microscopist (*CH*). × 22,000.

FIGS. 2 and 3. Sections through somatic interphase nuclei in the anther of *Lilium*. Methacrylate removed, uranium-shadowed. Sections through large numbers of twisted microfibrils are visible. End views of the cut fibrils show a dark center (arrows), indicating that the fibrils have a less electron-dense core. × 80,000.

PLATE 131

FIG. 4. Section through a late prophase chromosome of a microspore mother cell (*Lilium*). Methacrylate removed, uranium-shadowed. The 200 A fibrils occur in pairs (brackets) corresponding to the 500 A units seen in Fig. 1. In end views a darker core is visible in each fibril (circle). × 49,000.

FIG. 5. Section through a somatic interphase nucleus of a *Tradescantia* anther. Methacrylate removed, uranium-shadowed. Randomly cut segments of twisted microfibrils are visible (arrows). × 80,000.

FIG. 6. Section through a leptotene chromosome in a microspore mother cell of *Lilium*. Note the double lines and circles representing the more electron-dense outer region of the microfibrils (arrows). Compare with Text-fig. 1 *A*. × 80,000.

FIG. 7. Section through a chromosome in first meiotic metaphase of a microspore mother cell (*Lilium*). Sections through microfibrils appear as double lines and circles 200 A in diameter (arrows). × 80,000.

FIG. 8. Section through an oocyte nucleus of *Triturus*. The microfibrils of the lampbrush chromosomes appear again as double lines and circles (arrows). × 80,000.

FIG. 9. Section through a rat spermatid. Arrows indicate sections through chromosomal microfibrils. × 80,000.

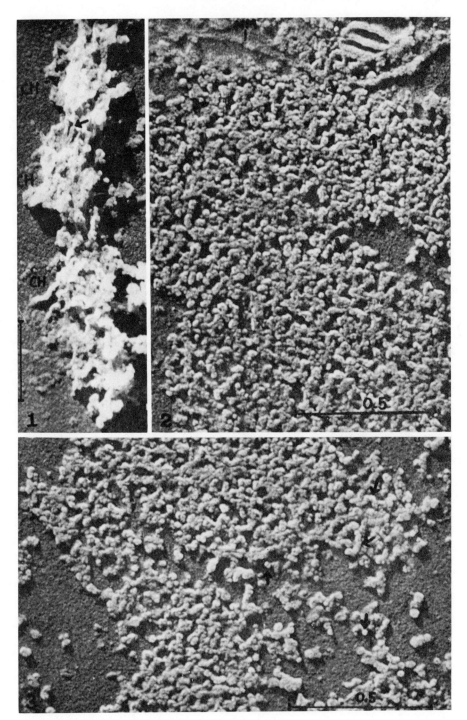

PLATE 130

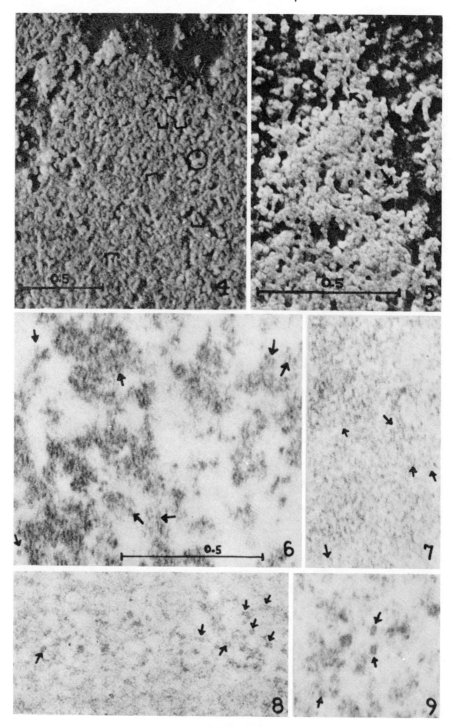

PLATE 131

6

Reprinted from *Science* **26**(657):151–152 (1907)

A PRELIMINARY NOTE ON THE CHROMOSOMES OF OENOTHERA LAMARCKIANA AND ONE OF ITS MUTANTS, O. GIGAS

Anne M. Lutz

*Station for Experimental Evolution,
Cold Spring Harbor, L.I.*

THE exceptional opportunities offered at this station for a study of inheritance as manifested in the germ cells of the *Œnotheras* led me to undertake a study of the chromosomes of *Œnothera Lamarckiana*, its mutants and hybrids.

The work was begun after the flowering season had passed, however; therefore only somatic cells from the growing root tips of potted plants in the rosette stage have so far been available for study; and it is the purpose of this note, pending the completion of a more general study of the *Œnotheras*, merely to call attention to a most unexpected contrast found in the number of chromosomes of *O. Lamarckiana* and one of its mutants, *O. gigas*, both pure bred.

Because of the smallness of the chromatic figures and the low percentage of figures studied in which the chromosomes could be counted with certainty, I do not at present feel justified in stating the exact number in either form; but I can state unreservedly what is of more interest, that in all the somatic cells of *O. gigas* arising from *O. Lamarckiana* in which the chromosomes could be counted with precision, *the number has become approximately double that of the parental form,* *O. Lamarckiana.* This result was unexpected, as a somewhat hasty survey of the tips of several other mutants previous to the study of *gigas* had indicated a number closely approaching or identical with that of the parental form. Gates, in his "Preliminary Note on Pollen Development in *Œnothera lata* de Vries and its Hybrids," published in SCIENCE, February 15, 1907, states that in a cross resulting from the pollination of *O. lata* by *O. Lamarckiana*, "the sporophyte count for the *O. Lamarckiana* side of the cross is at least twenty. The conclusion from this is that pure *O. Lamarckiana* itself must have over twenty chromosomes." In his paper on "Pollen Development in Hybrids of *Œnothera lata* × *O. Lamarckiana*, and its Relation to Mutation,"[1] he adds in a foot-note on page 109: "The inference that *O. Lamarckiana* itself has the same number of chromosomes as the dominant *O. Lamarckiana* hybrid is also apparently not borne out by the facts." From my own observations on all

Oenothera Lamarckiana.   Oenothera gigas.

[1] *Botanical Gazette*, February, 1907.

pure-bred *O. Lamarckiana* so far studied I have found no indication of the number ever approaching twenty; but from the evidence of repeated counts it seems to be fourteen or fifteen. I have at least eighteen good clear demonstrations of mitotic figures showing only fourteen chromosomes, all distinctly outlined and clearly defined—with no trace of a chromosome in a preceding or following section; on the other hand, I have encountered a sufficient number of less clearly defined figures, in which there seems to be but thirteen, and in others fifteen chromosomes, to make it necessary to state the number for the present with reserve. Chromosomes frequently lie in such positions as to make it impossible to distinguish between a long-looped form and two so placed as to give a similar appearance; also a looped chromosome may be sectioned at a point to give the two halves the appearance of distinct individuals.

The number of chromosomes characteristic of the somatic cells of *O. gigas* is probably twenty-eight or twenty-nine, although the difficulty in counting is here increased by the large number; however, I have six or seven excellent figures showing twenty-eight sharply-defined chromosomes, and as many more, not so clearly outlined, in which there is a strong indication of twenty-nine. It is hoped that the hundreds of new sections now in process of preparation for study will establish the facts, shortly.

Other points of interest are coming to light, particularly in connection with the hybridization of mutants, and will be mentioned in a later note.

7

Reprinted from *Natl. Acad. Sci. (U.S.A.) Proc.* **27**(11):507–512 (1941)

THE FREQUENCY OF POLYPLOIDY AND OTHER SPONTANEOUS ABERRATIONS OF CHROMOSOME NUMBER AMONG LARVAE OF THE NEWT, TRITURUS VIRIDESCENS

By Gerhard Fankhauser

Department of Biology, Princeton University

Communicated October 8, 1941

Spontaneous deviations from the normal somatic chromosome number have been investigated extensively in populations of plants, partly because of the ease with which the chromosome number of each individual may be determined in root-tip preparations. Comparatively little information is available concerning the range and frequency of such aberrations among animals. Recently, a procedure was described which permits the determination of the chromosome number of living amphibian larvae, in whole mounts of the amputated tip of the tail.[1] In young larvae the fin surrounding the tail is sufficiently transparent to allow the examination of mitotic figures in the large, flattened cells of the epidermis. Furthermore, reliable indirect evidence of polyploidy may be obtained from measurements of the size of the interkinetic nuclei of the epidermis as well as of other tissues of the tailtip which do not contain countable mitotic figures. This secondary evidence is particularly valuable because it reveals the chromosomal constitution of red blood cells and of cells of the lateral line sense organs both of which have moved into the tail from distant regions of the embryo, during earlier stages of development. The tailtip thus offers a truly representative picture of the embryo as a whole.

The tailtip method was originally devised to identify in life the triploid individuals which had been shown by several authors to occur among amphibian embryos in preserved material. The search for triploids among larvae of the newt, *Triturus viridescens*, was almost at once successful and provided material for a study of the effects of triploidy on development.[2,3] During the following three years many more larvae of this species were examined and classified with regard to their chromosome number. In addition to other triploid larvae, a single pentaploid larva was discovered in 1939.[4] The past season, 1940–41, was particularly productive in spontaneous chromosome aberrations. Not only did more triploid and pentaploid individuals appear, but also larvae with other chromosome numbers, both expected and unexpected. These included a haploid,[5] a tetraploid, a hyper-diploid individual with $2n + 2$ chromosomes, a haploid-diploid chimaera[5] and a larva containing both diploid and hyper-triploid cells.

All these individuals with deviating chromosome numbers appeared among larvae which developed from normal untreated eggs; they represent, therefore, natural or spontaneous aberrations. The term "spontaneous" is preferred for the reason that the larvae were raised in the laboratory from eggs which were laid by pituitary-stimulated females from November to May. Earlier investigations of Kaylor[6] had shown that such "pituitary eggs" behave in all respects like those laid during the breeding season by unstimulated females. This might be expected since the ovaries, following the normal breeding season in April and May, grow rapidly during the summer and, by late fall, contain full-grown oöcytes which are ready for laying and fertilization. However, the possibility still exists that the precocious release of the eggs may favor the production of chromosome abnormalities in some unspecific way. The importance of this factor will be discussed below (see table 2).

The search for aberrant individuals will be continued in the future until a sufficiently large number of larvae will have been investigated to give a reliable index of the range and frequency of spontaneous deviations in this species. However, in view of the considerable number of larvae

TABLE 1

Chromosome Numbers of Larvae of *Triturus viridescens* Examined from 1937 to 1941. Diploid Number = 22

Total number of larvae	1074
Diploid	1056
Triploid	10
Tetraploid	1
Pentaploid	3
Haploid	1
Hyper-diploid ($2n + 2$)	1
Haploid-diploid chimaera	1
Diploid-hyper-triploid chimaera	1

which have already been examined, a preliminary estimate of the frequencies of various types of numerical aberrations seems to be worth while.

A summary of the chromosome numbers of the 1074 larvae which were studied during the last four years is given in table 1. As was to be expected, triploidy is the commonest aberration, with a frequency of about 1%. The triploids must owe their origin to the fairly frequent formation of unreduced gametes. Theoretically, either a diploid egg fertilized by a haploid spermatozoon, or the reciprocal combination, would produce a triploid embryo. In plants, the extra set of chromosomes seems to be of maternal origin in most cases. An instance of paternal origin of two chromosome sets has been described in maize by Rhoades;[7] however, triploidy in this case probably resulted from the union of a haploid egg with two haploid sperms, since crosses between diploid and tetraploid plants have shown that diploid pollen grains, as a rule, do not function normally on a diploid stigma. It has not been possible so far to obtain evidence concerning the origin of the third chromosome set in triploid

Triturus larvae. Earlier observations on frogs by G. and P. Hertwig[8] and by Bataillon and Tchou-Su[9] indicate that, in amphibians also, triploidy arises in the majority of cases from the union of a diploid egg and a haploid sperm.

A low frequency of tetraploidy, less than 1 in 1000, would be expected if it owes its origin to the obviously rare occurrence of a union between two exceptional, unreduced gametes. The astonishingly high frequency of pentaploidy, if it is substantiated by future investigations, would indicate that pentaploidy could hardly be caused by the chance union of a diploid egg, a diploid sperm and a second, haploid sperm, as had been suggested before.[4] While polyspermy is a natural phenomenon in *Triturus viridescens*, all sperm nuclei remain separated from one another because each nucleus is associated with a large aster; and, furthermore, the supernumerary sperm nuclei degenerate early, during the first cleavage mitosis of the diploid fusion nucleus.[10] Another hypothesis seems at present more probable, namely, the occasional suppression of the formation of both polar bodies in an oöcyte; this would leave four sets of chromosomes in the egg and require nothing more to produce pentaploidy than the normal fertilization of such an egg by a haploid sperm.

The presence of one haploid and one partially haploid larva in this population is also of special interest. Haploidy, although easily induced in amphibians by various experimental techniques,[11] had not previously been known to occur spontaneously among untreated larvae. In less than a year, however, the first cases of spontaneous haploidy appeared in three different species of salamanders.[5,12] The hyper-diploid larva also seems to be the first individual of its type to be recorded among normal amphibian larvae. The morphological individuality of the chromosomes has not been studied sufficiently in this species to allow an identification of the two extra chromosomes that are present.

It has been mentioned already that all the larvae used in these investigations developed from eggs laid in the laboratory by pituitary-stimulated females, and that the more or less precocious ovulation of the eggs might conceivably increase their tendency towards chromosomal aberrations. It should be pointed out, however, that triploid amphibian embryos were first discovered in preserved material developed from eggs laid in the spring by unstimulated females, long before the pituitary technique had been introduced. Moreover, a single triploid adult newt of a related species, *Triton taeniatus*, was discovered in nature in Sweden by Böök.[13] Attempts will be made to determine the frequency of triploidy in natural populations of larval and adult salamanders of various species. The chances that the other, less viable types of aberrations could be discovered under natural conditions are very slight.

In view of the possible influence of artificially induced ovulation it is interesting to compare the frequencies with which polyploidy and other aberrations appeared among our larvae during early winter, late winter and spring. If the artificial release of the eggs is at all an important factor, it would be expected to exert more influence in the early part of the winter than in the spring, just before or during the normal breeding season when the eggs are certainly ready for laying. Table 2 shows, however, that during the season 1940–41 the incidence of chromosome abnormalities was almost as high in March, April and May as it was in November and December. It seems safe, therefore, to assume that the frequencies of different degrees of polyploidy given in table 1 are at least representative of, if not identical with, those present among young larvae in their natural habitat.

With the aid of the tailtip method the occurrence of polyploids has also been studied among larvae of two other species of salamanders, *Triturus*

TABLE 2

FREQUENCY OF NUMERICAL CHROMOSOME ABERRATIONS DURING DIFFERENT PERIODS OF WINTER AND SPRING 1940–41

	NOV.–DEC.	JAN.–FEB.	MARCH–MAY
Total number of larvae examined	162	27	195
Diploid	156	26	190
Total aberrant	6	1	5
Triploid	3	1	1
Tetraploid	1	—	—
Pentaploid	—	—	2
Haploid	1	—	—
$2n + 2$	1	—	—
Haploid-diploid	—	—	1
Diploid-hyper-triploid	—	—	1

pyrrhogaster and *Eurycea bislineata*. In the former species, three triploid individuals and one haploid were found among 273 larvae raised in the laboratory during the past two years.[14,12] In a population of 134 larvae of *Eurycea bislineata* which were investigated in 1938–39, thirteen triploid and two tetraploid individuals were present.[15] The conclusion that this high frequency of polyploidy may be typical for the species has not been substantiated so far by the current investigations of J. V. Michalski which, on the other hand, have demonstrated the occurrence of haploidy and partial haploidy. Apparently, "outbursts" of polyploidy may occur under certain conditions that are not yet understood aside from the generally accepted fact that some matings produce more polyploids than others and that genetic factors may therefore be involved.

In this connection Skalińska's[16] observations on the origin of polyploidy in *Aquilegia* are of interest. She found that repeated selfing of a strain decreased both vigor and fertility of the plants which began to produce many abortive pollen grains, as well as diploid pollen grains and mega-spores in considerable numbers.   This increased tendency to polyploid mutations was probably caused by the generally abnormal and retarded development of the plants of this weakened or "senile" strain.   It is not improbable that similar physiological factors may also play a rôle in the occasional production of unusually large numbers of polyploid embryos in amphibians.

Little information is available regarding the frequency of spontaneous numerical aberrations in other animals and in plants.   In *Drosophila melanogaster*, the frequency of triploidy is probably much less than one per cent.   In *Oenothera* cultures, triploidy is not uncommon; the frequency varies between different families and may reach a maximum of 12%.[17]   In maize, triploid plants appear with a frequency considerably below one per cent, while not a single spontaneous tetraploid has been recorded so far.[18]   On the other hand, haploids appear at a rate of approximately 1 in 2000.   The same frequency has been recorded for spontaneous haploidy in *Antirrhinum majus*.[19]   As a whole, these very incomplete figures show that numerical chromosome aberrations among salamander larvae are surprisingly abundant.

With the exception of the two chimaeras listed in table 1, the chromosome number seemed to be uniform in each larva studied, as far as could be judged from a comparison of nuclear and cell sizes in sections through different organs and regions of the body.   At least, no evidence has been found so far that diploid larvae, for instance, may contain sizable polyploid regions.   Partial polyploidy of a very limited extent has been recorded in diploid larvae of *Triturus pyrrhogaster*[14] and *Eurycea bislineata*.   In two tailtips of the former species, and in one of the latter, the epidermis of the fin included a small area of tetraploid cells.   However, it should be emphasized that the tailfin is a purely larval structure which disappears completely during metamorphosis.

[1] Fankhauser, G., *Proc. Am. Philos. Soc.*, **79**, 715–739 (1938).

[2] Fankhauser, G., *Anat. Rec.*, **77**, 227–245 (1940).

[3] Fankhauser, G., *Jour. Morph.*, **68**, 161–177 (1941).

[4] Fankhauser, G., *Proc. Nat. Acad. Sci.*, **26**, 526–532 (1940).

[5] Fankhauser, G., and Crotta, R. (in press).

[6] Kaylor, C. T., *Jour. Exp. Zool.*, **76**, 375–394 (1937).

[7] Rhoades, M. M., *Jour. Genet.*, **33**, 355–357 (1936).

[8] Hertwig, G. and P., *Arch. mikr. Anat.*, **94**, 34–54 (1920).

[9] Bataillon, E., and Tchou-Su, *Roux' Arch.*, **115**, 779–824 (1929).

[10] Fankhauser, G., and Moore, C., *Jour. Morph.*, **68**, 347–385 (1941).

[11] Fankhauser, G., *Jour. Hered.*, **28**, 1–15 (1937).

[12] Fankhauser, G., and Crotta, R. *Physiol. Zool.* (in press).

[13] Böök, J. A., *Hereditas*, **26**, 107–114 (1940).

[14] Fankhauser, G., Crotta, R., and Perrot, M., *Jour. Exp. Zool.* (in press).

[16] Fankhauser, G., *Jour. Hered.*, **30**, 379–388 (1939).

[16] Skalińska, M., *Proc. Seventh Internat. Genet. Congr.*, Edinburgh, pp. 265–266 (1941).

[17] Shull, G. H., *Proc. Nat. Acad. Sci.*, **15**, 268–274 (1929).

[18] Randolph, L. F., personal communication.

[19] Knapp, E., *Ber. deutsch. bot. Ges.*, **57**, 371–379 (1939).

8

Reprinted from *Science* **86**(2235):408 (1937)

METHODS OF INDUCING CHROMOSOME DOUBLING
IN PLANTS BY TREATMENT WITH COLCHICINE

A. F. Blakeslee and A. G. Avery

A number of chemicals have been tested with Datura and other plants in an effort to induce hereditary mutations. Narcotics had previously been found effective in inducing doubling of chromosomes in roots, but chloral hydrate and nicotine were found ineffective in inducing chromosome doubling in stems which alone bear seeds and thus might lead to production of 4n races. The alkaloid colchicine we have found will induce an abundant production of branches with doubled chromosome number. When seeds are heavily treated, all the seedlings may be affected. The stem becomes swollen while the growth of root and plumule is checked; buds are abnormally arranged leading to sectors with roughened leaves characteristic of mixed 4n and 2n tissue like spontaneous 4n sectorial mutations. Normal 2n tissue tends to outgrow the mutated 4n tissue, but the latter may include the whole shoot. Between one half and one third of plants from treated seeds have produced 4n flowers. Apparently sectors of 8n tissue have been secured by treating 4n plants as well as by heavy treatments of 2n individuals. Tetraploid tissue involving the flower may readily be determined by examination of pollen. In Datura and Portulaca the determination by pollen size has been checked by chromosome counts. Doubling in adult tissue has been induced by immersion of twigs in solutions and in agar, by treatment of buds with mixtures of colchicine and lanolin and by spraying with solutions. By use of colchicine changes have been induced which are interpreted as due to doubling of chromosomes in the following genera: Datura (several species), Portulaca (2 species), Cosmos (2 species), Phlox, Stellaria, Nicotiana, Digitalis, Mirabilis, Tropaeolum, Cheiranthus, Raphanus, Cucurbita, Trifolium, Medicago and Allium. If control of chromosome doubling by chemical means proves of general application, as seems to be the case, the plant breeder will be able to work with greater precision in his efforts to control the evolution of economic forms both of plants propagated vegetatively and of those reproduced by seed. For example, it should be possible, starting with a sterile hybrid, to synthesize a pure-breeding double diploid which would have hybrid vigor and the desirable characteristics brought about by tetraploidy. This we have apparently succeeded in doing with a species hybrid in Nicotiana. Doubling chromosome number would give enlarged flowers and fruits to the horticulturist and through triploids would be the basis of a wide range of 2n+1 types. Tetraploidy and presence of unbalanced extra chromosomes are known to have been factors in the origin of a large proportion of our most desirable varieties of fruits and flowers. In addition to increase in size of organs of the plant, tetraploidy has changed a self-sterile to a self-fertile form, a dioecious to a hermaphroditic race, an annual to a perennial, and has increased winter hardiness. The ability to induce chromosome doubling, therefore, is of importance to practical as well as to theoretical genetics.

Part II

CHROMOSOME REPLICATION

HISTORICAL PERSPECTIVES

In 1953 Watson and Crick described the structure of DNA. Replication occurs from the separation of the two strands of the double helix, each strand then directing the synthesis of a complementary strand resulting in two molecules both containing an original parental strand and a newly synthesized daughter strand. This mode of replication, referred to as *semiconservative replication*, has become one of the basic principles of cytogenetics. The two extreme alternatives to this mode of replication would be (1) conservative replication where a double helix directs the synthesis of a new double helix that contains both daughter strands; and (2) dispersive replication where the parental double helix becomes fragmented and both strands of the two DNA double helices produced would contain segments of the original parental strand and newly synthesized DNA.

The first experimental evidence as to the mode of replication was the report by Taylor, Woods, and Hughes (Paper 9). They used tritium, an isotope that gives resolution in photographic emulsions in elements as small as individual chromosomes, to label thymidine in the chromosome of *Vicia faba* and to follow the distribution of the label in the two succeeding cell generations. The use of colchicine to monitor chromosome numbers allowed them to determine the number of division cycles following exposure to the radioactive thymidine. The results indicated that daughter chromosomes consist of an old and a new DNA "unit" and that the units remain intact except for the occasional sister chromatid exchange.

77

The report by Taylor and coworkers was followed the next year by Meselson and Stahl's (1958) report that DNA extracted from *Escherichia coli* cells, grown first on a medium containing the ^{15}N isotope and then on an ^{14}N medium, was hybrid ^{15}N^{14}N. The use of tritium as a label for thymidine—a specific DNA precursor, as opposed to RNA—coupled with autoradiographic techniques have become powerful aids in cytogenetic studies.

We now know that chromosomes or parts of chromosomes of a particular nucleus do not all replicate at the same time in the S (DNA synthetic)-period. Using root-tip cells of *Crepis*, Taylor (1958) was the first to demonstrate that there was a progression of DNA synthesis from the chromosome ends to the centromeres. Asynchrony in replication between chromosomes was first noted in Lima-de-Faria's (1959) report that the sex chromosome of a grasshopper species replicated after the autosomes. Observations of this sort have been made on a wide variety of organisms. The first report of asynchronous replication in mammalian cells is Taylor's (1960) work with cultured cells of the Chinese hamster. He found that five or six chromosomes of the complement have segments that typically replicate late in the S-period.

Evidence from labeling studies suggested that the eukaryotic chromosome consists of several replicons (units of replication). In addition, considerations of the amount of DNA per nucleus, the rate of DNA synthesis, and the time spent in the S-period of the nuclear cycle indicated that there are several replicons per chromosome (Taylor 1968). The existence of numerous concurrently replicating replicons within a chromosome appeared likely for certain species in which the entire chromosome was autoradiographically labeled even with short pulses of ^{3}H-thymidine (i.e., the chromosomes were "continuously labeled"). In cases of discontinuously labeled chromosomes, i.e., when chromosomes are labeled only in certain segments, replication of the chromosomal replicons is under temporal control with both intra- and interchromosomal coordination. Howard and Plaut (1968) and Plaut (1969) showed such coordination in *Drosophila melanogaster* among several segments in three different chromosomes.

The polytene chromosomes of *Drosophila* have provided excellent materials for studying chromosome replication. For many years cytogeneticists recognized that the amount of heterochromatin in salivary-gland chromosomes was much less than predicted based on the chromosomes of other somatic cells of the same organism. Using Feulgen microspectrophotometry, Rudkin (1969) showed that the increase in DNA of salivary-gland chromosomes at various larval stages (representing different degrees of polyteny) followed a pattern suggesting that the DNA of the heterochromatin was not replicating while the euchromatin replicated several times. This information explained the relatively small amount of centric heterochromatin as compared with the amount of euchromatin found in *Drosophila* salivary-gland chromosomes. For other regions, intermediate degrees of replication may be occurring. Hennig and Meer (1971) have shown that the ribosomal DNA of *D. hydei* is about three steps behind the euchromatin in polytenization.

The polytene chromosomes of *Chironomus* have provided yet another dimension to considerations of chromosome replication. In the hybrid between two subspecies of *C. thummi*, the chromosomes have areas of loose pairing where the chromosomes of each species are separated and identifiable based on their characteristic widths. Although the sequence of bands in each chromosome is the same, certain bands are more prominent in one chromosome than the homologous band in the other chromosome. DNA can be measured in these bands by Feulgen microspectrophotometry. When differences in the bands exist, they follow a doubling series; that is, the proportions of DNA between homologous bands are 1:2, 1:4, 1:8, or 1:16 (Bauer and Keyl, 1964). This finding suggests that individual segments of a chromosome may increase by some doubling process; Bauer and Keyl believe the doublings are related to an abnormal replication process.

A striking cytogenetic demonstration of chromosome replication in a prokaryote was given by Cairns (Paper 10). The *E. coli* chromosome was labeled with tritiated thymidine and visualized through EM autoradiography. The chromosome was shown to be replicating as a circle with a single growing point. Huberman and Riggs (1968) showed that the DNA of mammalian chromosomes replicates bidirectionally from several initiation points. Eukaryotic DNA replication rate is up to 100 times faster than that of *E. coli*. Differences in S-phase lengths appear to be at least a function of the number of initiation sites and their time of activation (see Callan 1973).

Editors' Comments
on Papers 9 and 10

9   TAYLOR, WOODS, and HUGHES
The Organization and Duplication of Chromosomes as Revealed by Autoradiographic Studies Using Tritium-labelled Thymidine

10   CAIRNS
The Chromosome of Escherichia coli

Paper 9 reports the use of tritiated thymidine in the discovery of what has become one of the basic tenets of cytogenetics—that chromosomes replicate semiconservatively. The study provided not only a significant advance in understanding chromosome replication, but it also described techniques for studying eukaryotic chromosomes with normal structure. In this study sister-strand mitotic crossing-over was also readily observable. Previously, sister-strand crossing-over could be demonstrated only by using a ring or some other aberrant chromosome. This paper presents the theoretical basis for using tritiated thymidine which is a DNA precursor, rather than an RNA precursor. The low-energy beta particle released through tritium decay allows resolution to individual chromosome dimensions.

Paper 10, a landmark in the field, gave geneticists their first glimpse of the DNA of a chromosome in the process of replication. Using tritiated thymidine as a label and electron-microscope autoradiography, Cairns visually demonstrated that the *Escherichia coli* chromosome replicates as a circle from a single growing point. Autoradiographic studies of DNA replication in eukaryotes were an outgrowth of this observation.

REFERENCES

Bauer, H., and H. G. Keyl. 1964. Verdopplung des DNS-Gehalts kleiner Chromosomenabschnitte als Faktor der Evolution. *Naturwissenschaften* **51**:46–47.

Callan, H. G. 1973. DNA replication in the chromosomes of eukaryotes. *Cold Spring Harb. Symp. Quant. Biol.* **38**:195–203.

Hennig, W., and B. Meer. 1971. Reduced polyteny of ribosomal RNA cistrons in giant chromosomes of *Drosophila hydei*. *Nature New Biol.* **233**:70–72.

Howard, E. F., and W. Plaut. 1968. Chromosomal DNA synthesis in *Drosophila melanogaster*. *J. Cell Biol.* **39**:415–429.

Huberman, J. A., and A. D. Riggs. 1968. On the mechanism of DNA replication in mammalian chromosomes. *J. Mol. Biol.* **32**:327–341.

Lima-de-Faria, A. 1959. Differential uptake of tritiated thymidine into hetero- and euchromatin in *Melanoplus* and *Secale*. *J. Biosphys. Biochem. Cytol.* **6**:457–466.

Meselson, M., and F. W. Stahl. 1958. The replication of DNA in *Escherichia coli*. *Proc. Nat. Acad. Sci. U.S.A.* **44**:671–682.

Plaut, W. 1969. On ordered DNA replication in polytene chromosomes. Proc. Int. Symp. Nuclear Physiology and Differentiation. *Genetics* **61**, Sup. 1:239–244.

Rudkin, G. T. 1969. Non replicating DNA in *Drosophila*. *Genetics* **61**, Sup. 1:227–238.

Taylor, J. H. 1958. The mode of chromosome duplication in *Crepis capillaris*. *Exp. Cell Res.* **15**:350–357.

_____. 1960. Asynchronous duplication of chromosomes in cultured cells of Chinese hamster. *J. Biophys. Biochem. Cytol.* **7**:455–464.

_____. 1968. Rates of chain growth and units of replication in DNA of mammalian chromosomes. *J. Mol. Biol.* **31**:579–594.

Watson, J. D., and F. H. C. Crick. 1953. Genetical implications of the structure of deoxyribonucleic acid. *Nature* **171**:964–967.

9

Reprinted from *Natl. Acad. Sci. (U.S.A.) Proc.* **43**(1):122–128 (1957)

THE ORGANIZATION AND DUPLICATION OF CHROMOSOMES AS REVEALED BY AUTORADIOGRAPHIC STUDIES USING TRITIUM-LABELED THYMIDINE

By J. Herbert Taylor,* Philip S. Woods, and Walter L. Hughes

DEPARTMENT OF BOTANY, COLUMBIA UNIVERSITY; BIOLOGY DEPARTMENT AND MEDICAL DEPARTMENT, BROOKHAVEN NATIONAL LABORATORY

Communicated by Franz Schrader, October 26, 1956

Information on the macromolecular organization of chromosomes and their mode of duplication has been difficult to obtain in spite of numerous attempts. One point of attack, long recognized but until recently unattainable, was the selective labeling of some component of the chromosome, the distribution of which could be seen in succeeding cell divisions. Reichard and Estborn[1] demonstrated that N^{15}-labeled thymidine was a precursor of deoxyribonucleic acid (DNA) and that it was not diverted to the synthesis of ribonucleic acid. Recently Friedkin *et al.*[2] and Downing and Schweiger[3] have used C^{14}-labeled thymidine to study DNA synthesis. In chick embryos and *Lactobacillus* there was no appreciable diversion of the tracer to ribonucleic acid. In view of these findings, thymidine appeared to be the intermediate required for the experiment, but the labels so far employed have not been satisfactory for microscopic visualization by autoradiographic means. In order to determine whether an individual chromosome among several in a cell is radioactive, autoradiographs with resolution to chromosomal dimensions must be obtained. Resolution at this level is difficult if not impossible to obtain with most isotopes, since the range of their beta particles is relatively great. Theoretically tritium should provide the highest resolution obtainable, since the beta particles have a maximum energy of only 18 Kev, corresponding to a range of little more than a micron in photographic emulsions. Consequently, identification of this label in particles as small as individual chromosomes should be possible. With this in mind, tritium-labeled thymidine was prepared and used to label chromosomes and to follow their distribution in later divisions by the use of photographic emulsions.

Materials and Methods.—Tritium-labeled thymidine of high specific activity $(3 \times 10^3 \text{ mc/mM})$ was prepared by catalytic exchange of tritium from the carboxyl group of acetic acid to a carbon atom in the pyrimidine ring of thymidine (details of the method to be described elsewhere).

Seedlings of *Vicia faba* (English broad bean) were grown in a mineral nutrient solution containing 2–3 µg/ml of the radioactive thymidine. This plant was selected because it has 12 large chromosomes, one pair of which is morphologically distinct, and because the length of the division cycle and the time of DNA synhesisl in the cycle are known.[4] After growth of the seedlings in the isotope solution for the appropriate time, the roots were thoroughly washed with water and the seedlings were transferred to a nonradioactive mineral solution containing colchicine (500 µg/ml) for further growth. At appropriate intervals roots were fixed in ethanol–acetic acid (3:1), hydrolyzed 5 minutes in 1 N HCl, stained by the Feulgen reaction, and squashed on microscope slides. Stripping film was applied, and autoradiographs were prepared as described previously.[5]

Experimental Design and Results.—Roots remained in the isotope solution for 8 hours, which is approximately one-third of the division cycle.[4] Since about 8 hours intervene between DNA synthesis in interphase and the next anaphase, few if any nuclei which had incorporated the labeled thymidine should have passed through a division before the roots were transferred to the colchicine solution. In the presence of colchicine, chromosomes contract to the metaphase condition, and the sister chromatids (daughter chromosomes), which ordinarily lie parallel to each other, spread apart. The sister chromatids remain attached at the centromere region for a period of time, but they finally separate completely before transforming into an interphase nucleus. Because colchicine prevents anaphase movement and the formation of daughter cells, but does not prevent chromosomes from duplicating, the number of duplications following exposure to the isotope can be determined for any individual cell by observing the number of chromosomes. Cells without a duplication after transfer to colchicine will have the usual 12 chromosomes at metaphase (c-metaphase), each with the two halves (sister chromatids) spread apart but attached at the centromere. Cells with one intervening duplication will contain 24 chromosomes, and those with two duplications will contain 48 chromosomes.

Two groups of roots were fixed. The first group remained in the colchicine solution 10 hours. The second group remained in the colchicine for 34 hours. In the first group, cells at metaphase had only 12 chromosomes, which indicated that none of these had duplicated more than once during the experiment. The chromosomes in these cells were all labeled, and, furthermore, the two sister chromatids of each chromosome were equally and uniformly labeled (Fig. 1, *a* and *b*). The amount of radioactivity in the chromosomes varied from cell to cell, as would be expected in a nonsynchronized population of cells, but within a given cell the label in different chromosomes was remarkably uniform.

In the second group, cells contained either 12, 24, or 48 chromosomes. Those with 12 chromosomes usually were not labeled, but when labeling occurred, sister chromatids were uniformly labeled as in the first group. In cells with 24 chromosomes, all chromosomes were labeled; however, only one of the two sister chromatids of each was radioactive (Fig. 2, *a* and *b*). Evidently the pool of labeled precursor in the plant had been quickly depleted after the plant was removed from the isotope solution, and these cells with 24 chromosomes had gone through a second duplication in the absence of labeled thymidine.

In the few cells with 48 chromosomes, analysis of all 48 was not possible. However, in several cases where most of the chromosomes were well separated and flattened, approximately one-half of the chromosomes of a complement contained one labeled and one nonlabeled chromatid, while the remainder showed no label in either chromatid. The appearance of cells with 48 chromosomes in a 34-hour period in colchicine also indicates that there was some variation in the predicted 24-hour division cycle.

In cells with 24 and 48 chromosomes a few chromatids were labeled along only a part of their length, but in these cases the sister chromatids were labeled in complementary portions (Fig. 2, *b*, *arrow*). This is the expected situation following sister chromatid exchange and demonstrates that resolution is sufficient to see crossing over in cytological preparations. A careful search of numerous cells with 12 chromosomes failed to yield a decisive case of half-chromatid exchange, which

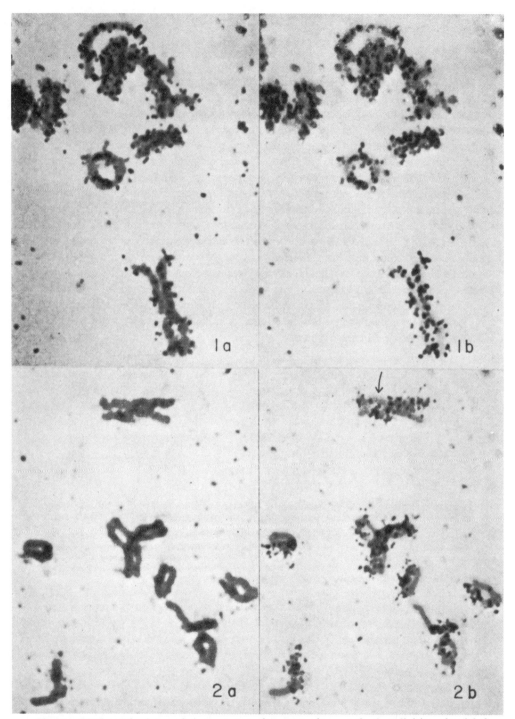

FIG. 1.—Photograph of several chromosomes of a c-metaphase at the first division after labeling occurred; *a*, chromosomes with the chromatids spread apart but still attached at the centromere; *b*, grains in the emulsion above the chromosomes. ×2,200.

FIG. 2.—Photograph of several chromosomes after labeling and one replication in the absence of labeled precursor; *a*, several of the chromosomes from a cell containing 24 chromosomes with chromatids spread but attached at the centromere; *b*, grains in the emulsion above the chromosomes in *a*. ×2,200.

should produce a portion of a chromatid without a label at the first c-metaphase following incorporation of the isotope.

Interpretation and Discussion.—These results indicate (1) that the thymidine built into the DNA of a chromosome is part of a physical entity that remains intact during succeeding replications and nuclear divisions, except for an occasional chromatid exchange; (2) that a chromosome is composed of two such entities probably complementary to each other; and (3) that after replication of each to form a chromosome with four entities, the chromosome divides so that each chromatid (daughter chromosome) regularly receives an "original" and a "new" unit. These conclusions are made clearer by the diagrams in Figure 3.   Beginning with two complementary nonlabeled strands in a chromosome, the two strands separate and a complementary labeled strand is produced along each original strand.   At the succeeding metaphase each chromatid would appear labeled, although it contains both a labeled and a nonlabeled strand.   At a succeeding replication in the absence of labeled precursor, each strand would have a nonlabeled complementary strand produced along its length.   At the succeeding metaphase only one chromatid of each chromosome would appear labeled.   Following another replication, only one-half of the chromosomes would contain a labeled chromatid, as demonstrated in those cells with 48 chromosomes.

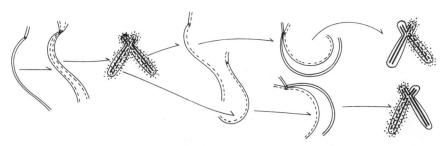

| Duplication with labeled thymidine | 1st c-metaphase after labeling | Duplication without labeled thymidine | 2nd c-metaphase after labeling |

FIG. 3.—Diagrammatic representation of proposed organization and mode of replication which would produce the result seen in the autoradiographs.   The two units necessary to explain the results are shown, although these were not resolved by microscopic examination.   Solid lines represent nonlabeled units, while those in dashed lines are labeled.   The dots represent grains in the autoradiographs.

It is immediately apparent that this pattern of replication is analogous to the replicating scheme proposed for DNA by Watson and Crick.[6]   We cannot be sure, of course, that separation of the two polynucleotide chains in the double helix is involved, for the chromosome is several orders of magnitude larger than the proposed double helix of DNA.

We know that these large metaphase chromosomes are coiled into at least one helix at the microscopic level and perhaps are twice coiled, a helix within a helix. That the chromosome could be a single supercoiled double helix of DNA is inconceivable when one considers the amount of DNA in a large chromosome.   Chromosomes are much more likely to be composed of multistranded units.   To explain their duplication as well as their mechanical properties at the microscopic level, they may be visualized as two complementary multistranded ribbons lying flat upon each other, as shown in Figures 4 and 5. Ribbons of this type with more flexible materials on their outer edges will have a tendency to coil. If the edges contract faster than the central strands when the chromosome begins to shorten, the ribbons fold, one within the other, so as to form a long, trough-shaped cylinder (Fig. 4), and with further contraction they assume the form of a helix. Continued differential contraction would produce a helix within a helix, but the mechanical properties of the model are outside the scope of this discussion.

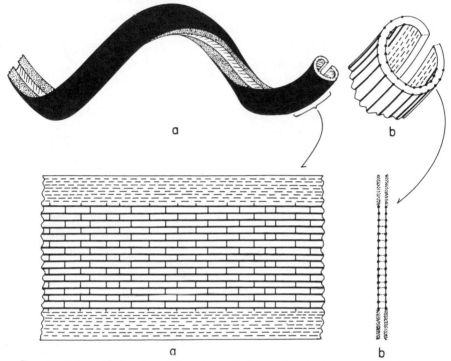

a   b

a   b

Fig. 4.—Schematic drawing of the proposed ribbon-shaped chromosome with the two multistranded units folded together and coiled; *a,* a single gyre from the coiled chromosome; *b,* detail in cross-section.

Fig. 5.—Diagrammatic sketch of the multistranded units uncoiled and flattened; *a,* short portion in longisection; *b,* cross-section. The number and size of strands shown have no special significance. Although the assumption is made that the strands contain DNA, they do not necessarily correspond to Watson-Crick double helices.

Although the chromosome model is provisional and may require considerable modification and refinement, it has many of the features necessary for duplication and the known stability of genetic materials. The large surface area exposed when the two complementary ribbons are extended would facilitate their rapid duplication. A double-stranded unit with two complementary faces has a high stability,[7] and if the two complementary units are composed of multiple, identical strands cross-bonded, the stability of the larger units should be even greater. Such large units would have a high probability of being transmitted as physical entities. If separation of the complementary faces involves the separation of intertwined double helices of DNA, unwinding presents a problem, but perhaps not an impossible one.[8]

The findings reported here are consistent with those recently reported for the distribution of P[32]-labeled phage DNA by Levinthal.[9] His data indicated that about 40 per cent of phage DNA is contained in one piece which is divided equally in the formation of two daughter particles, but undergoes no further distribution during the production of about 150 particles that result from the infection of a bacterium. A phage particle would be analogous to a chromosome before duplication, and the first two daughter particles, each of which would be labeled, would be analogous to the two sister chromatids. Since the chromosome is much larger and contains much more DNA than the phage, it is remarkable that they behave in a similar manner in distribution of their DNA during replication.

Our findings are at variance with the report by Mazia and Plaut[10] which was based on the analysis of the anaphase distribution of chromosomes labeled with C[14]-thymidine. Their data, obtained by the estimation of number of grains over pairs of anaphase or telophase nuclei, indicated a segregation of labeled and non-labeled units at the first division following the incorporation of the isotope. It is entirely possible that in their experiment more than one division occurred between the time of incorporation and the time the telophase nuclei were analyzed. If this had been the case in the present experiment, unequal distribution of activity in sister chromatids would have been observed in diploid cells.

Summary.—Tritium-labeled thymidine was prepared and used for labeling chromosomes during their duplication. Analysis of autoradiographs showed that both daughter chromosomes resulting from duplication in the presence of labeled thymidine appeared equally and uniformly labeled. After an ensuing duplication in the absence of the labeled DNA precursor, the label appeared in only one of each two chromatids (daughter chromosomes). These findings indicate that DNA is synthesized as a unit which extends throughout the length of the chromosome. The units remain intact through succeeding replications and nuclear divisions, except for occasional chromatid exchanges. Each chromosome is composed of two such units, probably complementary to each other. After each replication the four resulting units separate, so that each daughter chromosome always contains an "original" and a "new" unit. To explain the results, a model with two complementary units and a scheme of replication analogous to the Watson-Crick model of DNA is proposed.

* This work was initiated and the original experiments carried out while the senior author was a research collaborator in the Biology Department, Brookhaven National Laboratory. It has been continued at Columbia University under Contract AT (30-1)-1304 with the Atomic Energy Commission.

[1] P. Reichard and B. Estborn, *J. Biol. Chem.*, **188**, 839, 1951.

[2] M. Freidkin, D. Tilson, and D. Roberts, *J. Biol. Chem.*, **220**, 627, 1956.

[3] M. Downing and B. S. Schweigert, *J. Biol. Chem.*, **220**, 521, 1956.

[4] A. Howard and S. R. Pelc, *Exptl. Cell Research*, **2**, 178, 1951.

[5] J. H. Taylor and R. D. McMaster, *Chromosoma*, **6**, 489, 1954.

[6] J. D. Watson and F. H. C. Crick, *Nature*, **171**, 964, 1953; *Cold Spring Harbor Symposia Quant. Biol.*, **18**, 123, 1953.

[7] H. Kacser, *Science*, **124**, 151, 1956.

[8] N. Ardley, *Nature*, **176**, 465, 1955; M. Delbrück, these PROCEEDINGS, **40**, 783, 1955; G. Gamow, these PROCEEDINGS, **41**, 7, 1955; J. R. Platt, these PROCEEDINGS, **41**, 181, 1955; D. P. Bloch, these PROCEEDINGS, **41**, 1058, 1955; C. Levinthal and H. R. Crane, these PROCEEDINGS, **42**, 436, 1956.

[9] C. Levinthal, these PROCEEDINGS, **42**, 394, 1956.

[10] D. Mazia and W. Plaut, *Biol. Bull.*, **109**, 335, 1955.

10

Reprinted from *Cold Spring Harbor Symp. Quant. Biol.* **28**:43–46 (1963)

The Chromosome of *Escherichia coli*

JOHN CAIRNS

Cold Spring Harbor Laboratory of Quantitative Biology,
Cold Spring Harbor, New York

INTRODUCTION

Autoradiography of *Escherichia coli*, labeled with tritiated thymidine and lysed with duponol, has shown that the bacterial chromosome comprises a single piece of DNA which is probably duplicated at a single growing point (Cairns, 1963). Further, it seemed likely from the variety of structures seen that this DNA is in the form of a circle while it is being replicated, even though no intact replicating circles had, at that time, been found.

There is no immediate prospect of proving that the bacterial chromosome is simply a continuous DNA double helix; the existence, for example, of protein linkers scattered along the chromosome (Freese, 1958) could be disproved only by a degree of purification of the intact chromosome that, at the moment, is technically impossible. So, rather than attempt any further purification, an effort was made to extract the chromosome as an intact but replicating circle by lysing labeled bacteria with lysozyme instead of duponol; extraction with lysozyme, it was thought, might leave the DNA complexed with basic proteins and polyamines and therefore less liable to breakage by turbulence (Kaiser, Tabor and Tabor, 1963) or by tritium decay (Hershey, unpublished). Whether or not this reasoning is correct, intact and replicating circles have now been found.

RESULTS

E. coli K12 #3000 Hfr thy⁻ was labeled by growth for about two generations in medium containing H³-thymidine (10 C/mM), as described previously (Cairns, 1963), and was lysed at a concentration of 10⁴/ml by incubation at 37°C, in the usual dialysis chamber, in a medium containing 1.5 M sucrose, 0.01 M KCN, 0.01 M EDTA (pH 8), 5 µg/ml calf thymus DNA, and 200 µg/ml lysozyme. Following incubation for 6 hr, the lysed bacteria were dialyzed against repeated changes of 0.005 M EDTA for 18 hr at room temperature. Finally the dialysis chambers were drained and the membranes (VM Millipore Filters) were subjected to autoradiography.

This procedure displays up to 1% of the chromosomes as more or less tangled circles which, when

fully extended, have a circumference of 1100–1400 µ. Usually these circles are seen to be engaged in duplication. Here, one such example will be described in considerable detail. First, however, it is simplest to describe the model of chromosome duplication to which it gives rise.

Figure 1 shows diagrammatically two rounds of duplication of a circular chromosome, following the

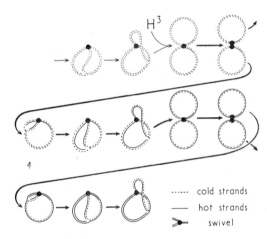

FIGURE 1. A diagrammatic representation of the replication of a circle, based on the assumption that each round of replication begins at the same place and proceeds in the same direction.

introduction of a labeled precursor at some arbitrary point in the cycle. The diagram is based on the assumption that duplication always starts at the same point (in this case, at 12 o'clock) and always advances in the same direction (in this case, counter-clockwise). To make the diagram in a sense complete, some provision must be made for free rotation of the unduplicated part of the circle with respect to the rest, so that the parental double helix can unwind as it is duplicated; this provision, which we may noncommittally refer to as a swivel, has been placed at the junction of starting and finishing point and is itself marked as being duplicated just before the completed daughter chromosomes separate and begin the next round of duplication. Aside from any question of the swivel,

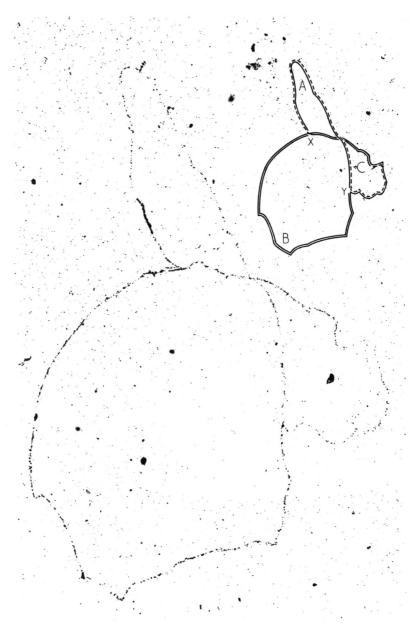

FIGURE 2. Autoradiograph of the chromosome of *E. coli* K12 Hfr, labeled with tritiated thymidine for two generations and extracted with lysozyme. Exposure time two months. The scale shows 100 μ. Inset, the same structure is shown diagrammatically and divided into three sections (A, B, and C) that arise at the two forks (X and Y).

we see that each daughter chromosome (and two of the four granddaughters) shows the stage in the duplication cycle at which label was originally introduced.

Figure 2 shows the autoradiograph of an unbroken replicating circular chromosome that is almost entirely untangled. For the purpose of grain counts and length measurements it was divided, according to the inset diagram, into three sections (A, B, and C) which all meet at two forks (X and Y). The grain counts and length measurements are given in Table 1.

We see here a predominately half-hot chromosome that has completed about two-thirds of the process of duplication. Part of the still-unduplicated section is half-hot (from Y to C) and part is hot-hot

TABLE 1. GRAIN COUNTS AND LENGTH MEASUREMENTS FOR THE THREE SECTIONS THAT UNITE THE FORKS, X AND Y, OF FIG. 2

Section		Grains	Length (μ)	Grains/μ
A		714	670	1.1
B		1298	680	1.9
C	Y to C	213	215	1.0
	C to X	359	205	1.8

(from C to X); as shown in Fig. 1, this situation arises when the moment of introduction of the label does not coincide with the start of a round of replication. And it is for this reason that one can, in this instance, identify one of the forks (X) as the starting and finishing point of duplication and the other (Y) as the growing point. Thus the history of this chromosome is taken to be as pictured in Fig. 1.

Discounting the excess due to replication, the total length of the chromosome seen here is 1100 μ (420 μ plus 670–680 μ) or about 22 times the length of T2 DNA—i.e., equivalent to about 2.8×10^9 daltons of DNA. This value is slightly higher than that reported earlier (Cairns, 1963) and agrees well with the maximum value of 23 T2-equivalents, obtained when the reported total DNA content of 32 T2-equivalents (Hershey and Melechen, 1957) is multiplied by ln 2 to correct for continuous duplication.

The process of duplication portrayed here is, in most respects, merely the physical embodiment of the conclusions of others. For it has become clear from quite unrelated experiments that the bacterial chromosome is duplicated at a single growing point (Bonhoeffer, 1963) which, at least in *E. coli* Hfr and in certain strains of *B. subtilis*, always starts at the same place and moves in the same direction (Nagata, 1963; Yoshikawa and Sueoka, 1963).

More problematical than the process of DNA synthesis itself, about which there is such satisfactory agreement, are the processes occurring between one round of duplication and the next. These have been represented diagrammatically in Fig. 1 as a separate stage in the duplication of the chromosome during which the swivel is supposed to be duplicated. Presumably, it is at this time that RNA and protein synthesis become obligatory (Maaløe and Hanawalt, 1961), that the Hfr chromosome becomes available for transfer during mating (Bouck and Adelberg, 1963), and that thymineless death may be consummated.

At first sight it seemed surprising to find that the chromosome is physically in the form of a circle even while it is being replicated, for this arrangement demands that somewhere in the circle there must be something that acts as a swivel. However, in view of the apparent importance of the structure that unites the ends of the chromosome and so completes the circle, one must now consider the possibility that the structure actively drives DNA replication by rotating one end of the chromosome relative to the other; in this way, single-stranded DNA might be continually produced at the replicating fork to act as primer for the polymerase. In short, it now seems conceivable that rapid DNA synthesis is possible only for circles.

ACKNOWLEDGMENTS

I am greatly indebted to V. J. Paral and R. Westen, of the Australian National University, who brought their technical skill to bear on the problem of photographing the autoradiographs of DNA; without their help such pictures could not have been taken.

REFERENCES

BONHOEFFER, F. 1963. Personal communication.

BOUCK, N., and E. A. ADELBERG. 1963. The relationship between DNA synthesis and conjugation in *Escherichia coli*. Biochem. Biophys. Res. Commun., *11*: 24–27.

CAIRNS, J. 1963. The bacterial chromosome and its manner of replication as seen by autoradiography. J. Mol. Biol., *6*: 208–213.

FREESE, E. 1958. The arrangement of DNA in the chromosome. Cold Spring Harbor Symp. Quant. Biol., *23*: 13–18.

HERSHEY, A. D., and N. E. MELECHEN. 1957. Synthesis of phage-precursor nucleic acid in the presence of chloramphenicol. Virology, *3*: 207–236.

KAISER, D., H. TABOR, and C. W. TABOR. 1963. Spermine protection of coliphage λ DNA against breakage by hydrodynamic shear. J. Mol. Biol., *6*: 141–147.

MAALØE, O., and P. C. HANAWALT. 1961. Thymine deficiency and the normal DNA replication cycle, I. J. Mol. Biol., *3*: 144–155.

NAGATA, T. 1963. The molecular synchrony and sequential replication of DNA in *Escherichia coli*. Proc. Natl. Acad. Sci., *49*: 551–559.

YOSHIKAWA, H., and N. SUEOKA. 1963. Sequential replication of *Bacillus subtilis* chromosome. 1. Comparison of marker frequencies in exponential and stationary growth phases. Proc. Natl. Acad. Sci., *49*: 559–566.

DISCUSSION

BUTLER: I should like to mention an idea put forward at a meeting of the British Biophysical Society December 1962 in a paper by Godson, Barr, and myself (see Fig. 1). The DNA polymerase is pictured as a disc with two holes or slots, through each of which one of the strands of the DNA passes. As it passes up the primer, this disc is pictured as rotating relative to it and so separating and unwinding the strands. The energy required for this could probably be provided by the energy of condensation of the triphosphates in the condensation process (about 8 kcals per nucleotide pair). The two new double fibers of DNA would initially be loosely wound round each other, as is often seen in chromosomes, but thermal agitation in the cell would tend to unwind them. The idea suggested by Dr. Cairns that the primer strand itself may rotate would get over the need of the two new fibers of DNA to be wound round each other.

Part III

CHROMOSOME PAIRING

HISTORICAL PERSPECTIVES

Blakeslee's discovery in *Datura* of the globe and other morphological mutants that showed exceptional breeding behavior and Belling's finding that each such mutant had an extra chromosome that was different in different mutants laid the foundation for advances in many areas of cytogenetics (Blakeslee, Belling, and Farnham, Paper 29). Belling recognized the experimental value of plants with extra chromosomes in the analysis of meiosis. Blakeslee recognized the importance of chromosome balance in relation to its effect on the phenotype. Comparisons of the various chromosome configurations and their frequencies at meiosis led Belling to conclude that the associations at the first division of meiosis were determined by homologies and hence could be used to identify chromosome ends (Belling and Blakeslee 1924 and Paper 11).

Belling (1925) reported that the results of his studies of semisterility in crosses between species A of *Stizolobium* and several Asiatic species or varieties

> were those that would follow if a small segment of each of two non-homologous chromosomes had been interchanged in the ancestry of species A. Some similar translocation of a segment between non-homologous chromosomes has now been proved genetically to occur in *Drosophila*, and segmental interchange has apparently been shown by the microscope to occur also in certain *Datura* mutants.

His cytological observations in *Datura* referred to above were on the class of trisomics which Blakeslee had termed *tertiary*, a third type of trisomic whose offspring included two different primary trisomics in contrast to primary and secondary trisomics whose offspring included only one. In the tertiary trisomics that arose only in the crosses with a diploid race called "B-white," the extra chromosome in the trivalent formed at meiosis was a smaller one and always attached by one end only with the bivalent (Belling and Blakeslee, Paper 11). To explain these observations, they hypothesized that there had been segmental interchange between two nonhomologous chromosomes in the ancestry of the B-white race. In quinquevalents observed in another tertiary trisomic line, the extra chromosome formed a connecting link between the two nonhomologous bivalents. The configurations in each case would be explainable if the extra chromosome were an interchange chromosome.

The interchange hypothesis was the final step needed to integrate the cytological and breeding information that had been accumulating on the Oenotheras and to aid in explaining the numerous problems the various races presented. A circle of chromosomes had been observed even in the early studies of *Oenothera* (Gates 1908). Cleland (Paper 12) had concluded from his observations on large circles of chromosomes in *Oenothera* that the order of the chromosomes in a particular race or hybrid was constant and that the alternate disjunction led to the formation mainly of parental chromosome combinations. Since the chromosomes were associated end to end, *Oenothera* was still assumed to have a continuous spireme in the early meiotic stages followed by segmentation into pairs and therefore to be an example of telosynapsis, even after this had been disproved in many other species. Circle formation had been assumed by many to be a genetic character. Renner, in a long series of well-planned and executed genetic experiments beginning in 1914, had discovered several types of lethals that were the basis for much of the unusual genetic behavior (Renner 1946).

Sturtevant's (1926a) review of Renner's studies called attention to that particular work and made the information readily available. Muller (1917) had published a paper entitled "An *Oenothera*-like case in *Drosophila*" in which, by using a balanced lethal situation, he was able to duplicate certain unusual breeding behaviors observed in *Oenothera*. Sterling Emerson (1928), following a suggestion of Belling's, was one of the first to apply the segmental-interchange hypothesis to circle formation in *Oenothera*. He joined the staff at the California Institute of Technology the same year and continued his studies there in collaboration with Sturtevant. Håkansson (1928) described the rings in *Oenothera* and discussed the interchange hypothesis as it might apply, citing Belling's 1925 paper. Papers by Blakeslee and Cleland (1930, presented in April 1929 before the American Philosophical Society), Darlington (1929), Sturtevant and Dobzhansky (1930), and Sturtevant (1931) all utilized the interchange hypothesis to interpret circle formation and the breeding behavior in *Oenothera*.

Brink had discovered semisterility in maize (1927) and reported its breeding behavior in 1929 (Brink and Burnham). Cytological observations on it and two other naturally occurring cases were reported by Burnham (Paper 13) together with genetic linkage data and observations on trisomics that occurred among the progeny. McClintock used one of these interchanges, T8–9a, first designated semisterile-2, (Paper 2) in the first pachytene studies to locate the breakpoints in the chromosomes. Creighton and McClintock also used the interchange (Paper 21) to demonstrate the fact that a cytological exchange accompanies genetic recombination. Translocations produced by X-rays were reported also in *Drosophila* (Dobzhansky 1929), together with their genetics and cytology, the latter from metaphase in oogonial cells.

. With the discovery that interchanges and other chromosomal aberrations could be produced experimentally by X-rays, other radiations, and other agents, it soon became apparent that collections of large numbers of them in a particular species would have many uses. Examples of such collections and sources of information on them are: *Drosophila* (Lindsley and Grell 1967); maize (Longley 1961, listed by Lambert 1969); barley (Ramage, Burnham, and Hagberg 1961); and *Neurospora* (listed by Barratt and Ogata, 1974, 1975).

One of the remarkable facts about chromosome behavior is that the chromosomes usually associate in pairs only in preparation for reduction at meiosis. Observations at the early stages of meiosis in plants with multivalents showed that only two chromosomes were paired at any one point (Newton and Darlington, Paper 14, using triploid and tetraploid tulips, and Darlington, 1929, using aneuploid hyacinths). At a later stage of meiosis, when the chromosomes became visibly double, one pair of chromatids repelled the homologous pair of chromatids except at certain points at which there appeared to be a change of partner, a chiasma. The nature of the chiasma will be discussed later. This 2-by-2 pairing led Darlington to ask the question: How does one account for the failure of chromosomes to pair at somatic mitosis and then to pair at meiosis? He proposed the following explanation. The beginning of meiosis occurs early, before the chromosomes have had time to become double. Since the threads are single, they then associate to form pairs. Darlington referred to this as the "precocity theory of meiosis." Contrasted with this, when somatic mitosis begins, each chromosome is already visibly double and thus is prevented from pairing with its homologue which is also double. Darlington stated that "the affinities of chromosomes are satisfied by association in twos" and thus accounted for the difference in pairing between somatic mitosis and meiosis. Huskins (1932) used the same basic reasoning but interpreted meiosis as being initiated by a delayed splitting of the chromosomes. His simple statement (1933) was that "at all stages of mitosis and meiosis, chromosome threads are attracted in pairs and pairs of pairs are repulsed." Darlington believed that in mitosis the chromosomes were single during the preceding resting stage, whereas cytological observations by Sharp (1929), by Huskins and Smith (1932) and by Huskins (1933) showed that they were visibly double at anaphase of the preceding division. Hence Huskins stated that meiosis was initiated by suppression of the chromosome doubling at the preceding anaphase. The differences in Darlington's and Huskins' interpretations led to a heated discussion at a late-afternoon session of the Genetics Congress in Ithaca, New York, in 1932. In agreement with the general principle that pairing of chromosomes is in twos were Huskins' and Smith's (1934) observations in *Fritillaria meleagris*, a species with localized chiasmata, that the chromosomes at leptonema were double in some regions, single in others, and that at zygonema pairing was in unsplit regions of the chromosomes.

Data from genetic studies in *Drosophila* using attached-X chromosomes (Anderson 1925) and in *Neurospora* using ascospores dissected in order (Lindegren 1933) were explainable if the first division of meiosis were reductional at the centromere. An explanation Darlington offered for this behavior was that division of the centromere was postponed until the second division of meiosis (Darlington 1937, p. 134).

Additional evidence on this point was obtained by Huskins and Spier (1934) from their work with three plants of *Triticum vulgare*, each with a different heteromorphic pair, analyzed at metaphase I and anaphase I of meiosis. In one plant one member of the pair had a median centromere; the other member had a terminal centromere as a result of losing an entire arm. Although equational separations at division I would have been easy to recognize, none were found (56 nuclei were analyzed, but many more were observed in which this would have been recognized if present). In the other two plants one member of the pair was shorter than the other. Crossing-over in the unequal arm would have resulted in equational separation at division I with a normal and a shorter chromatid passing to each pole if the first division were reductional at the centromere. This was seen in 4 of 39 nuclei for one plant and in 11 of 48 in the other. Hence the observations on the three plants were not in disagreement with the conclusion that the first division is reductional at the centromere.

The tendency for 2-by-2 pairing extends to pairing that is not between homologous segments. In maize whenever a chromosome is present as a univalent at pachynema, as in trisomic and in monosomic plants, that chromosome often pairs with itself. The foldback occurs at different points but with a strong tendency for the ends to be paired (McClintock, Paper 16). The association appears to be as intimate as in pairing between homologues. Some derivative types of chromosomes that McClintock described might have resulted from exchanges in such paired regions. Nonhomologous association is found also in interchange heterozygotes in maize. In most of these the position of the pachytene cross is variable. Since there may be intimate associations at all positions, it is obvious that in all but one (the true) position, there must have been nonhomologous pairing. Darlington has interpreted this as due to torsion and not true pairing.

In experiments designed to furnish situations in which two or more genetically marked chromosomes or fragments had not crossed over with their homologues (and therefore assumed not to be paired), Grell (1962) reported data showing that these chromosomes disjoined from each other with relatively high frequencies, yet recombination values were not affected. She postulated that in *Drosophila* at meiosis there is pairing for exchange (exchange pairing); after that, the chromosomes that have not crossed over are available for 2-by-2 pairing without regard to homology (distributive pairing). Extensive studies have all been given the same explanation.

The basic argument Grell gave in favor of two kinds of pairing in stages following zygonema was that recombination values were normal in individuals showing high frequencies of distributive pairing. To obtain information related to this theory, behavior of nonhomologous univalents in maize plants that were doubly trisomic ($2n + 1 + 1 = 20 + 1 + 1$) was studied cytologically at pachynema and later stages of meiosis (Michel and Burnham 1969). In 13 different doubly trisomic combinations, there were two unpaired univalents at the pachytene stage in some cells, and in others these two chromosomes were paired. Occasionally 11 pairs were observed at metaphase I in the eight plants examined at that stage. There was no evidence for a separate stage of distributive pairing that followed exchange pairing.

Weber (1969) used plants with two extra chromosomes, in most cases either chromosome 4 or 6 plus a "B" chromosome, but he observed no cells that definitely had 11 bivalents (2 were questionable) at diakinesis or metaphase. Since the 11-11 and 10-12 disjunctions at anaphase I were about equally frequent, he concluded that distributive pairing is either not present in maize or is operating at a much lower level than in *Drosophila*.

Experimental data in *Drosophila* reported by Novitski (1975) strongly suggest that the associations that lead to the disjunction of nonhomologues are set up prior to the time of crossing over and not following exchange pairing.

As mentioned earlier, McClintock (Paper 16) had noted a strong tendency for pairing to begin at the ends in foldbacks of univalents. Conclusive evidence in maize that pairing at meiosis is not initiated at the centromere but at or near the ends was furnished by crosses between chromosomal interchange lines that involved the same two chromosomes but with the breaks in the second interchange in the other arm of each chromosome (Burnham, et al. 1972). At pachynema the "pairs" that were formed in that type of intercross were associated at the interchanged ends, but the intercalary segment of one "pair" including its centromere was not homologous with the intercalary segment of the other chromosome with its centromere. If pairing had been initiated at the centromere, the homologous centromere-bearing intercalary segments would have been paired.

The reduced crossing-over observed in certain strains of *Drosophila* was shown genetically to be the result of heterozygosity for an inversion (Sturtevant 1926b). Bridges and Li (1937) reported a salivary-gland analysis of this inversion (In(3R)C). The first report of reverse pairing at pachynema in such an inversion heterozygote was in maize (McClintock 1931). Painter (1934) showed similar configurations in the polytene salivary-gland chromosomes of an inversion heterozygote in *Drosophila*.

Model for Chromosome Pairing

The views of Darlington and Huskins on chromosome pairing, namely, that "chromosome threads are attracted in pairs, and pairs of pairs are repulsed," seemed difficult to reconcile once it was learned that the chromosomes had actually doubled during interphase preceding the stages that followed in mitosis or meiosis. However, their line of reasoning is accommodated in a current model for chromosome pairing.

Chromosome pairing, as interpreted today, may involve the following processes. The chromosomes are attached to the nuclear membrane at their ends and at certain other places. Homologous chromosomes presumably have membrane-attachment sites that are not widely separated or somehow come into close proximity preceding pairing. During interphase before meiosis, most of the DNA is synthesized, perhaps 99.7 percent (Hotta, Ito, and Stern 1966). The 0.3 percent (determined for *Lilium*) that is not synthesized is represented by short segments scattered along all chromosomes. As the cell enlarges in preparation for meiosis, the homologous chromosomes, which have their ends associated, are brought into closer proximity and pair loosely through associations between the unreplicated segments of homologous chromosomes; thus Darlington's and Huskins' views may still be correct at least in part. The synaptonemal complex (Moses, Paper 17) is then synthesized and holds the four chromatids of the bivalent together in a rigid form. In the process of synaptonemal complex formation during zygonema, the unreplicated regions complete their replication. The synaptonemal complex is not associated with the entire length of the chromatin, only at certain sites. Thus the chromosome pairing observed at pachynema and stabilized by the synaptonemal complex does not represent precise chromatin alignment in a base-for-base fashion. Although the chromatin may be closely aligned due to the pairing of the segments left unreplicated at the premeiotic S-phase, it is not paired point for point throughout the chromosome.

The synaptonemal complex appears to hold the homologues together in order for crossing-over to occur in an efficient manner, as opposed to the inefficient occurrence of crossing-over in mitosis (see Part IV). After crossing-over is completed, the synaptonemal complex is released from the chromosomes in diplonema, except in regions where chiasmata exist (the synaptonemal complex in these regions is released eventually). Subsequently, the chromosomes enter metaphase and then disjoin.

Genes Affecting Meiotic and Postmeiotic Events

Genes affecting meiosis have been found in many different species. The first one reported was a recessive gene (c(3)G) located in chromosome III of *Drosophila*, discovered in 1917. In homozygous females there was complete linkage in the two chromosomes tested, in chromosome II and the X (Gowen and Gowen 1922). Since triploids and various intersex types appeared among the progeny, it probably was an asynaptic gene. It did not affect meiosis in males. Gowen (1933) discussed similarities to the behavior in asynaptic maize, but there is no published account of the cytological observations referred to but not described in the abstract (Gowen 1932). Meyer (1964) reported that the synaptonemal complex core structures characteristic of electron-microscopic preparations of normal oocyte nuclei were absent in females homozygous for this gene.

A frequent recessive mutant in maize cultures is male-sterility. Emerson at Cornell University had accumulated a number of these in the course of his genetic experiments. Beadle (1932a) undertook a genetic study of these and other male sterile mutants from various sources. Most of them showed normal behavior in megasporogenesis, as shown by normal seed set. A few of these were examined cytologically at microsporogenesis and showed normal behavior through meiosis. The breakdown that resulted in shriveled anthers occurred after those stages. A few had poor seed set. For one of those with poor seed set, there was at diakinesis and metaphase I during microsporogenesis a lack of synapsis (asynapsis, designated *as*) in many or most of the chromosomes (Beadle and McClintock, Paper 18). This was the first published cytological analysis of a genic disturbance of meiosis. It was followed by a more extensive study of this mutant (Beadle 1930) and others affecting meiosis, e.g., polymitotic, characterized by additional cytokineses without chromosome duplication immediately following the second meiotic division (Beadle 1932b). Genes for asynapsis, sometimes referred to as "desynapsis" since the chromosomes are paired in early prophase stages, have been reported in many different species.

An important question concerning the behavior of polyploids is: How has almost complete bivalent pairing evolved in species that arose probably by spontaneous chromosome doubling (autopolyploidy) or by crossing of different related species followed by doubling? The clue to the answer to this question was furnished by Riley and Chapman (Paper 19) and at about the same time by Sears and Okamoto (1958) in studies of common bread wheat (*Triticum vulgare*, renamed *T. aestivum*). This species is a hexaploid ($2n = 42$) in which the genomes of three different diploid ($2n = 14$) species are combined. Although there is considerable homology between the chromosomes of these three species, usually only bivalents are formed in *T. aestivum*; and even in the haploid there is only an occasional association. In a particular 20-chromosome haploid (a nullihaploid), i.e., lacking one chromosome, later identified as chromosome 5B, there was extensive pairing between different chromosomes which was in marked contrast to the very low frequency in a 21-chromosome haploid with the complete genome. The interpretation was that the missing chromosome in the nullihaploid carried a gene or genes that suppressed pairing between chromosomes with only partial homology. A similar genetic situation has been reported in hexaploid oats (*Avena sativa*) (Gauthier and McGinnis 1968). Hence an amphiploid, and presumably also an autopolyploid, could evolve from a situation with frequent multivalents and resultant sterility to one with mostly bivalents and nearly complete fertility by a single mutational event. This discovery and the extensive subsequent studies by Feldman (1966) represent an important advance in our knowledge about polyploids and have sparked new thinking and approaches in plant breeding. Genes have been found in *Neurospora* that increase recombination (Jessop and Catcheside 1965; Smith 1966). As noted earlier, closeness of pairing can be affected by certain morphological features, an example being the effect of the abnormal chromosome 10 in maize in bringing about a more intimate pairing of the chromosomes that can lead to increased recombination.

Editors' Comments
on Papers 11 Through 19

Paper 11 reports the typical configuration that Belling observed in primary and secondary trisomics in *Datura* which led him to conclude the associations at division I of meiosis were determined by homologies and therefore could be used to identify chromosome ends. The paper also includes cytological observations and genetic results for tertiary trisomics. One tertiary trisomic formed chains (strings)-of-five chromosomes in which the extra chromosome was the connecting link between the two nonhomologous bivalents. These observations led Belling and Blakeslee to propose the interchange hypothesis to explain the configurations and also to explain Belling's earlier results on the breeding behavior of semisterility in *Stizolobium* (velvet bean). The extra chromosome in tertiary trisomics is an interchange chromosome. This hypothesis has been important in many areas of research, including cytogenetic studies in man. Rarely has one hypothesis had such far-reaching applications to so many different areas in biological science.

Paper 12 established the fact that *Oenothera* chromosomes in the large circles at meiosis were in a definite order and that the usual alternate type of segregation produced only the two parental combinations of the chromosomes that comprised the ring. This report also set the stage for the application of Belling's interchange hypothesis to the analysis of the unusual cytogenetic behavior in that species.

Burnham's (1930) paper (Paper 13) reports the first genetical and cytological analysis of naturally occurring interchanges in maize, including crosses between them.

The excerpts from the reports by Newton and Darlington (Paper 14) and Darlington (Paper 15) document two fundamental generalizations Darlington advanced. The first was that at prophase of meiosis only two chromosomes are associated at any one point, this being true not only for bivalents but also for trivalents and quadrivalents. This generalization led to his precocity hypothesis of meiosis, that is, that the reason homologues pair at meiosis is that meiotic pairing begins too early, before the chromosomes have become visibly double. Hence homologues pair because they are single stranded. Homologues do not pair at somatic mitosis because they are already visibly double. The second generalization was that in the later stages of meiosis after the chromosomes have become visibly double, the chromosome pairs are held together only by the changes of partner (chiasmata) that occur at certain points (see Part IV for Darlington's paper demonstrating cytologically that these are points at which a crossover had already occurred).

These papers, together with the publication of Darlington's book *Recent Advances in Cytology* (first published in 1932, revised in 1937, and a supplement added in 1964), had a remarkable impact on cytology and genetics. As discussed by Schrader (1948), Darlington brought this about by making "an analysis of the needs of the geneticist and the difficulties of the cytologist" and then saying in effect, "Let us cut the Gordian knot and reduce this mass of cytological information to what is essential, always keeping in mind the basic conclusions that have been reached through genetic procedure." Darlington's generalizations mentioned above were the result. Again, quoting Schrader:

> Under the impact of this attack, the unwieldy and undigested mass of cytological data seemed to dissolve and reprecipitate as a usable system. Viewing this change, the geneticist gained a new sense of power and once more began to look on cytology as an aid to his researches and something that he himself could employ.

Paper 16, excerpts from McClintock's full report, shows that the tendency for 2-by-2 pairing is so strong that associations may occur between nonhomologous chromosomes or segments of chromosomes. McClintock reported observations of nonhomologous associations of supernumerary B-chromosomes, univalents in monosomic, trisomic, monoploid, and asynaptic plants and also in inversion, deficiency, and translocation heterozygotes as well as in normal diploids. Her conclusions regarding chromosome pairing and the consequences of nonhomologous association represent important contributions to cytogenetic knowledge.

Often the significance of a cytological structure is not generally appreciated until it is observed in a variety of organisms. This was true of the synaptonemal complex that was discovered by Moses and independently by Fawcett but not widely known by cytogeneticists until the early sixties. Moses' paper (Paper 17) describes the appearance of a distinctive chromosomal structure at meiotic prophase as viewed with the electron microscope. Hundreds of papers have followed in recent years describing the appearance of the synaptonemal complex (a term first used by Moses in 1958) in meiotically paired chromosomes of all chiasma-forming organisms. Various observations have led to the interpretation that the complex provides the chromosomal stability and configuration for the high levels of crossing-over that normally occur in meiosis as compared with mitosis. Although Moses did not have sufficient information in 1956 to suggest the structure's probable function, he realized its possible significance as indicated in his concluding paragraph:

> It is realized that our present knowledge of chromosomes based on observations with the light microscope does not lead

us to expect the structures described. However, they may well represent either a special or transient form of a fundamental structure whose role in chromosome function is less obscure.

Paper 18 constitutes the first cytological documentation of a gene (asynaptic) affecting meiosis. This gene in maize was studied in detail by Beadle (1933) and later by Miller (1963). That meiosis is under genetic control was emphasized by discovery of this mutant. The wide array of meiotic mutants discovered in various organisms has advanced greatly our knowledge of the meiotic process.

Paper 19 by Riley and Chapman represents a very important advance in our understanding of the essentially diploid behavior in allopolyploids. The cytogenetic results Sears obtained in hexaploid wheat had indicated considerable homology between corresponding chromosomes in the three genomes. Paper 19 demonstrates that the lack of pairing between chromosomes that are at least partially homologous is the result of a gene or genes on a single chromosome pair. Thus it is possible that frequent multivalent formation and its accompanying irregular breeding behavior and sterility can be changed to the normal behavior seen in a diploid by a single gene mutation. Sears and Okamoto reached similar conclusions in the same year (1958).

REFERENCES

Anderson, E. G. 1925. Crossing over in a case of attached X chromosomes in *Drosophila melanogaster. Genetics* **10**:403–417.

Barratt, R. W. and W. N. Ogata. 1974. *Neurospora* Stock List. Seventh Revision (June 1974) Part V. Aberration Stocks. *Neurospora Newsletter* **21**:82–87.

_____. 1975. First supplement to *Neurospora* Stock List Part V. Aberration Stocks. *Neurospora Newsletter* **22**:26–27.

Beadle, G. W. 1930. Genetical and cytological studies of Mendelian asynapsis in *Zea mays. Cornell Univ. Agric. Exp. Stn. Mem. 129.*

_____. 1932a. Genes in maize for pollen sterility. *Genetics* **17**:413–431.

_____. 1932b. A gene in *Zea mays* for failure of cytokinesis during meiosis. *Cytologia* **3**:142–155.

_____. 1933. Further studies of asynaptic maize. *Cytologia R.C.* **4**: 269–287.

Belling, J. 1925. A unique result of certain species crosses. *Z. Indukt. Abstamm. –Vererbungsl.* **39**:286–288.

_____ and A. F. Blakeslee. 1924. The configurations and sizes of the chromosomes in the trivalents of 25-chromosome *Daturas. Proc. Nat. Acad. Sci. U.S.A.* **10**:116–120.

Blakeslee, A. F., and R. E. Cleland. 1930. Circle formation in *Datura* and *Oenothera. Proc. Nat. Acad. Sci. U.S.A.* **16**:177–183.

Bridges, C. B., and J. C. Li. 1937. (p. 301) in Morgan, T. H., C. B. Bridges, and J. Schultz. Constitution of the germinal material in relation to heredity. *Carnegie Inst. Washington Yearb.* **36**:298–305.

Brink, R. A. 1927. The occurrence of semi-sterility in maize. *J. Hered.* **18**:266–270.

———— and C. R. Burnham. 1929. Inheritance of semi-sterility in maize. *Am. Nat.* **63**:301–316.

Burnham, C. R., J. T. Stout, W. H. Weinheimer, R. V. Kowles, and R. L. Phillips. 1972. Chromosome pairing in maize. *Genetics* **71**:111–126.

Darlington, C. D. 1929. Ring-formation in *Oenothera* and other genera. *J. Genet.* **20**:345–363.

————. 1937. *Recent Advances in Cytology*. P. Blakiston's Son & Co., Philadelphia.

Dobzhansky, T. 1929. Genetical and cytological proof of translocations involving the third and the fourth chromosomes of *Drosophila melanogaster*. *Biol. Zentralbl.* **49**:408–419.

Emerson, S. H. 1928. The mechanism of inheritance in *Oenothera*. Ph.D. Thesis. Univ. of Michigan, Ann Arbor.

Feldman, M. 1966. The effect of chromosomes 5B, 5D, and 5A on chromosomal pairing in *Triticum aestivum*. *Proc. Nat. Acad. Sci. U.S.A.* **55**:1447–1453.

Gates, R. R. 1908. A study of reduction in *Oenothera rubrinervis*. *Bot. Gaz.* **46**:1–34.

Gauthier, F. M., and R. C. McGinnis. 1968. The meiotic behavior of a nulli-haploid plant in *Avena sativa* L. *Can. J. Genet. Cytol.* **10**:186–189.

Gowen, J. W. 1932. Meiosis as a genetic character (Abstr.). *Proc. Sixth. Int. Congr. Genet.* **2**:69–70.

————. 1933. Meiosis as a genetic character in *Drosophila melanogaster*. *J. Exp. Zool.* **65**:83–106.

Gowen, M. S., and J. W. Gowen. 1922. Complete linkage in *Drosophila melanogaster*. *Am. Nat.* **56**:286–288.

Grell, R. F. 1962. A new hypothesis on the nature and sequence of meiotic events in the female of *Drosophila melanogaster*. *Proc. Nat. Acad. Sci. U.S.A.* **48**:165–172.

Håkansson, A. 1928. Die Reduktionsteilung in den Samenanlagen einiger Oenotheren. *Hereditas* **11**:129–181.

Hotta, Y., M. Ito, and H. Stern. 1966. Synthesis of DNA during meiosis. *Proc. Nat. Acad. Sci. U.S.A.* **56**:1184–1191.

Huskins, C. L. 1932. Factors affecting chromosome structure and pairing. *Proc. and Trans. R. Soc. Can.* **26**, Ser. 3, Sect. V:17–28.

————. 1933. Mitosis and meiosis. *Nature* **132**:62–63.

———— and S. G. Smith. 1932. Observations bearing on the mechanism of meiosis and crossing over. *Proc. Sixth Int. Congr. Genet.* **2**:95–96.

————. 1934. Chromosome division and pairing in *Fritillaria meleagris*: the mechanism of meiosis. *J. Genet.* **28**:397–406.

Huskins, C. L., and J. D. Spier. 1934. The segregation of heteromorphic homologous chromosomes in pollen-mother cells of *Triticum vulgare*. *Cytologia* **5**:269–277.

Jessop, A. P., and D. G. Catcheside. 1965. Interallelic recombination at the *his-I* locus in *Neurospora crassa* and its genetic control. *Heredity* **20**:237–256.

Lambert, R. J. 1969. List of reciprocal translocation stocks maintained by the Maize Genetics Cooperative. *Maize Genet. Coop. News Letter* **43**:216–230.

Lindegren, C. C. 1933. The genetics of *Neurospora* III. Pure bred stocks and crossing-over in *N. crassa. Bull. Torrey Bot. Club* **60**:133–154.

Lindsley, D. L., and E. H. Grell. 1967. Genetic variations of *Drosophila melanogaster. Carnegie Inst. Washington Publ. No. 627*

Longley, A. E. 1961. Breakage points for four corn translocation series and other corn chromosome aberrations. U.S. Dep. Agric., *Agric. Res. Serv. ARS 34–16:*1–40.

McClintock, B. 1931. Cytological observations of deficiencies involving known genes, translocations and an inversion in *Zea mays. Missouri Agric. Exp. Stn. Res. Bull.* **163**:1–30.

Meyer, G. F. 1964. A possible correlation between the submicroscopic structure of meiotic chromosomes and crossing over. *Proc. 3rd Eur. Reg. Conf. Electron Microsc.* **B**:461–462.

Michel, K. E., and C. R. Burnham. 1969. The behavior of nonhomologous univalents in double trisomics in maize. *Genetics* **63**:851–864.

Miller, O. L. 1963. Cytological studies in asynaptic maize. *Genetics* **48**:1445–1466.

Muller, H. J. 1917. An *Oenothera*-like case in *Drosophila. Proc. Nat. Acad. Sci. U.S.A.* **3**:619–626.

Novitski, E. 1975. Evidence for the single phase pairing theory of meiosis. *Genetics* **79**:63–71.

Painter, T. S. 1934. A new method for the study of chromosome aberrations and the plotting of chromosome maps in *Drosophila melanogaster. Genetics* **19**:175–188.

Ramage, R. T., C. R. Burnham, and A. Hagberg. 1961. A summary of translocation studies in barley. *Crop Sci.* **1**:277–279.

Renner, O. 1946. Artbildung in der Gattung *Oenothera. Naturwissenschaften* **33**:211–218.

Schrader, F. 1948. Three quarter-centuries of Cytology. *Science* **107**:155–159.

Sears, E. R., and M. Okamoto. 1958. Intergenomic chromosome relationships in hexaploid wheat. *Proc. X Int. Congr. Genet.* **2**:258–259.

Sharp, L. W. 1929. Structure of large somatic chromosomes. *Bot. Gaz.* **88**:349–382.

Smith, B. R. 1966. Genetic controls of recombination I. The *recombination-2* gene of *Neurospora crassa. Heredity* **21**:481–498.

Sturtevant, A. H. 1926a. Renner's studies on the genetics of *Oenothera. Q. Rev. Biol.* **1**:283–288.

———. 1926b. A crossover reducer in *Drosophila melanogaster* due to inversion of a section of the third chromosome. *Biol. Zentralbl.* **46**:697–702.

———. 1931. Genetic and cytological studies on *Oenothera*. I. *Nobska, Oakesiana, Ostreae, Shulliana*, and the inheritance of old-gold flower-color. *Z. Indukt. Abstamm.-Vererbungsl.* **59**:367–380.

——— and T. Dobzhansky. 1930. Reciprocal translocations in *Drosophila* and their bearing on *Oenothera* cytology and genetics. *Proc. Nat. Acad. Sci. U.S.A.* **16**:533–536.

Weber, D. F. 1969. A test of distributive pairing in *Zea mays. Chromosoma* **27**:354–370.

11

Reprinted from *Natl. Acad. Sci. (U.S.A.) Proc.* **12**(1):7–11 (1926)

ON THE ATTACHMENT OF NON-HOMOLOGOUS CHROMOSOMES AT THE REDUCTION DIVISION IN CERTAIN 25-CHROMOSOME DATURAS

By John Belling and A. F. Blakeslee

Carnegie Institution of Washington

Communicated November 17, 1925

There are now at least three classes of 25-chromosome Daturas which have been more or less thoroughly studied, both by breeding experiments and by microscopical examination of their chromosomes. Of these, the main class, the *primaries*, come directly from triploids, or from non-disjunction in diploids; they give normally only diploids and primaries like themselves in their progenies; the different groupings of the chromosomes in their trivalents (Fig. 1) are those which would result in random assortment of three homologous chromosomes; they show triploid in-

FIGURE 1	FIGURE 2
Three trivalents of the primary $2n + 1$ mutant, *Rl*, with the extra chromosome in set I. The configurations are: (1) chain of three, (2) open V, and (3) ring and rod. These were drawn with the camera from iron-acetocarmine preparations.	Three trivalents of the secondary $2n + 1$ mutant, *Sg*, with the altered chromosome in set I. The configurations are: (1) closed V, (2) ring and rod, and (3) another closed V.

heritance of genes in the trivalents, in the two or more cases which have been extensively investigated; non-disjunction is apparently no more abundant than in diploids; and non-homologous chromosomes have not been found attached to one another at the maturation divisions.

The second class of 25-chromosome Daturas, the *secondaries*, occur, as exceptions, in the offspring of the corresponding primaries (and of diploids and, of course, other primaries), they give normal diploids, secondaries like themselves, and the corresponding primaries, in their progeny; the groupings of the chromosomes in their trivalents (Fig. 2) point to one chromosome of the three having two homologous ends (and consequently being probably composed of two homologous halves), so that assortment is not at random; diploid inheritance of genes in the trivalent occurred in the one case investigated;[1] non-disjunction is no commoner than in diploids; and non-homologous chromosomes have not been seen connected.[2] Notwithstanding that one of the three chromosomes of the trivalent differs from the other two, measurements show that the chromosomes of the secondaries do not differ appreciably in size from those of their primaries.

In the third class of 25-chromosome Daturas, only one, *Wy*, has been as yet sufficiently investigated, both as to its breeding and its chromosomes, to allow deductions to be drawn (although some significant data have also been obtained with regard to *Hg*). The following statements then relate, primarily at least, only to *Wy*. This 25-chromosome mutant of the third class occurs in the offspring of a primary, *Pn* (which came from a *Pn* after being crossed with the *B* race of *Datura Stramonium*,[3]); *Wy* gives in its progeny, diploids, *Wy* and *Pn;* in the groupings of the three chromosomes which belong to set IX one of the three is often (in that form of *Wy* which was examined with the microscope) attached to one of the large chromosomes of bivalent I; and thus would not give free assortment nor triploid inheritance; notwithstanding this, the form of *Wy* tested by breeding showed triploid inheritance; non-disjunction was not more abundant than in diploids (but was common in *Nb*); and, as already stated, the non-homologous chromosomes, I and IX, were frequently united.

Hence there are apparently two forms of *Wy;* one which gives triploid inheritance, and one which should give approximately diploid inheritance.

The *Wy* trivalent was, in all cases seen in metaphase, an open V (although one ring and rod was seen in late prophase), but the ring and rod trivalent has been seen in the more or less parallel mutant *Hg*, which also had M-shaped chains of 4 or 5 chromosomes, groupings not yet identified with certainty in *Wy*, though they probably do occur. The trivalents observed and measured in *Wy* were 13 cases of two of the largest chromosomes, set I, forming an unequal V with one of the *Pn* chromosomes of set IX, which latter is only half the size of chromosomes I. This small chromosome was always at one free end of the V, never in the middle. Also there were two cases of an open V of three chromosomes of set IX. Lastly one case was measured where a V consisted of two small chromosomes of set IX, and one large chromosome of set I at a free end of the V.

From the measurements which have been made, it can be concluded that the chromosomes of sets I and IX in *Pn*, *Rl*, and *Wy*, are of the same sizes.

A working hypothesis which may be seen to fit most or all of the facts so far discovered is that in the ancestry of the B diploids there has been segmental interchange between two non-homologous chromosomes. In plants, such segmental interchange probably occurred between two non-homologous chromosomes in the ancestry of *one species* of Stizolobium (Mucuna).[5,6]   In crosses between this species and three allied species and varieties, dwarf mutants with peculiar leaves appeared in F$_2$, in the proportion of about 0.5 per cent.[7,8]   Similarly *Wy* might appear in the F$_2$ of crosses of normal and B diploids.   The F$_1$ plants of the Stizolobium crosses were semisterile, as were also the F$_1$ plants of the crosses of normal and B diploids in Datura.

The following are the chief points that must be included in any working hypothesis:

(1) The tertiary mutants *Wy* and *Hg* have 25 chromosomes, as counted in the second metaphase.

(2) They show commonly 11 bivalents and one trivalent at the first metaphase.

(3) Non-homologous chromosomes in two particular sets (I and IX in *Wy*) are often or usually combined into trivalents (quadrivalents or quinquevalents).   (See Figs. 3 and 4.)

FIGURE 3	FIGURE 4
Three trivalents from the tertiary 2*n* + 1 mutant, *Wy*. The large chromosomes are from set I, while the small one comes from set IX, and is the altered chromosome (9).	Three configurations from the tertiary 2*n* + 1 mutant, *Hg*. The large chromosomes belong to set I, and the smaller ones (shown in outline) perhaps to set VI.

(4) In such trivalents the two chromosomes of one set are always together (that is, not separated by a non-homologous chromosome).

(5) Thus the single chromosome of the other set is connected by one end only with the non-homologous bivalent.   (Fig. 3.)

(6) Sometimes the three chromosomes of the one set (IX in *Wy*) form a trivalent.

(7) Only one chromosome of the three is ever connected with non-homologous chromosomes.

(8) The open V is the common configuration of the trivalents, but the ring and rod also occurs.

(9) The triangle or closed V does not occur.

(10) The three chromosomes (in set IX) are equal in size to the three chromosomes of the primary *Pn* (set IX).

(11) There is no perceptible difference in size or shape between the three chromosomes of set IX in *Wy*.

(12) Thus one of the chromosomes of set IX, in *Wy*, has one end normal, and attracting its fellow of set IX; while the other end is changed so that it attracts one of the ends of chromosome I.

(13) This could have been brought about by segmental interchange, at the constriction of chromosome IX; one-half of this chromosome interchanging with an equal length of chromosome I, in the ancestry of the B diploids.

In addition to these facts, the following data from the pedigreed cultures have to be reckoned with by any hypothesis.

(1) *Pn* may give *Wy* after crossing with the B strain, but without such crossing does not.

(2) Some *Wy* plants throw *Pn*.

(3) Some *Wy* plants give triploid ratios for genes in the chromosomes of set IX.

(4) *Pn* crossed twice by B diploids may give ratios, for genes in the trivalents, which show increase in the number of recessives.

(5) *Rl* after crossing with B diploids, may give ratios for genes in the IX bivalent, which are different to the ordinary diploid ratios.

(6) Crosses between the normal and B strains have given F_1 plants with 50 per cent of the pollen and ovules aborted.

A detailed working out of the hypothesis has shown that both the chromosomal and the cultural data fit the hypothesis tolerably well in most points, though there remain a few seeming exceptions (see note at end of paper). Since this hypothesis fits these facts, it may well be employed until a better is found.

Of the other $2n + 1$ mutants which apparently more or less resemble *Wy*, in the configurations of their chromosomes, viz., *Hg, Nb, Ph, Dv* and *X*; *Hg* has been partially studied (Fig. 4). Its largest chromosomes are larger than No. I. The meaning of this is as yet unknown.

NOTE.—All our primaries as well as the majority of our secondaries have been obtained in a purple-flowered race secured from Washington, D. C., which has been designated Line 1. When these mutants have been rendered heterozygous for certain white lines (called A whites), secured from various sources, they have all thrown disomic ratios; except *Pn* which has thrown trisomic ratios, since the genes for purple and white are located in the *Pn* chromosome. When these same Line 1 mutants are rendered heterozygous for certain other white lines (called B whites), abnormal ratios are thrown by *Rl* and by *Pn* if the latter appears to have the constitution Pp_2. The mutant *Wy* has occurred in the offspring of parents heterozygous for Line 1 and B whites; but never in the offspring of Line 1, of A whites, nor of parents heterozygous for Line 1 and A whites. Crosses between Line 1 and three different B whites have given F_1 plants with normal pollen and ovules.

[1] Blakeslee, A. F., 1924. "Distinction between Primary and Secondary Mutants," these Proceedings, **10**, 109–116.

[2] Belling, J. and A. F. Blakeslee, 1924. "The Configurations and Sizes of the Chromosomes in the Trivalents of 25-Chromosome Daturas," these Proceedings, **10**, 116–120.

[3] Blakeslee, A. F., 1924. Quoted in *Yearbook of Carnegie Institution of Washington*, No. **23**, 1924, pp. 24–27.

[4] Belling, J., 1924. Quoted in *Yearbook of Carnegie Institution of Washington*, No. **23**, 1924, pp. 28–30.

[5] Belling, J., 1914. "On the Inheritance of Semi-sterility in Certain Hybrid Plants." *Zeitsch. ind. Abst. Vererbungslehre*, **9**, Heft 5, 303–342.

[6] Belling, J., 1915. *Rept. Fla. Agr. Exp. Sta.* for 1914. Pp. 81–105.

[7] Belling, J., 1925. "A Unique Result in Certain Species Crosses." *Zeitsch. ind. Abst. Vererbungslehre.* (In press.)

[8] Belling, J., 1925. "On the Origin of Species in Flowering Plants." *Nature*, **116**, 279. (Aug. 22).

12

Reprinted from *Am. Nat.* **57**(653):562–566 (1923)

CHROMOSOME ARRANGEMENTS DURING MEIOSIS IN CERTAIN OENOTHERAS

Ralph E. Cleland

DURING the past two years I have been studying the pollen mother cells of pedigreed strains of the following Oenothera species: *Oe. franciscana sulfurea*,[1] *Oe. biennis, Oe. biennis sulfurea, Oe. muricata* and *Oe. oblonga*. The following is a brief account of certain meiotic stages in these plants. Fuller descriptions I hope to present in the near future.

I have recently described the reduction process in the anthers of *Oe. franciscana*.[2] Late heterotypic prophase stages in this plant (corresponding to "diakinesis" in most plants) reveal an interesting situation, characterized by (1) a marked tendency on the part of the homologous chromosomes to become paired, a condition quite in contrast to that in most of the Oenotheras so far studied; (2) a constant and definite arrangement of the chromosomes in the nucleus; (3) the association of certain of the chromosomes end to end throughout the period, in such a way as to form a closed circle.

In regard to the first of these characteristics, we find that the species in question differ markedly from one another. On the one hand, *Oe. oblonga* shows a large amount of pairing of homologous chromosomes. *Oe. franciscana sulfurea*, on the other hand, succeeds in forming but one pair; and in *biennis, biennis sulfurea* and *muricata*, there is no pairing whatever. With respect to the other two characteristics, however, the plants show a striking uniformity. All present at this stage an arrangement of chromosomes in the nucleus which is constant for the species; and all display at least one large closed circle, consisting of a definite number of chromosomes united end to end. The arrangement of the chromosomes at this stage may be most briefly portrayed in a series of diagrams (see Figs. 1–7).

In the case of *muricata*, the arrangement shown in the figure persists unchanged until metaphase. In *franciscana sulfurea*, the single ring usually breaks away from the large circle, just previous to metaphase; otherwise, the arrangement persists. In *biennis* and *biennis sulfurea*, the typical arrangement (Fig. 2) may become somewhat modified by a separation of the two rings, or by an attachment of the two in such a way as to form a figure

[1] A fuller account of this species is soon to appear in the *Botanical Gazette*.

[2] *Amer. Jour. Bot.*, 9: 391–413. 1923.

113

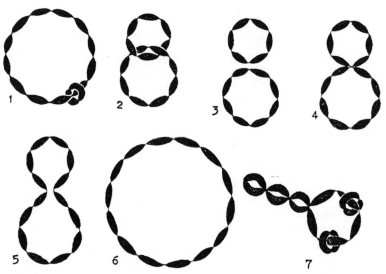

FIGS. 1–8. Diagrams of chromosome arrangements in late heterotypic prophase. 1, *Oenothera franciscana sulfurea;* 2, *Oe. biennis,* typical; 3, 4, 5, modifications of fig. 2: 6, *Oe. muricata;* 7, *Oe. oblonga,* probable complete arrangement.

8, or rarely, by an opening of the latter figure at the point of attachment of the two circles, so that the whole becomes one large circle of 14.

The situation in *oblonga,* a 15 chromosome form, is not so clear, owing to the fact that the chromosomes begin to break apart soon after the second contraction period has past. The complete grouping being rarely seen, therefore, the evidence for a definite arrangement of the chromosomes is not so complete as in the case of the other species. However, I think that the evidence is clear enough to make it fairly certain that the arrangement herewith presented is the correct one (Fig. 7). When the chromosomes separate, and the figure breaks up, the end result is not always the same. It appears that three sets of homologous chromosomes are always formed into definite pairs. In the case of the other chromosomes, however, the affinity between homologues appears to be weaker, and they sometimes succeed in pairing, and sometimes do not. The largest number of distinct pairs that I have observed in a nucleus is six. Such a large number of separate pairs can only be formed by the breaking up of the circle of 5 chromosomes, two of the components pairing. In most cases, however, the five chromosomes which form the circle remain attached throughout this period, and the number of pairs formed is less than six. When the number of pairs formed is less than five, the chromosomes which failed to pair are found still attached to the circle of five, which breaks open, so that chains of seven or nine chromosomes are frequently seen (see Figs. 11–14). In all these species, therefore, there seems to be a definite arrangement of chromosomes in late heterotypic prophase, the arrangements differing in different species, but being constant in each. There is also in each species at least one closed circle, the number of chromosomes composing it varying from four in *franciscana,* to the whole group of fourteen in *muricata.*

Turning now to the heterotypic metaphase, we find that the chromosomes are arranged in the equatorial region as illustrated in the figures (Figs. 8–14). Chromosomes that were attached together in circles or chains in late prophase, are still attached, and those that were separated into pairs are still separated. But the most noticeable feature of this stage is the regular way in which the chromosomes forming the large circles are arranged. With but a small percentage of irregularity, contiguous chromosomes become attached to spindle fibres leading to opposite poles, alternate ones to fibers leading to the same pole. Thus it is brought about that in spite of the fact that the chromosomes fail

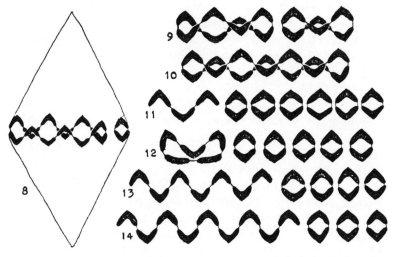

Figs. 8–14. Diagrams of chromosome arrangements in heterotypic metaphase, lateral view. Outline of the spindle omitted except in fig. 8; 8, *Oe. franciscana sulfurea;* 9, *Oe. biennis;* 10, *Oe. muricata;* 11–14, *Oe. oblonga.*

in large measure to pair, but instead, remain attached to one another end to end, they are for the most part distributed evenly and normally to the poles.

The chromosomes in all the Oenotheras that I have studied are so similar in size and appearance that it is impossible to distinguish one from another, and so, by direct observation, determine the exact position of each chromosome pair. However, the uniform linking of groups of chromosomes in each species, and the regular way in which they are arranged in metaphase, and distributed in anaphase, suggests very strongly that the chromosomes making up the circles are not arranged by chance, but have positions in these circles which are definitely fixed. If this were not the case, there would be very little chance that the two members of each of the homologous pairs in the circles would be disjoined in anaphase, by the process employed in these plants. It is more than likely, therefore, that the chromosomes are placed in such a way in the circles, that homologous ones are separated to opposite poles. Bearing in mind the condition found in metaphase, the most natural arrangement allowing of such a result is an end to end relation between the homologous chromosomes, and this I believe is in all likelihood the actual condition. If we assume the probability that within the larger groups of chromosomes there is a constant arrangement of the homologous chromosomes into pairs, it is but an easy step to the further assumption that the chromosomes are not only arranged into pairs, but have definite positions within the pairs also. Thus the arrangement AA' BB' CC' (representing three adjacent pairs of homologous chromosomes) should not, except through irregularity, become AA' B'B CC'.

Supposing, then, that the chromosomes are thus fixed in their position in the circle, it follows clearly that the group as a whole constitutes a unit, both structurally and in behavior, acting very much like a single chromosome pair. Chromosomes ABCD normally go to one pole, and A'B'C'D' to the other. Chromosomes ABCD are then as much linked in effect as though they were merely parts of a single chromosome, the homologue of which is composed of A'B'C'D'.

The possible bearing of this situation upon the experimental work of Shull[3] in which he finds evidence of a large amount of

[3] *Eugenics, Genetics and the Family,* Vol. I, pp. 86–99, 1923; *Genetics* 8: 154–167, 1923.

character linkage in the Oenotheras, need only be mentioned. Shull finds that most of the characters that he has studied so far seem to belong to the same linkage group. He believes, therefore, that the genes for all of these characters are carried in the same chromosome pair. In view of the facts here presented, it seems to me that the possible correspondence of larger units composed of several chromosomes, rather than of individual chromosome pairs, to character linkage groups such as Shull has discovered, should be kept in mind.

13

Reprinted from *Natl. Acad. Sci. (U.S.A.) Proc.* **16**(4):269–277 (1930)

GENETICAL AND CYTOLOGICAL STUDIES OF SEMISTERILITY AND RELATED PHENOMENA IN MAIZE†

By C. R. Burnham*

Bussey Institution, Harvard University

Communicated March 12, 1930

This paper is a preliminary report of studies on two new partially sterile lines of maize derived from a pedigreed strain of genetic material grown at the University of Wisconsin in 1927. The parent line is unrelated to the stock in which semisterile-1, reported by Brink,[4] arose. In semisterile-1, approximately 50 per cent of the pollen and ovules abort and, when selfed or crossed with normals, the plants throw a ratio of 1 normal : 1 semisterile. In order to account for these results and for the origin of semisterility, Brink[4] and Brink and Burnham[6] advanced the hypothesis that a chromosomal change involving non-homologues had occurred. Simple translocation of a portion of one chromosome to a non-homologue would give a plant in which, with random distribution in the two pairs, 50 per cent of the spores either would be deficient or would possess a portion in duplicate. This unbalance was assumed to be sufficient in both cases to cause abortion. From a semi-sterile plant selfed a new normal type was isolated which, when crossed with standard normals, gives all semisterile plants in F_1.[6] The new normal, which should be homozygous for the translocation, was designated x-normal, as contrasted with the standard or o-normal. It has been found also that semisterile-1 shows linkage with factors in two different groups, $P\ br$ and $B\ lg$.[5]

To explain somewhat similar phenomena in Datura, Belling and Blakeslee[2] advanced the hypothesis of segmental interchange between two non-homologous chromosomes. Semisterility in Stizolobium, originally explained as due to two factors, has been reinterpreted more recently by Belling[1] on the basis of such interchange. According to this hypothesis, with random distribution, 50 per cent of the spores are deficient in chromatin material and are assumed to abort. A diagram of the assumed chromosomal constitution is given in figure 1. The combinations $A_1\ B_2$ and $A_2\ B_1$ are assumed to abort, while $A_1\ B_1$ and $A_2\ B_2$ form apparently normal spores. A plant with the constitution $A_2\ A_2\ B_2\ B_2$ would correspond to the x-normal type in maize. Both the interchange hypothesis and the simple translocation hypothesis fit the breeding data thus far obtained in maize. Although duplication of chromatin material may cause abortion in certain cases, as in the simple trisomic Datura called Echinus where about half of the pollen is devoid of starch,[3] deficiency seems a more probable cause of abortion. This favors the idea of segmental interchange.

In maize it should be possible to obtain critical genetic evidence as to the type of change in different semisterile lines.

Classification for different degrees of sterility is based on estimates of the relative amounts of normal and aborted pollen. The two classes are easily distinguishable. The aborted grains are either entirely devoid of starch or nearly so, while the normal ones are well filled. Counts for sterility on the ear are made most easily from two to three weeks after pollination. For cytological study of meiosis in microsporocytes, the aceto-carmine smear method was used.

At diakinesis of the first meiotic division in normal plants ten bivalent chromosomes are found regularly. At the corresponding stage in semisterile-1 only eight bivalents occur, plus a group of four chromosomes arranged either in a ring or in a chain. In occasional cells there are ten bivalents. Apparently some chromosomal change has occurred which causes the association of non-homologues in this semisterile race. With the conjugation of homologous ends, ring configuration might be expected if segmental interchange had occurred; although, with crossing-over in the "four-strand" stage, rings might occur as a result of simple translocation. Referring to figure 1, which represents segmental interchange, the order of the chromosomes in the ring would be $A_1 A_2$ $B_1 B_2$. Only when alternate chromosomes in the ring go to the same pole would normal spores result. When adjacent non-homologues go to the same pole, i.e., A_2 with B_1 and A_1 with B_2, aborted spores result. More study is necessary to determine the detailed behavior of the altered chromosomes. The longest chromosome pair appears to be included in the group of four. Since semisterile-1 is linked with the *B lg* and *P br* groups, and since *B lg* is the third from the largest chromosome,[10] the longest one in maize must carry the factors in the *P br* group.

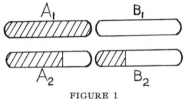

FIGURE 1

Diagrammatic representation of the assumed chromosomal constitution of a semisterile plant on the basis of segmental interchange. One end of chromosome A has interchanged with an end of chromosome B.

The new partially sterile lines were derived from sibs in a progeny of twenty-six plants from a homozygous waxy ear. Thirty-four plants from the same ear have been grown since, but all were normal. Crosses were obtained with only two of the partially sterile plants.

The first partially sterile line to be considered shows the typical breeding behavior of a semisterile. An average for eight plants showed 56.9 per cent pollen abortion. Although more than 50 per cent sterile, it breeds as a semisterile and will be referred to hereafter as semisterile-2. The cross of semisterile-2 with *x*-normal-1, i.e., the *x*-normal of semisterile-1, gives all partially sterile plants in F_1. About half of these are semisterile. The other half are more than 75 per cent sterile, and will be referred to as the 75+ per cent sterile class. Pollen counts on eleven plants in the latter class gave an average of 78.5 per cent sterility. Kernel counts on eight of the same plants gave an average of 83.7 per cent ovule abortion. These results in F_1 indicate that semisteriles-1 and -2 are different. Had they been similar, there would have been a ratio of 1 normal : 1 semisterile. As far as tested, the semisterile F_1 plants are of the semisterile-1 type. Plants in the 75+ per cent sterile class were crossed with standard normals. The results from these crosses, given in the last line of table 1, show a close approximation to a ratio of one normal: one 75+ per cent sterile : two semisterile. These are the results expected if the two semisteriles are entirely independent of each other. The intermediate classes will be considered later.

TABLE 1

SEGREGATION FOR WAXY AND PARTIAL STERILITY IN THE F_1 FROM THE CROSS OF NORMAL WITH (75+ PER CENT STERILE FROM: SEMISTERILE-2 $wx\ wx$ × SEMISTERILE-1 $Wx\ Wx$)

| | | | | | | | | INTERMEDIATE | | | |
		NORMAL		75+ PER CENT STERILE		SEMISTERILE		30± PER CENT STERILE		65± PER CENT STERILE	
		$Wx\ Wx$	$Wx\ wx$	$Wx\ Wx$	$Wx\ wx$	$Wx\ Wx$	$Wx\ wx$	$Wx\ Wx$	$Wx\ wx$	$Wx\ Wx$	$Wx\ wx$
Cross A, Normal = $Wx\ Wx$		74	10	18	61	86	80	1	..	1	9
		$Wx\ wx$	$wx\ wx$	$Wx\ wx$	$wx\ wx$	$Wx\ wx$	$wx\ wx$	$Wx\ wx$	$wx\ wx$	$Wx\ wx$	$wx\ wx$
Cross B, Normal = $wx\ wx$		8	1	3	6	11	10	1	2	..	1
Total		82	11	21	67	97	90	2	2	1	10
Total		93		88		187		4		11	

The above conclusion is supported also by cytological evidence. At diakinesis in microsporocytes of the 75+ per cent sterile plants, there are six bivalent chromosomes plus two separate groups of four chromosomes each. Two separate groups would be expected if semisterile-2 arose as the result of an interchange or a translocation involving two non-homologous chromosomes, neither of which is involved in semisterile-1. In semisterile-2, rings are formed frequently, but chain configurations are more frequent than in semisterile-1. Possibly the size of the piece involved in the change may affect the relative frequency of chains and rings. Chains might be more frequent where the piece was very small in relation to the size of the chromosomes concerned. Two of the smaller chromosome pairs apparently are involved in semisterile-2. The deviation from 50 per cent sterility may be due to a lack of random distribution in the two pairs.

The segregation for waxy in the crosses of 75+ per cent sterile with standard normals is shown also in table 1. Linkage of semisterile-2 with wx should give, in cross A with $Wx\ Wx$, an excess of the two parental types: normal $Wx\ Wx$, and 75+ per cent $Wx\ wx$. In cross B with $wx\ wx$, the reverse should be true. A 1 : 1 ratio is expected in the semisterile class with or without linkage. Although the numbers are small, apparently waxy is linked with semisterile-2, the calculated cross-over value being 17.7 per cent. The amount of recombination in the two classes is very different, but possibly this is not significant.

The plants showing intermediate degrees of sterility (table 1) have not been tested thoroughly. Cytological evidence and abnormal genetic ratios thus far obtained indicate that they are 21-chromosome plants. I am indebted to Dr. McClintock for suggesting the hypothesis that these intermediate types have an extra chromosome of one of the kinds composing the ring. The presence of an extra chromosome in a spore enables certain deficient combinations to survive which otherwise would abort, thus markedly reducing the amount of sterility. The four plants belonging to the class intermediate between normal and 50 per cent sterile showed 27.1, 32.9, 33.1 and 39.8 per cent pollen sterility, respectively. Only two of these were examined cytologically. Both were found to have 21 chromosomes. At diakinesis in the one showing 32.9 per cent aborted pollen there is usually a group of five chromosomes in addition to eight bivalents, making a total of twenty-one. In certain cells, the group forms a chain; in others a single chromosome is attached at its ends to two bivalents, showing that the extra chromosome belongs to one of the groups involved in semi-sterility. If, in partially sterile 20-chromosome plants, three chromosomes of the quadripartite ring go to the same pole, spores with $n + 1$ chromosomes will be formed. Anaphases have been counted in which there were eleven chromosomes at one pole and only nine at the other. The amount of sterility in the resulting 21-chromosome plant depends on which three chromosomes in the ring went to the same pole to form the $n + 1$ spore. Referring to figure 1, the two types of combinations possible in the $n + 1$ spore would be $A_1\ A_2$ with B_2 and $A_1\ A_2$ with B_1. The amount of sterility also depends on whether or not there is random pairing in the set containing the extra chromosome. There might be preferential pairing of the two normal chromosomes. The frequency with which the extra chromosome lags at meiosis and is eliminated also will affect the degree of sterility.

From 75+ per cent sterile plants, there should be two classes of intermediates: those falling between semisterile and normal, and those between semisterile and 75+ per cent sterile. The former class is distinct; but in the latter there is probably some overlapping with 20-chromosome semisterile-2 plants which show an average of about 57 per cent abortion. The upper class of intermediates has not been examined cytologically.

Data from two 21-chromosome plants backcrossed to a $c\ sh\ wx$ stock are given in table 2. On the basis of interchange as represented in figure 1, both plants are assumed to have arisen from an $n + 1$ egg of the type $A_1 A_2 B_2$. The extra chromosome in B11–13 belongs probably to the *sh wx* set. In the crosses with both plants, the ratios for *sh* and *c* deviate widely from the normal 1 : 1 ratios obtained from semisterile sibs. Close fits to the observed numbers and also to the degree of sterility may be obtained by assuming the proper amount of preferential pairing and the proper frequency of lagging univalents in the trisomic groups. The assumption that the break in the *sh wx* chromosome occurred on the side of *wx* away from *sh*, placing *sh* about 38 units from the break, fits the observed data. Further study is necessary to determine the amounts of preferential pairing and lagging, and to determine the frequency of crossing over between *sh* and semisterile-2.

TABLE 2

SUMMARY OF BACKCROSS DATA FROM TWO EXCEPTIONAL 21-CHROMOSOME PLANTS IN THE PROGENY OF THE CROSS: [75+ PER CENT STERILE FROM ($C\ Sh\ wx \times ?\ Sh\ Wx$)] × NORMAL $c\ sh\ wx$

CROSS	PER CENT POLLEN ABORTION	C				c				TOTAL
		Sh		*sh*		*Sh*		*sh*		
		Wx	*wx*	*Wx*	*wx*	*Wx*	*wx*	*Wx*	*wx*	
B11–13 × c sh wx	32.9	94	73	..	2	1	..	9	38	217
c sh wx × B11–2	27.1	..	189	..	11	..	1	..	365	566

From B11–2, the *Sh* : *sh* ratio through the pollen is approximately 1 : 2. This might appear to be a normal trisomic ratio, since in maize $n + 1$ pollen functions only occasionally.[10] However, the progeny in which plant B11–2 was found came from the cross: 75+ per cent sterile *Sh Sh* × normal *sh sh*. Since the female parent in which the $n + 1$ spore probably arose was *Sh Sh*, *Sh* seeds would have been in excess in the backcross given in table 2 if the extra chromosome were the one bearing this factor. The extra chromosome is proably the other one involved in semisterile-2. With random pairing in the trisome in a plant of the type assumed for B11–2 and B11–13, it would be possible to get apparent trisomic ratios for factors which were not in the trisome, provided they were closely linked with semisterility. In this case, where *sh* is assumed not to be closely linked with sterility, the assumption of preferential pairing in the trisome gives a close fit to the observed numbers and also to the degree of sterility.

Plant B11–2 was crossed also on two *Pr pr* plants, giving 233 *Pr*:156 *pr* seeds. The deviation from the expected 3 : 1 ratio is over ten times its probable error, and suggests that the *pr* v_2 chromosome is the other one involved in semisterile-2. This is not certain, however, since ratios for *pr* often deviate widely from the expected (Fraser, unpublished data from Cornell).

In two other exceptional plants, abnormal ratios for *waxy* were obtained in the pollen. One plant which showed 37.7 per cent aborted pollen had only 11.9 per cent waxy grains. Another plant showing 9.7 per cent pollen sterility had 16.3 per cent waxy grains. In semisterile plants heterozygous for waxy, 50 per cent of the pollen is waxy.

The abnormal shrunken and waxy ratios furnish additional evidence that the *sh wx* chromosome is one of those involved in semisterile-2.

A partially sterile sib in the culture in which semisterile-2 was found was crossed with x-normal-1. All but one of the forty F_1 plants were partially sterile. The exception was a normal which may have resulted from a stray pollination. About half of the F_1 plants were semisterile, while the other half were somewhat less than 75 per cent sterile (referred to as 75− per cent). Pollen counts on nine plants in the latter class showed an average of 68.7 per cent sterility, while kernel counts on the same plants gave an average of 71.3 per cent abortion. These results indicate that the new sterile is different from semisterile-1. Since the new class from the corresponding cross with semisterile-2 showed 78.5

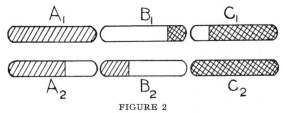

FIGURE 2

Diagram of the assumed chromosomal constitution of a 75 per cent sterile plant from the cross of semisterile-1 with semisterile-3, according to the segmental interchange hypothesis. One member of the B pair has interchanged with an A chromosome while the other member of the B pair has interchanged with a C chromosome. If homologous ends associate, a ring of six chromosomes might be expected at diakinesis.

per cent pollen abortion, the new type is also different, apparently, from semisterile-2. Plants belonging to the 75− per cent sterile class were crossed with standard normals. Out of 97 plants from these crosses, 90 were semisterile, one was normal, one was 81.5 per cent sterile and five belonged to a distinct class intermediate between normal and 50 per cent sterile. Practically every functional spore from the 75− per cent sterile plants must have been a semisterile producer. This suggests that the new type, semisterile-3, also involves interchange or translocation between non-homologous chromosomes, but that one of them is the same as is involved in semisterile-1. A diagram of the assumed chromosomal constitution of a 75− per cent sterile plant, based on segmental interchange is given in figure 2. In the pair common to both steriles, the B_1 member has interchanged with chromosome C_1, while the B_2 member has interchanged with A_2. Separation at meiosis of the members of this common pair would give only two kinds of normal spores, $A_1 B_1 C_1$ and $A_2 B_2$ C_2. In crosses with standard normals, the first would give semisterile-3 and the other would give semisterile-1.

On the above hypothesis, configurations involving six chromosomes should be found at meiosis in 75− per cent sterile plants. Cytological examination of microsporocytes at diakinesis does show a group of six chromosomes plus seven bivalents. The group is either a large closed ring or, occasionally, an open one. On the basis of Belling's interchange hypothesis illustrated in figure 2, the order of the chromosomes in the ring of six would be $A_1 A_2 B_1 C_2 C_1 B_2$. Only the combinations of alternate chromosomes in the group would give normal spores. The deviation from 75 per cent sterility is probably due to lack of independent distribution of the three pairs involved. Semisterile-3 may be considerably less than 50 per cent sterile; or the association of the three pairs in a ring or chain may prevent their independent distribution.

If opposite ends of the two members of the common chromosome pair are involved in semisteriles-1 and -3, as illustrated in figure 2, cross-overs between them might occur. In crosses with standard normals, these would give normal and 75≑ per cent sterile plants. One normal plant and one with 81.5 per cent sterility were found. They are being tested to determine whether they are the result of crossing-over or stray pollinations.

Two of the three pairs in the ring of six chromosomes must be $B\ lg$ and $P\ br$, respectively, since these pairs are involved in semisterile-1. Either the $B\ lg$ or the $P\ br$ chromosome, therefore, must be concerned in semisterile-3. Certain configurations at diakinesis indicate that $B\ lg$ is the one involved. Genetic tests are in progress.

The five plants belonging to the class intermediate between normal and 50 per cent sterile have been examined cytologically using either root tip or microsporocyte material. All proved to have 21 chromosomes. Where microsporocyte material was available, groups of five chromosomes were found in addition to eight bivalents, showing that the extra chromosome belongs to the group involved in semisterility. The frequency of these plants is very high, over six per cent having been found following crosses where the 75− per cent sterile plant was the female parent. The 20-chromosome, partially sterile plants, thus furnish another source for the isolation of certain simple trisomics in maize.

In Pisum, an unexpected linkage of two factors which are ordinarily independent has been found by Hammarlund.[9] Håkansson[8] examined plants from this cross cytologically and found a ring of four chromosomes at meiosis. No mention of sterility was made in these papers; but by comparing the number of plants per progeny from the normal line with that from the cross which shows linkage, I find that the latter progenies are about half as large.   In the normal line the average for seven progenies is 115.6.   In the cross showing linkage, the average for 28 progenies is only 53.0.   This suggests that Hammarlund's double dominant line K, which shows linkage of the two factors in its crosses, may be an x-normal type.   Since only the parental combinations of chromosomes in the two pairs involved in semisterility are viable, two factors closely linked with sterility but in different chromosome pairs would show apparent linkage with each other.   This would be true whether or not either or both were in the interchanged or translocated piece.   Unless both were linked with sterility to about the same degree, the recombination classes should not be of equal size.   Examination of the data shows that the two classes are approximately the same size.   Another case of ring-formation in Pisum was reported recently by Miss Richardson[12] in a cross showing 50 per cent sterility.

Ring formation in Oenothera has been interpreted by Darlington[7] on the basis of segmental interchange between non-homologous chromosomes. One of the cases described in maize in the present paper shows how configurations involving more than two pairs of chromosomes may be built up by combining the proper semisterile types.

Müntzing[11] has studied partial sterility in crosses of Galeopsis species. Although the complete story has not been published, the breeding behavior suggests a situation similar to that in maize and in Datura.   Cytological examination revealed no abnormalities, but only one diakinesis preparation was examined.

Summary.—1.   In semisterile-1 and in two new types, semisteriles-2 and -3, there is an association of non-homologous chromosomes at meiosis. A group of four chromosomes occurs plus eight bivalents.

2.   Sufficient evidence is not available to decide definitely whether this is the result of segmental interchange or simple translocation.

3.   The combination of semisteriles-1 and -2 gives a new class which is somewhat more than 75 per cent sterile, and in which two separate groups of four chromosomes occur plus six bivalents.

4.   Neither of the two pairs involved in semisterile-2 is concerned in semisterile-1.   The *sh wx* pair is one of those involved in semisterile-2.

5.   The combination of semisteriles-1 and -3 gives a new class which is a little less than 75 per cent sterile, and in which there is a group of six chromosomes plus seven bivalents.

6.   One of the pairs involved in semisterile-3 is probably the same as is concerned in semisterile-1, and, therefore, must be either *P br* or *B lg*.

7.   Exceptional 21-chromosome plants showing intermediate degrees of sterility have arisen from partially sterile 20-chromosome plants.   The extra chromosome belongs to the group involved in semisterility.

† This work was begun in the Department of Genetics at the University of Wisconsin.

The greater part of the analysis was made at Cornell University, certain parts being finished at the Bussey Institution.

I am indebted to Dr. R. A. Brink for giving me the semisterile stocks, to Dr. R. A. Emerson for making it possible to grow the material at Cornell University in 1929, and to Drs. L. W. Sharp and R. A. Brink for helpful suggestions in the preparation of this manuscript. To Dr. Barbara McClintock of Cornell University I wish to express my sincere thanks for her aid and encouragement in attacking the cytological aspects of the problem.

* NATIONAL RESEARCH COUNCIL FELLOW at Cornell University and Bussey Institution, Harvard University.

[1] Belling, J., *Zeit. indukt. Abstamm. Vererb.*, **39,** 286–288 (1925).

[2] Belling, J., and A. F. Blakeslee, *Proc. Nat. Acad. Sci.*, **12,** 7–11 (1926).

[3] Blakeslee, A. F., *Ibid.*, **10,** 109–116 (1924).

[4] Brink, R. A., *J. Hered.*, **18,** 266–270 (1927).

[5] Brink, R. A., (Abstract) *Anat. Rec.*, **44,** 280 (1929).

[6] Brink, R. A., and C. R. Burnham, *Amer. Nat.*, **63,** 301–316 (1929).

[7] Darlington, C. D., *J. Gen.*, **20,** 345–363 (1928–29).

[8] Håkansson, A., *Hereditas*, **12,** 1–10 (1929).

[9] Hammarlund, C., *Ibid.*, **4,** 235–238 (1923); **12,** 210–216 (1929).

[10] McClintock, B., and H. E. Hill, (Abstract) *Anat. Rec.*, **44,** 291 (1929).

[11] Müntzing, A., *Hereditas*, **12,** 297–319 (1929).

[12] Richardson, Eva, *Nature*, **124,** No. 3128 (1929).

14

Reprinted from pp. 1–2, 4, 12–13 of *J. Genet.* **21**(1):1–15 (1929)

MEIOSIS IN POLYPLOIDS.

PART I. TRIPLOID AND PENTAPLOID TULIPS.

By W. C. F. NEWTON and C. D. DARLINGTON.

(*John Innes Horticultural Institution.*)

1. INTRODUCTION.

MEIOSIS is ordinarily looked upon as a process by which the corresponding or homologous elements of the nucleus, contributed by the maternal and paternal gametes, are enabled to segregate from one another to give a new generation of cells with the reduced gametic complement of these elements. Now in certain polyploids, instead of each maternal element being capable of pairing with, and separating from, a particular paternal element, more than one of these corresponding elements are derived from one or both sides, so that at the first maturation division several such elements may be associated. It is naturally desirable that the special process of pairing and separation involved in these forms should be studied in the most favourable material, and for this purpose the triploid garden tulips and hyacinths seem to be marked out.

Belling has studied the multiple association of chromosomes at metaphase extensively, but no attempt has been made to correlate these structures with early prophase behaviour except in the case of triploid tomatoes (M. M. Lesley, 1926). It has been the object of the authors to describe the essential features of the process in detail.

Owing to the illness which preceded his death in December 1927, Newton was unable to complete these studies, and the junior author is responsible for the illustrations and for the details of interpretation.

[*Editors' Note:* Material has been omitted at this point.]

127

3. MATERIAL.

In the tulips of the section Leiostemones, two circumstances will account for the occurrence of triploids. First, while doubling has frequently been observed at the chalazal end of the embryo-sac in many liliaceous species, in *Tulipa* (Newton, 1927) this doubling may occasionally occur at the micropylar end and diploid egg-cells result. These on fertilization would give rise to triploids. From their analogous constitution and behaviour it would appear that the triploid hyacinths have arisen in the same way. Secondly, they need suffer no drawback from sterility, for their vegetative means of propagation (like the apogamy of *Hieracium*) makes them independent of ordinary seed-production. Triploids have been produced experimentally by hybridisation of distinct diploid and tetraploid species, as well as of diploid and tetraploid individuals of the same species. Either way they are as a rule infertile, and their accidental appearance without hybridisation amongst diploid seedlings of short-lived plants would not lead to the production of a new race.

Amongst these bulb genera on the other hand the greater vigour of a triploid might give it, although not an interspecific hybrid, a definite advantage over its diploid sister seedlings. The cytological evidence seems to support the view that these triploids occupy an unusual position amongst cultivated plants in being auto-polyploid; that is to say, they have three sets of chromosomes which, as far as observation can show, are similar and indifferently attracted to one another.

4. TRIPLOID TULIPS.

A. Somatic Chromosomes.

De Mol (1925) and Newton (1927) have found triploid forms among the garden tulips. Three of these are the subject of the present study. They have, so far as we can tell, similar chromosome complements, numbering 36. Most of the chromosomes have relatively subterminal constrictions, except one rather short type and one longer type (Text-fig. 1). The types can only be illustrated rather unsatisfactorily from polar views of metaphase owing to the longer distal segments of the chromosomes never lying flat.

The three varieties examined are: Keizerskroon, Massenet, Pink Beauty.

The description of meiosis is taken chiefly from the first, but the others have been studied sufficiently to show that they obey the same rules as Keizerskroon and differ from it in no recognisable detail.

B. *Pollen Mother-cell Divisions.*

(i) *Pairing of chromosomes.* The earliest stages that have been examined show the fine threads ("leptotene") only here and there coming together to form thicker threads. At this stage threads can be followed running parallel with one another for long distances in twos and threes (Pl. I, figs. 1, 3 *a–b*) without apparently coming into actual association. The first sign of pairing is the intimate association along a part of their length of two of the threads—the corresponding part of the third being always unaffected (Pl. I, figs. 2, 3 *c–m*). It seems as though pairing, once initiated between two at a particular point, passes along the chromosome until this association runs into another association. This frequently happens to be between a different pair of the three lying together, so that an exchange of partners occurs at this point. Thus we come to the condition (Pl. I, fig. 3 *d, e, l, o*) where a pair appears to be touched at one point by a third thread. This is a typical result of two associations meeting, for an exchange of partners amongst the three chromosomes can usually be seen to take place at the point of contact, and the thread that leaves the association is not the same one that entered into it.

This pairing, which is obviously between two of the three threads, and never between all three at any one point, takes place over a prolonged period. Longitudinal contraction and thickening of the chromosomes appears to follow pairing as a consequence, for double threads often appear shorter than the corresponding portions of single threads. It is probably also as a result of this contraction, following pairing, that paired threads in the same nucleus, having paired at different times, appear of different thicknesses. There is no polarisation of the nucleus in *Tulipa*, and apparently association can begin at many different points along the chromosomes. Occasionally it has been seen to begin at the ends (Pl. I, fig. 3 *c*).

[*Editors' Note:* Material has been omitted at this point.]

6. Discussion.

In recapitulation it may be worth while to describe the fundamental facts of conjugation and reduction in the tulips in terms of the principles which the behaviour of the triploid forms seems to demonstrate.

First, the three homologous chromosomes consisting of single threads come to lie parallel with one another, probably throughout their length. This proves Newton's parasynaptic interpretation in the diploid (1927) to have been correct, for three pairing threads cannot be taken to be half chromosomes as Digby has suggested in *Osmunda* (1919).

Secondly, parts of these chromosomes associate intimately in pairs, leaving the odd chromosomes unassociated for the time being, a phenomenon apparently also observed in triploid *Solanum* by M. M. Lesley (1926). The same chromosome need not be unassociated throughout its length; on the contrary, exchanges take place between the partners at intervals. While the earlier attraction may have been between chromosomes as wholes, this association is evidently between the smallest elements, the chromomeres. Each chromomere acts independently of its neighbour.

Thirdly, when pairing is as complete as possible—it can never be complete where an odd number is involved—each pair is seen to consist of four threads which are probably sometimes equally associated with one another, sometimes more definitely attracted to one another in pairs. This is therefore a state of attraction between four threads. Pairing appears to lead to linear contraction, unpaired threads being often longer in the early stages than the corresponding parts of paired threads. This stage lasts unchanged for a long time in the diploid, but in the triploid it seems probable that the odd chromosome may displace one of the partners, or, more properly, two of the chromatids, so that readjustments will occur.

Fourthly, at various points in the paired chromosomes splits appear (diplotene) which extend along the threads until they meet. From this stage there is a state of attraction between two chromatids. At the points at which the loops meet, chiasmata, it can frequently be seen that splits have evidently occurred between different pairs; two threads thus appear to exchange partners with two others. These chiasmata appear at random and vary in number.

Fifthly, without any essential change in their relationships the threads now contract to the diakinesis condition and then arrange themselves on the equatorial plate preparatory to division. In the course of this process there is probably a reduction in the number of chiasmata. Owing to the state of attraction being only between the threads, and not between the chromosomes themselves, the chiasmata form the only bond between associated chromosomes. Thus varying numbers of odd chromosomes are left unassociated.

[*Editors' Note:* Material has been omitted at this point.]

REFERENCES

Digby, L. 1919. On the archesporial and meiotic mitoses in *Osmunda*. *Ann. Bot.* **33**:135–172.

Lesley, M. M. 1926. Maturation in diploid and triploid tomatoes. *Genetics* **11**:267–279.

Newton, W. C. F. 1927. Chromosome studies in *Tulipa* and some related genera. *J. Linn. Soc. (Bot)* **47**:339–354.

PLATE I

Figure 1.   Pairing of chromosomes taking place (zygotene) in Massenet. x2250.

Figure 2.   The same stage in Keizerskroon. x2250.

Figure 3.   a–o Successive stages of association of particular threes of chromosomes in Keizerskroon and Massenet. c, k and 1 x3200, the rest x2250.

[*Editors' Note:* The references and legends for Plate I are from Darlington, C. D. 1929. Part II. Aneuploid hyacinths. *J. Genet.* **21**:17–56.]

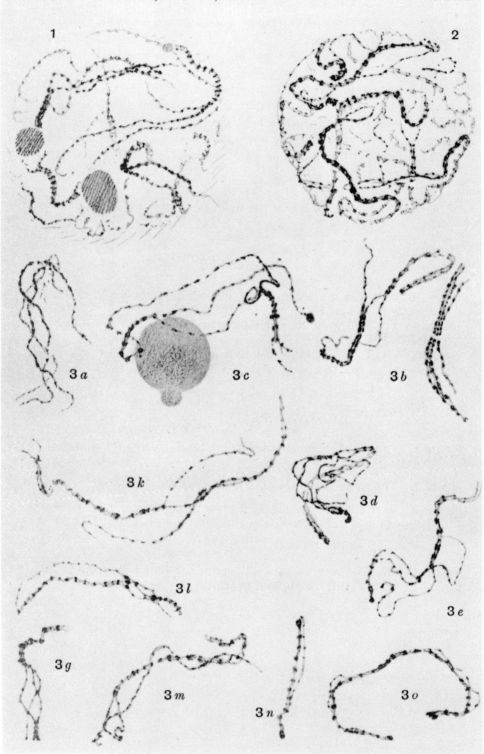

15

Reprinted from pp. 221, 233–236, 256–257 of *Biol. Rev.* **6**(1):221–264 (1931)

MEIOSIS

By C. D. DARLINGTON.

(John Innes Horticultural Institution, Merton.)

(*Received* August 28, 1930.)

[*Editors' Note:* In the original, material precedes this excerpt.]

SECTION 3. THE CYTOLOGICAL THEORY OF MEIOSIS.

ITS RELATION TO MITOSIS.

[*Editors' Note:* Material has been omitted at this point.]

Those who have held that chromosomes paired side by side have hitherto looked upon pairing as directly determined by an "affinity" of pairs of similar chromosomes; as, in fact, a mere exaggeration of the attraction sometimes observed at somatic mitoses. Such an affinity must be specific to the parts of the chromosomes. This fourth hypothesis is rather a description in genetic terms of the observed conditions in most diploids. It applies to the great body of observations of pure forms and undefined "hybrids." But it does not meet the requirements of observations on certain structural and numerical hybrids (Section 4 (*b*)), for in these a failure of pairing is observed which has no simple relation to likeness (Section 4 (*c*)). Nor does it explain the failure of pairing at meiosis in diploid parthenogenesis (Section 5). Finally it does not explain the observations of multivalent formation in non-hybrid polyploids, for here the chromosomes do not regularly associate *either* in pairs *or* up to the limit allowed by their known affinity (Section 2 (*b*)). "Affinity" is one of the conditions of pairing, but it is not the immediate one. Chromosome configurations observed between diplotene and metaphase are, in fact, symptomatic of repulsion rather than of attraction.

It is now possible to examine the relationship of mitosis to meiosis in the light of recent observations. And since meiosis is evidently an abnormality of mitosis it is only by such a comparison that the other conditions of pairing can be determined and a serious theoretical advance made.

Let us first recall the essential and (apart from certain Protozoa such as *Aggregata*, Bĕlař, 1926 b), we may say, universal, facts of mitosis. Mitosis consists in a succession of chromosome changes in the following order:

(1) Resting stage; longitudinal division of each chromosome into two halves.

(2) Prophase; gradual longitudinal contraction of the chromosomes, the halves of each remaining associated, *i.e.* in pairs. The closeness of the association varies, but it is uniformly preserved at the spindle attachment.

(3) Metaphase-anaphase; division and separation of the paired spindle attachments leading to the separation of the half-chromosomes.

Meiosis presents an essential difference at the earliest stage. The contraction of the chromosomes which renders them first visible, and produces the condition of prophase, *anticipates* their longitudinal division. They are in a condition in which chromatin threads are normally associated in pairs. In mitosis these threads are derived by division, but they are none the less potentially separate and independent, as is shown by their easy separation at anaphase. It is therefore possible to regard the pairing of undivided homologous chromosomes at the prophase of meiosis as the re-establishment of conditions upset by their precocious contraction. This pairing is probably not a cause of further contraction, for single unpaired chromosomes contract before the paired ones, *e.g.* X-chromosomes in Orthoptera and unequal pairs (McClung, 1927 a; *v.* Bĕlař, 1928) and idiochromosomes in Hemiptera, (Wilson, 1905 b). The ring chromosomes in *Mecostethus*, which must be supposed to have paired over a greater length than where the chiasmata are strictly localised, are on the other hand relatively compact (McClung, 1927 a). It should be noted that the attachment of each chromosome does not appear to divide, or if divided, behaves as one; each pair has therefore only two effective attachments at metaphase.

The later division of the paired chromosomes to give four threads, the chromatids, in association, will upset the mitotic conditions again and require the separation of chromatids in pairs. Then the chromosomes are said to be paired, but they are not paired by virtue of any present affinity of one chromosome for another. Their pairing depends immediately on two conditions:

(1) A universal mitotic affinity of half-chromosomes for one another in pairs.

(2) The occurrence of exchanges of partner (chiasmata) between pairs of pairs.

Before going into the evidence for and against this theory of meiosis in detail, certain general considerations in its favour may be summarised.

(*a*) The prophase is often preceded by a curtailed or incomplete resting stage (*v.* Section 2 (*a*)). This is evidence that it is the prophase contraction which is precocious, rather than the division of the chromosomes which is delayed.

(*b*) The same physical relationship between pairs of threads is seen at every stage of the prophase of meiosis as at mitosis but the process is relatively *out of step*.

(*c*) A pre-metaphase repulsion is seen at diakinesis, especially after terminalisation, and is comparable perhaps less to the repulsion seen at anaphase of mitosis

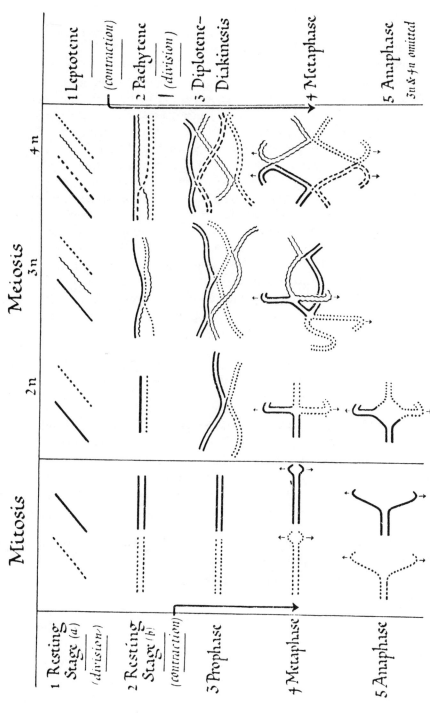

FIG. 4. Comparative diagram of the course of chromatid change between prophase and anaphase in mitosis and meiosis. Figures correspond in relation to the *division* of the chromosomes into two associated halves. Meiosis is therefore shown as beginning prophase earlier. The distribution of chiasmata is represented as random. The metaphase configuration in the triploid does not correspond to that illustrated at diakinesis. Association after diplotene is considered as taking place only between two of the four chromatids following crossing-over at each chiasma (partial chiasmatypy) (v. Section 7). (Original.)

NOTE. In the prophase of mitosis, contraction of the chromosomes follows division so that the condition is of the association of identical elements of chromatin in pairs. In meiosis, contraction precedes any apparent division of the chromosomes. If, then, the normal condition of mitosis is to be fulfilled, identical chromosomes must associate in pairs. They do so (pachytene) until the chromosomes have divided. They then fall apart, giving the normal mitotic condition of association of identical chromatids in pairs (diplotene). These chromatids are however differently paired at different points through crossing-over, so that the exchanges of partner (chiasmata) hold all four chromatids (two chromosomes) together.

135

ordinarily, than to the repulsion between whole chromosomes which leads to their even distribution on the metaphase plate.

(*d*) The precocious prophase is prolonged and the chromosomes therefore contract further than in mitosis (cf. Section 2 (*a*)).

(*e*) Similar changes in viscosity take place in the nucleus between the resting stage and metaphase in meiosis and in mitosis, but, since these changes are more prolonged in meiosis, the chromosomes might be expected to pass through different stages of viscosity in relation to their medium. Hence characteristic distortions might be observed after treatment, especially in plants. Such are the so-called "contraction" or synaptic stages of prophase (cf. Gates, 1928, and Section 2 (*a*)).

(*f*) The configurations that appear at diakinesis and metaphase in polyploids must be dependent on two factors: (i) association between pairs of chromosomes—by the formation of chiasmata; (ii) prophase changes of partner amongst the pairing chromosomes. The result is neither simple pairing nor the maximum satisfaction of "affinity," and quadrivalents should occur more frequently amongst the long chromosomes of a tetraploid than amongst the short ones. This is found to be the case (*v*. Section 2 (*b*)). But since pairing at metaphase depends upon two variable factors, there should be a wider range of variation in the occurrence of pairing in a tetraploid than in a corresponding diploid, and unpaired chromosomes might be expected, as in *Primula sinensis* (D. 1931), in the one although never found in the other.

[*Editors' Note:* Material has been omitted at this point.]

SECTION 8. CONCLUSIONS.

The aim of the present account has been to show that a uniformity of principle underlies the external diversity of meiosis. This diversity is due to the occurrence of differences in detail, which for the most part have some mechanical and hence genetic significance. Such differences do not affect the universal principles on which the hereditary mechanism works, but rather provide critical tests of their validity.

To sum up:

1. Meiosis may be supposed to have originated as an abnormality of mitosis in a diploid (zygote) nucleus in which prophase contraction has *anticipated* the division of each chromosome into two threads. (Section 3.)

2. At mitosis, in this stage of contraction, the split halves of chromosomes are constantly associated side-by-side in pairs. This relationship is restored in meiosis by the association of whole undivided chromosomes in pairs (zygotene). (Section 3.)

3. The later division of the chromosomes (at pachytene) upsets their relationship, which is restored again by their separation (at diplotene). (Sections 2 and 3.)

4. But, probably at the moment of their division, the new halves exchange segments (genetic crossing-over). The separation along the original line of cleavage therefore involves exchange of partners—"chiasmata"—more or less at random along the four paired half-chromosomes. (Sections 3 and 7.)

5. The chromosomes do not therefore pair at metaphase as a result of any present affinity between them. Rather are four half-chromosomes, associated in pairs (as in mitosis), held together by exchanges of partner amongst them. (Sections 2, 3 and 4.)

6. Combining these two hypotheses we may say that "genetic crossing-over" determines the pairing of chromosomes. (Sections 3 and 7.)

7. Therefore, while the structure of the chromosomes at mitosis is constant, their structure at meiosis cannot be constant except where crossing-over is absent or apparently absent. (Sections 2, 4 and 7.)

8. Pairing of chromosomes may fail as a result of: (i) insufficient precocity (preventing full pachytene pairing, as in diploid parthenogenesis); (ii) insufficient pachytene pairing (owing to fragmentation, hybridity or competition); (iii) genetic, developmental or external conditions leading to insufficient chiasma formation. (Sections 4 and 5.)

9. The separation of pairs of half-chromosomes at anaphase precipitates a second division, for these half-chromosomes are in a prophase relationship. (Section 2.)

10. The succession of two cell-divisions without an intervening division of the chromosomes gives numerical reduction and genetic segregation. Segregation may occur at a particular locus at the first division, at the second, or, in polyploids, at both. (Sections 2 and 7.)

11. The occurrence of meiosis has made possible the recurrence of fertilisation. It has inaugurated sexual reproduction as a self-repeating process or habit. (Section 1.)

[*Editors' Note:* Material has been omitted at this point. The references from these excerpts appear on the following pages.]

C. D. DARLINGTON

REFERENCES

Bělăr, K. 1926b. Zur Zytologie von *Aggregata Eberthi*. *Arch*. *Protist*. **53**:312–325.

———. 1928. Die cytologische Grundlagen der Vererbung. *Handb*. *Vererbungswiss*. 1B, 412.

Darlington, C. D. 1931. Meiosis in diploid and tetraploid *Primula sinensis*. *J*. *Genet*. **24**:65–96.

Gates, R. R. 1928. The cytology of *Oenothera*. *Bibliogr*. *Genet*. **4**:401–492.

McClung, C. E. 1927a. Synapsis and related phenomena in *Mecostethus* and *Leptysma* (Orthoptera). *J*. *Morph*. **43**:181–265.

Wilson, E. B. 1905b. Studies on chromosomes. II. The paired microchromosomes, idiochromosomes and heterotropic chromosomes in *Hemiptera*. *J*. *Exp*. *Zool*. **2**:507–545.

16

Reprinted from pp. 191, 192–193, 196–201, 213–214, 221–224, 232, of
Z. Zellforsch. Mikrosk. Anat. **19**(2): 191–237 (1933)

THE ASSOCIATION OF NON-HOMOLOGOUS PARTS OF CHROMOSOMES IN THE MID-PROPHASE OF MEIOSIS IN ZEA MAYS.

BARBARA MCCLINTOCK. [1]

[Editors' Note: Plates VII–XII, consisting of photographs 1 through 62, and text figures 2, 3, 16–32, and 35–51 are not reproduced here.]

I. Introduction.

The ten chromosomes composing the monoploid complement of *Zea mays* are all morphologically distinguishable. The distinctive features of the members of the complement, such as relative size, position of spindle fiber attachment regions, and knobs are given in text figure 1.

In a normal diploid plant similar chromosomes usually associate in the prophase of meiosis to produce a side-by-side alignment of homologous parts. However, when unbalanced chromosomal complements are present, as in monosomic or trisomic plants or plants with deficiencies, or when structurally dissimilar complements are present, as in plants heterozygous for an inversion or a translocation, non-homologous associations involving the chromosomes responsible for the unbalance or structural dissimilarity frequently take place. Non-homologous

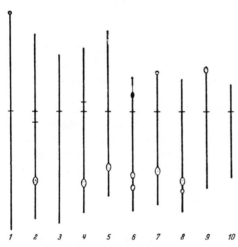

Fig. 1. Diagram of the ten chromosomes composing the monoploid complement of *Zea mays*. The positions of the spindle fiber attachment regions are indicated by cross lines. In chromosomes 2 and 4 two positions for this region have been found in different cultures of maize. In chromosome 4 this difference in position has been correlated with an inversion. The positions of the knobs which are most likely to appear in the different cultures of maize are indicated. In any one culture a particular knob may or may not be present.

associations are not limited to such complements but may occur occasionally in sporocytes of normal diploids.

At pachytene, association of chromosomes is two-by-two whether or not the parts associated are homologous. There are two apparent exceptions, spindle fiber attachment regions and knobs. This appears to be due to a stickiness which these parts possess at this stage of meiosis. Knobs of one bivalent may stick to the knobs of another bivalent. Similarly, the spindle fiber attachment regions of one bivalent

[1] National Research Council Fellow in the biological Sciences.

may appear fused with the spindle fiber attachment regions of another bivalent. Therefore, the association of three knobs or three attachment regions in a trivalent prophase figure is not considered as evidence for three-by-three synapsis. Such association of knobs and spindle fiber attachment regions does not continue into the diakinesis stage.

The evidence for non-homologous association of parts of chromosomes has been obtained from monoploids, diploids, monosomics, trisomics, deficiencies, inversions, translocations, ring-shaped chromosomes, "asynaptic" plants and so-called "B-type" chromosomes. This non-homologous association can be just as intimate, so far as the eye can see, as that between homologous parts of chromosomes.

[*Editors' Note:* Material has been omitted at this point.]

II. Non-homologous association within the B-type chromosome.

The monoploid complement of *Zea mays* consists of ten readily distinguishable chromosomes each of which has been associated with a particular linkage group. In certain strains of maize there is, in addition to the normal complement (A-type chromosomes, RANDOLPH 1928), a type of chromosome with a very peculiar but definite morphology. This has been referred to as the B-type chromosome (RANDOLPH 1928). In morphological features, synaptic relations and transmission behavior it shows no evidence of relationship with any member of the monoploid complement. Dr. RANDOLPH has succeeded through successive crosses in accumulating more than twenty-five of this one type of chromosome in a single plant in addition to the normal diploid complement of twenty. This chromosome has shown no indication of affecting the morphological characteristics of the plant, carries no known genes and is easily transmitted through pollen and eggs (RANDOLPH 1928).

[*Editors' Note:* Material has been omitted at this point.]

2n plus three B-type chromosomes.

Many types of configurations have been observed in plants with twenty normal chromosomes plus three B-type chromosomes. Sometimes

Fig. 4. The association of three B-type chromosomes. In this figure only homologous association has occurred.   Plant 536-2. Mag. × 1650.

Fig. 5. The association of three B-type chromosomes to produce a T configuration. Homologous association produced the two side arms. The upper arm resulted from non-homologous association. Plant S 31-314-2b. Mag. × 1650. See photomicrograph of same, photo. 8, Plate I.

Fig. 6. The association of three B-type chromosomes to produce an asymmetrical T configuration. The 2-by-2 association in the arm to the right is non-homologous. Plant S 31-314-2b Mag. × 1650.

Fig. 7. The association of three B-type chromosomes. Only homologous parts are associated 2-by-2. The univalent threads to the lower right and left have remained unassociated. Plant 536-2. Mag. × 1650. See photomicrograph of same, photo. 12, Plate VII.

Fig. 8. The association of three B-type chromosomes. The chromosomes are numbered 1, 2 and 3. Chromosome 1 associated at the spindle fiber attachement end with chromosome 2. Chromosome 1 associated at the opposite end with chromosome 3. The rest of chromosome 2 is in the form of a foldback. Such figures are comparatively frequent. Plant 536-2. Mag. × 1650. See photomicrograph of same, photo 10, Plate VII.

Fig. 9. The association of three B-type chromosomes to form a complex configuration. Such complex trivalent configurations have been observed occasionally in plants trisomic for a member of the monoploid complement. Plant 536-2. Mag.×1650. See photomicrograph of same, photo. 11, Plate VII.

only homologous associations occur, text figure 4. Among the figures showing non-homologous association there are two simple types which will be considered first.

If the three B-type chromosomes are numbered 1, 2 and 3, chromosome 1 may associate with chromosome 2 to form a normal bivalent.

Chromosome 3, then, folds upon itself, photos. 6 and 7. In photo. 6, arrow 1 points to the bivalent, arrow 2 to the foldback univalent. In photo. 7, arrow 1 points to the foldback univalent, arrow 2 to the bivalent.

The second simple type of configuration is that of a T. The figures suggest the following interpretation. Chromosome 1 synapses with the homologous part of chromosome 2 at one end and with the homologous part of chromosome 3 at the other end. The 2-by-2 association travels along the chromosome until chromosome 2 meets chromosome 3. At the junction point, 2-by-2 association continues bringing the non-homologous sections of chromosome 2 and 3 together to form the third arm of the T. T-figures are shown in text figure 5 and photos. 5 and 8. If the junction is not in the middle of the chromosome, modified Ts will be produced, text figure 6. It is rare that non-homologous parts do not associate to form the third arm of the T. However, occasionally this occurs. An illustration of this is given in text figure 7 and photograph of same, photo. 12. Simple T configurations at pachytene are frequent but a ring of three chromosomes at diakinesis has not been observed.

There are other more complex types of configurations, only two of which will be illustrated, text figure 8 and 9 and photographs of same, photos. 10 and 11.

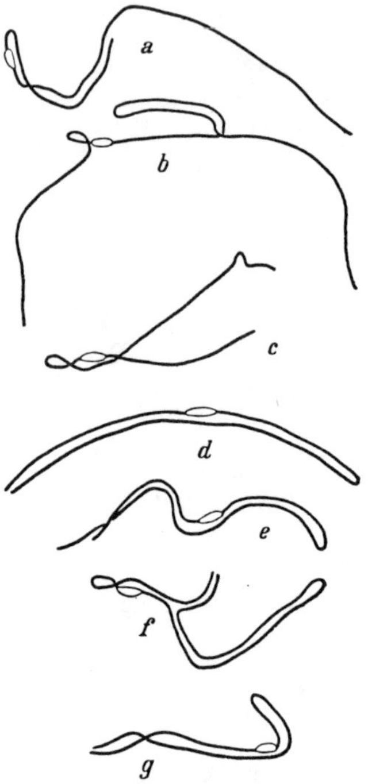

Fig. 10. Outline sketch of pachytene configurations of the univalent in a plant monosomic for chromosome 10. Association of the two ends of the univalent, such as in d, f and g, were frequently observed. Plant 541-9. Mag. ×1450.

III. Non-homologous association within the univalent in monosomic plants.

The tendency for parts of chromosomes to be associated 2-by-2 in the prophase of meiosis is well illustrated by the behavior of the univalent chromosome in monosomic plants. A plant monosomic for chromosome 10 will be used as an example. In the mid-prophase of meiosis in this plant there were nine homologously associated bivalents composed of chromosomes 1 to 9. Each of the bivalents was readily distinguishable by morphological features illustrated in text figure 1. The univalent chromosome 10 stood out in marked contrast to the bivalent chromosomes. Its appearance was variable depending upon the degree of non-homologous association which took place within this chromosome. Text figure 10 illustrates this variability. Although occasional sporocytes were observed in which the univalent was completely unassociated, the usual appearance involved some form of folding of the chromosome to bring non-homologous parts into intimate association. The folding frequently commenced at or close to the spindle fiber attachment region, text figure 10, *d*, *e* and *g*. In some sporocytes the univalent possessed several foldback points, *b* and *f*, text figure 10. The figures of univalent foldbacks suggest that the two ends of the chromosome have a tendency to become associated. In a large number of cases the two ends of a univalent were associated regardless of the type of configuration involving the rest of the chromosome, compare *f* with *d* and *g*, text figure 10. Four illustrations of foldbacks in univalent chromosome 2 are given in photos. 41, 42, 43 and 44 [1].

Although some form of non-homologous association between parts of the univalent characterized most prophase figures, diakinesis nearly always showed the univalent as an open rod-shaped chromosome.

IV. Non-homologous association of parts of the three homologues in trisomic plants.

The synaptic configurations of the three homologous chromosomes in trisomic plants is essentiall y similar to those of the three B-type chromo-

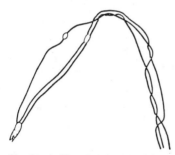

Fig. 11. Outline sketch of the 2-by-2 association of three homologous chromosomes (chromosome 9) in mid-prophase of meiosis. In this figure only homologous associations occurred. Note that the three knobs are together. Plant 539-4. Mag. × 1450.

somes. Since the chromosomes of the normal complement are much longer than the B-type chromosome, there is more chance for complexity. Configurations have been examined in plants trisomic for chromosomes 7, 8, 9 and 10 and in addition 2n plus 1 plants possessing *I* and *i* (two chromosomes resulting from an interchange between chromosomes 8 and 9, see text figure 12) as the extra chromosome. Chromosome 9, with a terminal knob at the end of the short arm, and chromosome *i* will be used as examples.

In a plant trisomic for chromosome 9, 2-by-2 association commonly takes place only between homologous parts of the three chromosomes, text figure 11. In this case it should be noted that all three knobs are together. The formation of a bivalent and a foldback univalent is less common than in B-types. Nevertheless, it occurs with sufficient

---

[1] Besides those in monosomics, foldback configurations of univalents of chromosomes 6, 7, 8 and 9 and *I* and *i* (two chromosomes resulting from an interchange between chromosomes 8 and 9, see section IV) have been observed.

frequency to be readily observed. Such figures can give rise to ten bivalents plus a univalent at diakinesis.

Usually the three chromosomes are associated to form some complex figure which shows a combination of homologous and non-homologous association. Three such configurations will be illustrated. Photo. 13 is a comparatively simple T configuration with non-homologous association along one arm of the T. All three knobs are together. Photo. 14 is a slightly more complex T configuration. The double thread in the

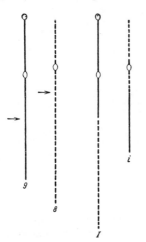

Fig. 12. Diagram of chromosomes 8 and 9 which were involved in an interchange at the approximate positions indicated by the arrows to form the two chromosomes, *I* and *i*. In the sketches which follow chromosome 9 and its components will be drawn as a solid line, chromosome 8 and its components as a broken line.

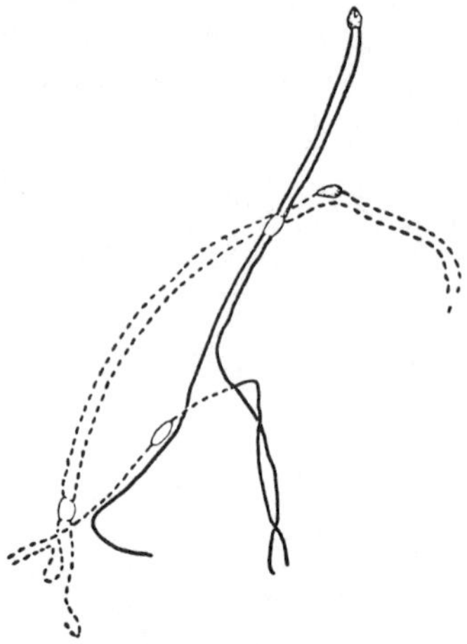

Fig. 13. Outline sketch of pachytene configuration in a 2n + 1 plant having the interchanged chromosome *i* (Fig. 12) as the extra chromosome. Homologus associations have occurred with the exception of a part of the short arm of chromosome *i* which is associated with a section toward the end of the long arm of one chromosome 9. Plant 533C-1. Mag. × 1450.

middle arm of the T represents the non-homologous association. In this case, also, all three knobs are together. Photo. 15 represents another common type of configuration. The non-homologous association is mainly in the foldback region of part of one of the chromosomes.

Excellent examples of non-homologous association in $2n + I$ plants have been obtained from individuals having an interchanged chromosome as the extra chromosome. The interchange occured between chromosomes 8 and 9 in the approximate position given by the arrows in text figure 12, to form the two interchanged chromosomes I (longer) and i (shorter). Figures from sporocytes of $2n + i$ plants will be considered. A variety of synaptic configurations have been seen in these prophases. Only a few of the more simple ones will be described. The common types of figures can be classified as follows: (1) ten bivalents plus a foldback i, (2) nine bivalents plus a T configuration (8-8-i or 9-9-i), (3) eight bivalents plus a configuration involving all five chromosomes, (4) nine bivalents plus either chromosome 8 or 9 with i, i being a partial foldback.

Text figure 13 illustrates the association of chromosome i with chromosomes 8 and 9. In this figure homologous parts of chromosomes are associated with the exception of a short region in chromosome i which is associated with a non-homologous section of the long arm of one chromosome 9. Such associations can give rise to chains of five in diakinesis. A second example of this type of association is shown in photo. 58. In this case, the full chromosome complement is visible.

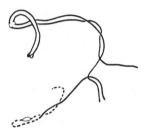

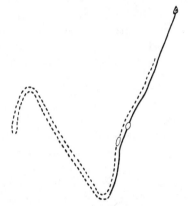

Fig. 14. Outline sketch of pachytene configuration in a 2n + 1 plant having the interchanged chromosome I as the extra chromsome. Chromosome i is homologously associated with one chromsome 9 for a short distance. The rest of chromosome i is in the form of a partial foldback. Plant 533C-1. Mag. ×1450.

Fig. 15. Outline sketch of the association of chromosome I (Fig. 12) with one chromosome 8 in a 2n + 1 plant having I as the extra chromosome. The seond chromosome 8 formed a totally unassociated univalent. Plant 538-4. Mag. ×1450.

In 2n + i plants many excellent examples of T configurations were found. Photos. 17, 18 and 20 show Ts formed by the association of chromosome i with chromosome 8. Photos. 16 and 19 show T configurations of chromosome i with chromosome 9. In all cases, with the exception of photo. 20, the region of non-homologous association is approximately that portion represented in the sketches by the association of the solid with the broken line. In photo. 20, the non-homologous association is represented by the right arm of the T. In all of these cases, diakinesis could show nine bivalents plus a trivalent, the trivalent being composed of 8-8-i or 9-9-i. By using the large knob on chromosome 9 as a marker, it has been possible to determine that with few exceptions, only homologous parts of chromosomes are associated at diakinesis in a trivalent involving this chromosome with chromosome i.

Text figure 14 illustrates a fourth type of configuration which frequently occurs. A section of chromosome i associates with an homologous section of either 8 or 9. The remaining portion of chromosome *i* folds upon itself.

Plants possessing *I* as the extra chromosome gave similar figures to those shown for *i*. In addition to the types already described, one exceptional but highly interesting case was found in a 2n + *I* plant. Chromosome *I* associated with one chromosome 8 throughout their homologous regions. The 2-by-2 association continued beyond the region of homology until all of the chromosome 8 was involved, text figure 15. This left the second chromosome 8 as a univalent. The occurrence of such figures probably is correlated with the presence of an inversion which involved a large portion of the short arm of one chromosome 8, page 201. Within the region of the inversion, homologous parts sometimes remain completely unassociated. Difficulties in bringing about homologous associations in this region probably increase the chances of associations such as shown in text figure 15.

From the observations of trivalent synapsis and the illustrations given here, it seems clear that non-homologous associations can be expressed in many ways. It is not without semblance of order. In general, the configurations of the three chromosomes were similar in the four normal trisomes, the three B-type chromosomes, and to some extent, the two reciprocal interchanged trisomes. The type of synaptic configuration probably depends upon the position the chromosomes occupy in the nucleus and thus the distance a chromosome must travel in order to reach an homologous part. If there is competitive synapsis, homologous parts of the nearest two will associate, the homologous part of the third will remain unassociated or participate in a non-homologous association. This judgment is based upon the fact that related cells very often showed similar configurations. Since associated cells are often related in division, it is not improbable that the chromosomes in their nuclei occupy relatively similar positions with respect to one another. On the basis of competitive 2-by-2 association, such a condition predisposes the chromosomes to similar synaptic configurations.

[*Editors' Note:* Material has been omitted at this point.]

2. Non-homologous association in deficiencies involving inner segments of chromosomes.

Deficiencies involving an internal segment of a chromosome have been obtained through X-ray treatment of pollen. When this pollen

Fig. 33. Pachytene configuration produced by the association of a normal chromosome 3 and its homologue, deficient for an inner segment in the long arm. Note the fold in the normal chromosome which compensates for the deficient region in the homologue. Compare with Fig. 34. Plant S 32-717-113. Mag. × 1450.

Fig. 34. Pachytene configuration produced by the association of the normal chromosome 3 and the one deficient for a segment within the long arm. Note the displacement of the fold in the normal chromosome. This displacement indicates that there is a relatively large section of non-homologous association. Compare with fig. 33. Plant S 32-717-113. Mag. × 1450.

is placed on silks of an untreated plant, the deficiencies are detectable at pachytene in the resulting individuals by a buckle in the normal chromosome produced by the association of the normal with the deficient chromosome. In most cases, the part representing the deficiency folds upon itself to satisfy the 2-by-2 tendency of chromosome association at the prophase of meiosis. Occasionally, however, a true loop is present. The position of the fold or loop is relatively constant when a large number of sporocytes are considered. Nevertheless, variations in the position of the loop or fold in individual sporocytes are encountered.

The usual position of the loop or fold in the sporocytes of a plant with a deficiency in one chromosome 3 is given in text figure 33. From this figure it is clear that the deficiency involved a section of the long arm. Variation in position of the fold or loop ranged along a short distance on either side of the spindle fiber attachment regions. An example of this is given in text figure 34. It is possible that 2-by-2 association which commenced between homologous regions of the two chromosomes, continued to travel along the chromosomes to the position of the deficiency. When this point was reached, the 2-by-2 association continued since there was no counter force of homologous attraction acting upon the region of the normal chromosome corresponding to the deficiency which would tend to prevent this association.

[*Editors' Note:* Material has been omitted at this point.]

X. Non-homologous associations in monoploids and "asynaptic" plants.

Monoploids.

Two monoploid plants from the cultures of Dr. L. J. STADLER have been examined at meiosis. From observations of monoploid meiotic prophases the following seems clear. Much of the doubleness of chromosomes in pachytene must be due to the association of non-homologous parts of chromosomes either within a chromosome (foldbacks) or between chromosomes. Typical foldback chromosomes and foldback regions within chromosomes were frequently observed. Associations between chromosomes also occurred. The complexity of the association in monoploid prophases made an analysis of all the chromosomes of the complement of a particular sporocyte exceedingly difficult. Since non-homologous association at early prophase usually leads to disassociation during diplotene the formation of ten univalents at metaphase would be expected. At metaphase *I* the sporocytes usually showed ten univalents. Frequently, however, two chromosomes were loosely associated to form a "bivalent". Occasionally there were several loosely associated "bivalents".

One feature of anaphase *I* in the monoploid deserves attention. In a number of figures an extra piece of chromatin was visible, ranging in size from small to large sections of a chromosome. Related cells did not show related conditions with regard to the presence or the size of these fragments. It is inferred that these fragments may represent the results of translocations between non-homologous parts of chromosomes which were associated in the prophase. Further evidence for translocations between non-homologous parts of chromosomes which are known to be associated in the mid-prophase of meiosis are given in C of the Discussion, page 226.

"Asynaptic" plants.

"Asynaptic" plants are characterized by variable numbers of univalents in diakinesis. Several plants from the progeny of X-rayed pollen showed a variable number of bivalents and univalents at diakinesis. Examination of the mid-prophase of meiosis suggested a reason for this. In some sporocytes homologous chromosomes were associated throughout their lengths. In other sporocytes homologous association occurred for a short distance. Each of the chromosomes then associated with some other chromosome to produce typical "translocation" configurations. Among the different figures there was no constancy of the chromosomes involved or the position at which a chromosome would start to associate with another non-homologous chromosome. Carrying this further would involve the non-homologous association of the whole chromosome. Since the association of non-homologous chromosomes rarely continues into diakinesis and metaphase, it is reasonable to assume that some of the univalent condition seen here is correlated with an earlier non-homologous association between chromosomes.

The author does not imply that all kinds of "asynapsis" are due to non-homologous associations. It is certain that in some cases there is a true asynapsis in pachytene. In some plants, also, early disjunction of normally synapsed chromosomes could account for the appearance of univalents at diakinesis. Nevertheless, one kind of "asynapsis" can be correlated with the presence of non-homologous associations.

XI. Discussion.

A. *Factors involved in synaptic association.*

From a study of synaptic configurations described in this paper the following conclusions are drawn with regard to the synaptic process in *Zea mays*.

1. There is a tendency for chromosomes to associate 2-by-2 in the prophase of meiosis whether or not the parts associated are homologous. NEWTON and DARLINGTON (1929) placed emphasis on the 2-by-2 association in the mid-prophase of meiosis of homologous parts of chromosomes in trivalents. In the present paper this emphasis is extended to include non-homologous associations.

If the two chromosome complements of a diploid plant are structurally similar, association is most frequently confined to homologous parts of chromosomes (for exceptions, see pages 220 and 221). When the two complements are structurally dissimilar (inversions and translocations) conflicting forces responsible for 2-by-2 synapsis may result in the association of non-homologous parts of chromosomes in those regions representing the structural dissimilarity.

In plants with unbalanced chromosomal complements (monosomics, trisomics, deficient rod-shaped chromosomes, deficient ring-shaped chromosomes) it is not possible to have all parts homologously associated 2-by-2. In these cases, a wide range of synaptic configurations results. Each type of unbalance produces its own range of synaptic associations. In general, it can be stated that there is a tendency for associations to occur between homologous parts leaving the parts without homologous sections with which to synapse to form non-homologous associations or to remain unassociated. Exceptions to this generalization have been considered in the sections on trisomics, ring-shaped chromosomes and deficiencies.

2. Association between non-homologous parts of chromosomes appears to be as intimate in many cases, as that between homologous parts, see photos. 4, 14, 20, 41, 48, 62, etc.

3. Association of non-homologous parts can proceed at the same time as homologous association and may even compete with it. See pages 199 and 201, also, text figures 5, 17, *b* and *d*, 24, 38, 40, 41, 48, and translocation, BURNHAM (1932 b).

4. Ends of chromosomes tend to associate in advance of other parts of chromosomes whether or not the ends are homologous. This is suggested by (1) the high frequency of T configurations in B-types and trisomics of small chromosomes (text figure 5, and photos. 5, 8, 13, 14, etc.), (2) by unusual configurations of B-type chromosomes and trivalents (photos. 10 and 11), (3) by the high frequency of end associations in foldbacks of univalents of the B-type chromosome and monosomics

150

(text figures 3 and 10; photos. 6, 10, 41, 43, 46), (4) by the association of the three terminal knobs in chromosome 9 trivalents (text figure 11; photos. 13 and 14) and (5) by occasional associations such as illustrated in text figure 29.

5. Although association tends to begin at the ends, it can start anywhere along the chromosome, as illustrated by (1) trivalent associations, text figure 11, (2) compound foldbacks in univalents, text figure 10, f, and photos. 42, 43 and 45, (3) foldbacks in monosomic regions of chromosomes, text figure 17, b, (4) and foldbacks in ordinary bivalents, text figure 49 and photo. 62.

[*Editors' Note:* Material has been omitted at this point.]

Summary.

1. Association of chromosomes at pachytene in Zea mays is 2-by-2 whether or not the parts associated are homologous.

2. Evidence for non-homologous association of parts of chromosomes has been obtained from monoploids, diploids, monosomics, trisomics, deficiencies, inversions, translocations, ring-shaped chromosomes, "asynaptic" plants and so-called "B-type" chromosomes.

3. In many cases, the non-homologous association at pachytene appears to be as intimate as homologous association.

4. Non-homologous association, present at pachytene, rarely continues into diakinesis.

5. Translocations probably result from the association of non-homologous parts of chromosomes.

REFERENCES

Burnham, C. R. 1932b. The association of non-homologous parts in a chromosomal interchange in maize. *Proc. 6th Int. Congr. Genet.* **2**:19–20.

Newton, W. C. F., and C. D. Darlington. 1929. Meiosis in polyploids. I. Triploid and pentaploid tulips. *J. Genet.* **21**:1–15.

Randolph, L. F. 1928. Types of supernumerary chromosomes in maize. (Abstract). *Anat. Rec.* **41**:102.

17

Reprinted from *J. Biophys. Biochem. Cytol.* **2**(2):215–218 (1956)

Chromosomal Structures in Crayfish Spermatocytes.* By Montrose J. Moses.‡

(From The Rockefeller Institute for Medical Research.)§

To date the electron microscope has yielded little about the orderly macromolecular structure of chromosomes that, from genetical and cytological evidence, might be expected. Indeed, even the grosser morphology long familiar to cytologists in the light microscope is harldy recognizable, partly for want of suitable electron stains and partly, undoubtedly, because of fixation and preparative effects and the complexity of interpreting structures in thin sections. It is true that refined methods of preparing tissues for study in the electron microscope—particularly those of buffered osmium tetroxide fixation, plastic embedding, and thin sectioning—have vastly improved the quality of material studied, but the cytology of the nucleus at the electron microscope level of resolution remains obscure.

Most consistent are recent reports of coiled structures about 500 A wide seen in isolated chromosomes (whole (1) and in sections (2)) and areas assumed to be occupied by them *in situ*. These have been reported as being comprised of paired strands about 200 A wide (3) but there is difficulty and hence disagreement in interpreting them as filaments or granules (4). While such entities may well prove to be morphological subunits of the chromosome, at the moment they present a disappointingly random aspect. Orderly organization intermediate to

them and the coarser details such as chromonemata (at the limit of light resolution) and gyres (best visible in specially treated material) seen in the light microscope is lacking.

Details of the latter structures are most apparent at stages in the cell cycle such as meiotic prophase where the chromosomes have emerged as easily discernible entities and have not yet undergone extensive coiling and condensation prior to metakinesis. Light microscopists have derived much of our present knowledge of chromosomes from studies of such stages. It is here also that we may expect to find structure with the electron microscope. Watson (5), for example, has reported fine paired filaments less than 400 A wide in primary spermatocyte prophases in the rat. These may be chromosomal elements, the counterparts of which we have observed in spermatocytes of grasshopper and crayfish. In the latter the condition is sufficiently striking to warrant a brief description.

Testes from mature *Cambarus clarkii*[1] were fixed for 1 hour in veronal-acetate buffered 1 per cent OsO_4 at pH 8 and 8.4. The meiotic chromosome cytology, which has been described by Fasten (6) was followed in the light microscope. The large number of chromosomes (*ca.* 100) together with their characteristic polarization and attachment to the nuclear periphery raises the probability of including aspects of various planes through chromosome axes in a thin section.

In the electron microscope, chromosomes are about 1 μ wide, depending on

* This work was supported by a Special Purpose grant (INSP-85A) from the American Cancer Society, Inc.

‡ The author gratefully acknowledges the generous collaboration of Dr. Keith R. Porter.

§ *Received for publication, January 30, 1956.*

[1] Material obtained in the spring from Carolina Biological Supply Co., Elon College, North Carolina.

stage and chromosome region. A central dense region (Fig. 1, *C*) about 150 mμ in width follows the chromosome axis and is clearly visible at low powers. Fig. 1 shows a characteristic part of a primary spermatocyte prophase chromosome (*CR*) with one end closely applied to the double nuclear envelope (*NE*). A second chromosome passes out of the section just below it. The chromosome profile is clear and of some interest, but will not be discussed here. Such structures are Feulgen-positive when studied by light microscopy in adjacent thick sections. The chromosome lies parallel to the plane of section for a good part of its length and has been cut so that the section passes through the central dense region. The latter can be seen to follow the bend in the chromosome and otherwise appears to be an integral part of it. A section of the same chromosome is shown in Fig. 3 at higher magnification. The interior of the dense region gives the appearance of a series of alternate parallel light and dark lines. They cannot be followed to the nuclear envelope, probably because of curving of the chromosome. Such parallel arrays in the dense region are commonly visible in prophase chromosomes of crayfish, though seldom at such length. Fig. 2 is a section from another cell that includes a similar longitudinal cut. It typifies our most consistent observation, that of a continuous dense central structure (*P*) less than 150 A wide bounded on either side by less dense areas about 250 A wide (*Q*). These are in turn bounded by one or more roughly parallel 150 *A* lines (*R*), frequently discontinuous and separated by varying distances. The characteristically non-homogeneous substance of the remainder of the chromosome blends with the dense area at its periphery, occasionally appearing to be arranged in parallel array perpendicular to the central apparatus. Fig. 4 presents three aspects of increasing degrees of obliquity (*a*, *b*, *c*), approaching the normal to the longitudinal axis of the chromosome in a thicker section. The appearance is consistent with the central structure or "core" (*C*, Figs. 1 and 2) being composed of a median dense rod embedded in a less dense material. This in turn is surrounded by varying numbers of concentric dense shells which become less distinct progressing radially. Fig. 5 is a very thin section showing cross (*X*) and longitudinal (*L*) sections of the "core." While the cross-section gives the impression of several concentric circles, it can be seen to be actually formed by imperfect aggregations of finely divided discontinuous material. The longitudinal section has a corresponding appearance with the rod and its shell being most nearly perfect. Almost every prophase chromosome observed to be cut through its axis has shown some evidence of this "core." Late prophase and metaphase chromosomes lack it entirely. We have not yet discovered its origin or followed it through sequential stages.

Chromosomal "cores" showing substantially the same detail have been occasionally encountered in other materials. However, our observations have so far been limited and it cannot be said if the structure reported here is common to all chromosomes. It is realized that our present knowledge of chromosomes based on observations with the light microscope does not lead us to expect the structures described. However, they may well represent either a special or transient form of a fundamental structure whose role in chromosome function is less obscure.

BIBLIOGRAPHY

1. Ris, H., *Genetics*, 1952, **37**, 619.
2. Gay, H., *Dissertation Abstr.*, 1955, **15**, 899.

3. Lafontaine, J., and Ris, H., *Rec. Gen. Soc. Am.*, 1955, No. 24, 579.

4. Gall, J., *Brookhaven Symp. Biol.*, 1956, No. 8, 17.

5. Watson, M., Spermatogenesis in the adult albino rat as revealed by tissue sections in the electron microscope, University of Rochester Atomic Energy Project Report UR-185, 1952.

6. Fasten, N., *J. Morphol.*, 1914, **25,** 587.

EXPLANATION OF PLATE 36

All chromosomes are from primary spermatocyte prophases of the crayfish. Bars denote approximately 100 mμ.

FIG. 1. Longitudinal section of chromosome (*CR*) cut along its axis. *C*, central dense region; *CY*, cytoplasm; *NE*, nuclear envelope; *AT*, region of chromsome association with nuclear envelope (membranes are cut obliquely at this point); *M*, discontinuous chromosomal material. Approximately × 25,000.

FIG. 2. Longitudinal section through axis of chromosome showing details of structure in central dense region (*C*). *P*, *Q*, *R*, parallel structures described in text. Approximately × 56,400.

FIG. 3. Higher magnification of portion of chromosome outlined in Fig. 1. Approximately × 77,000.

FIG. 4. Three oblique sections through chromosome central region, approaching the normal and the axis at *c*. Approximately × 56,400.

FIG. 5. Cross (*X*) and longitudinal (*L*) sections of chromosome "cores" seen in very thin section. Approximately × 56,400.

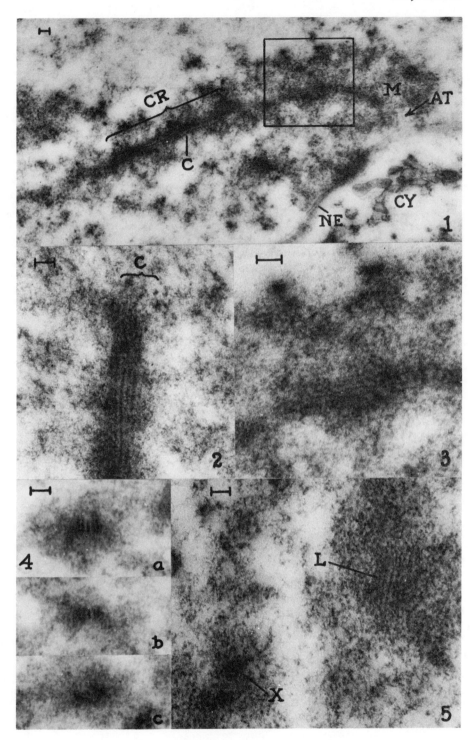

PLATE 36

18

Reprinted from *Science* **64**(68):433 (1928)

A GENIC DISTURBANCE OF MEIOSIS
IN ZEA MAYS

George W. Beadle
Barbara McClintock

Cornell University

DURING and following the summer of 1927, a collection of maize carrying factors for male sterility (Eyster)[1] was made for the purpose of genetical and cytological investigation. The occurrence of male-sterile plants in material from thirty or more unrelated cultures suggests the possibility of several genetic factors causing such sterility.

In segregating material obtained from I. F. Phipps, it has been found that sterility is due to a recessive mendelian factor causing irregular meiosis. In a count of 144 plants the observed ratio was 109 normal to 35 sterile plants, a deviation from the calculated ratio of but one plant.[2] The cytological behavior in these sterile plants has been determined by studies of the meiotic divisions in the microsporocytes.

Early stages of microsporogenesis in the male-sterile plants have not been extensively studied. During the stages just previous to and during diakinesis there is observed a partial or complete failure of synapsis. Because of this lack of synapsis and the consequent presence of a large number of univalents, metaphases are characteristically irregular. Microsporocytes most frequently show twenty univalents. Progressively fewer cells show one bivalent and eighteen univalents, two bivalents and sixteen univalents, and so on, cells with ten bivalents rarely being observed. Some anthers show a high percentage of sporocytes containing some bivalents while other anthers show a high percentage of sporocytes containing twenty univalents.

Irregularity in the appearance of metaphase *I* increases with an increase in the number of univalents. A microsporocyte with ten bivalents in metaphase *I* appears normal. When univalents are observed they do not always lie in one spindle. Usually there is one major spindle containing the several bivalents, when present, plus some of the univalents, and one to several minor spindles containing one or more

univalents (Fig. 1). In consequence of the presence of several spindles, the sporocyte is divided into a number of unequal cells after the first meiotic mitosis. Each cell contains one or more nuclei and each nucleus contains one or more chromosomes. These cells undergo a second division to form microspores. It is obvious that most of these microspores and the pollen grains formed from them do not contain a normal haploid set of chromosomes, and they are probably non-functional under ordinary conditions.

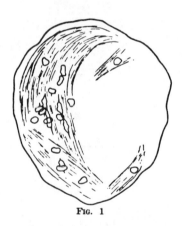

FIG. 1

This particular type of male sterility is accompanied by a certain amount of female sterility. Several pollinations have given only sparsely-filled ears. Female sterility has been observed in male steriles from a few of the other sources also. Megasporogenesis remains to be studied in these cases.

With regard to at least one male-sterile culture, it may be stated that male sterility is due to a simple mendelian gene affecting synapsis and consequent meiotic behavior, the result being the formation of gametes containing varying chromosomal complements, only a few of which are viable.

[1] L. A. Eyster, *Journal of Heredity*, 12: 138–141.
[2] Data partly from I. F. Phipps.

19

Reprinted from *Nature* **182**(4637):713–715 (1958)

GENETIC CONTROL OF THE CYTOLOGICALLY DIPLOID BEHAVIOUR OF HEXAPLOID WHEAT

By Dr. RALPH RILEY and VICTOR CHAPMAN

Plant Breeding Institute, Cambridge

COMMON wheat, *Triticum vulgare*, is a hexaploid species with 42 chromosomes in which the genomes of diploid wheat and of the diploid species *Aegilops speltoides* and *Aegilops squarrosa* are combined together[1-5]. There is considerable similarity between the chromosomes of these three diploids. Indeed, in hybrids between diploid wheat and *Aeg. squarrosa* and between diploid wheat and *Aeg. speltoides*, complete pairing at meiosis occasionally results in the formation of seven bivalents, although the mean bivalent frequency is somewhat less, being between three and four per cell[5,6]. Moreover, the attraction which results in bivalent formation in the hybrids is of such strength as to cause multivalent formation in tetraploids induced from them. However, each chromosome normally conjugates with its complete homologue at meiosis in *T. vulgare*, and only bivalents are formed. Thus, the pairing attractions between the equivalent, homoeologous, chromosomes of different genomes either no longer exists or is no longer expressed. Polyploid wheat has therefore evolved to behave cytologically as a diploid. Further, the shift has been of such efficiency that there is very little chromosome pairing even in 21-chromosome haploids of *T. vulgare*, in which intergenome pairing is not restricted by the affinity of complete homologues. The cytological distinctiveness of the equivalent chromosomes of different genomes is all the more striking in view of the close genetic relationships which have enabled Sears[7], by nullisomic-tetrasomic compensation, to recognize the seven homoeologous groups into which the complement of wheat can be arranged.

Recent results with various aneuploids have revealed something of the genetic control of the diploid behaviour of *T. vulgare*. The evidence centres on the derivatives of a 41-chromosome monosomic line, designated *HH*, which arose at Cambridge in the variety Holdfast. The formation of twenty bivalents and one univalent (Fig. 6) in *HH* monosomics is like the behaviour of all other monosomics of common wheat[8]. Meiosis is normal in the 42-chromosome segregants in the monosomic line, and in the 42-chromosome hybrids from crosses of *HH* monosomics to euploid plants of Holdfast. Therefore, the chromosomes of *HH* monosomics are structurally unaltered compared with those of Holdfast. One arm of the chromosome monosomic in the *HH* line is rather more than twice the length of the other, but the chromosome has not yet been positively identified relative to the numbering based on the variety Chinese Spring.

Monosomics of *HH* have been used in crosses with 44-chromosome plants which had the full complement of wheat chromosomes plus a single pair of rye chromosomes. From these crosses five 20-chromosome, nulli-haploid, plants have been obtained, which, by the nature of their origin, must have had the haploid complement of Holdfast minus the chromosome which is monosomic in the *HH* line. The nulli-haploids arose in (a) the F_1 of the cross *HH* monosomic × rye chromosome III disomic addition, and in (b) the F_2 of crosses of *HH* monosomic × rye II, or rye III, disomic additions[9]. All the F_1's which produced nulli-haploids in F_2 had twenty bivalents composed of wheat chromosomes and the *HH* chromosome and a rye chromosome as univalents. In each situation, therefore, the 20-chromosomes in the embryo sacs which functioned parthenogenetically must have been those which had segregated from bivalents—that is, all the wheat chromosomes except *HH*.

The meiotic pairing of the *HH* nulli-haploids has been compared with the pairing in 21-chromosome euhaploids of Holdfast. The meiotic behaviour of the euhaploid plant listed in Table 1 was similar to that of twelve other 21-chromosome plants of Holdfast examined over a period of four years. The number of bivalents in these plants never exceeded four, with a mean frequency of between 1·3 and 1·7 per cell, and trivalents were rare (Fig. 1). There was usually only one chiasma per bivalent.

By contrast, in the *HH* nulli-haploid enumerated in Table 1, which is typical of five such plants examined in two successive years, the mean bivalent frequency exceeded the maximum observed in euhaploids. Trivalents were quite frequent with as many as five in some cells (Fig. 3), although there was never more than one per cell in the euhaploids. Very many more chromosomes per cell, a mean of 11·0 compared with a mean of 2·9, were involved in chiasma-associations and a much higher frequency of closed bivalents and of 'pan-handle' trivalents resulted from more intimate pairing. As many as nineteen of the twenty chromosomes of the nulli-haploid have been observed in various associations simultaneously, although never more than nine were involved in associations in the same cell in euhaploids.

The most reasonable hypothesis to account for these observations was that the deficient chromosome carried a gene, or genes, which restricted

Type	Chromosome No.	Cells	Mean pairing			Proportion of bivalents as rings	Conjugated chromosomes per cell	
			Bivalents	Trivalents	Quadrivalents		Mean	Range
Euhaploid	21	100	1·38 ± 0·09	0·07	—	0·00	2·86 ± 0·23	0–9
Nulli-haploid	20	75	4·16 ± 0·12	0·86 ± 0·03	0·02	0·23	10·95 ± 0·33	4–19

intergenomic, homoeologous, pairing ; the alternative being that association between randomly distributed duplicate segments was normally inhibited.   On either view, there being no other independent mechanism, in the absence of the pairing restriction normally undetected affinities were expressed. Crucial evidence that the pairing was homoeologous rather than random was afforded by the high frequency of trivalents and the infrequency of quadrivalents.

If homoeologous pairing could take place in *HH* nulli-haploids, it was argued that 40-chromosome nullisomics deficient for the *HH* pair should deviate from the strictly bivalent-forming regime of *T. vulgare*. Consequently, three *HH* nullisomics have been examined.   All had large and frequent multivalents, and associated univalents, at meiosis.   No method has yet been found of consistently preparing well-spread pollen mother cells of this material, but the complexity of the behaviour is readily apparent (Fig. 4).   More than half the cells have at least one multivalent, usually a quadrivalent, and many have several. Associations of three, four, five and six are common, but no higher multivalents have been observed. However, the greater the number and size of multivalents the less is the chance of disentangling the snarl, so there is danger of observational bias. No doubt, also, there are homoeologous, or allosyndetic, bivalents which would be undetected unless markedly heteromorphic.   The magnitude of the meiotic disturbance is therefore hard to assess ; nevertheless, it is clear that, in the absence of the *HH* chromosome, *T. vulgare* ceases to be a stable bivalent former, a classical example of the autosyndetic pairing allohexaploid, and behaves as an intermediate, auto allopolyploid.

The diploidizing mechanism is effective in the hemizygous state in monosomics and in euhaploids. Furthermore, in a haploid plant which had twenty normal wheat chromosomes plus the iso-chromosome formed from the long arm of the *HH* chromosome (Fig. 5), meiotic pairing was the same as in euhaploids. Moreover, there was no multivalent formation in a 41-chromosome monosomic which had twenty normal pairs of Holdfast chromosomes and in which the *HH* chromosome was represented only by a single telocentric of the long arm.   The long arm alone is thus effective in prohibiting homoeologous association ; but no evidence is available on the effects of the short arm alone.   It may be that both arms have genes with equally effective control over the pairing behaviour, but this seems unlikely. The control of bivalent formation is probably restricted to one arm, and may indeed be effected by quite a localized region.

The implications of these results are both theoretical and practical.   First, there has been considerable discussion on the methods by which many polyploids have attained their cytologically diploid character, with its obvious advantages in fertility and genetic stability.   Several authors have favoured the theory that this situation has been achieved by the selective accumulation of many small changes of chromosome structure, which lead to the divergence of homoeologues.   However, the selection of the localized product of a single mutational step, such as seems likely to have happened in wheat, is very much simpler to envisage, and would clearly be a more rapid and efficient process.   To compare the frequencies of the genetical, or cytological, determination of diploid meiotic behaviour, other allopolyploid species should be examined for evidence of a mechanism similar to that in wheat.

Secondly, knowledge of the pairing restricting mechanism may be useful in wheat improvement.

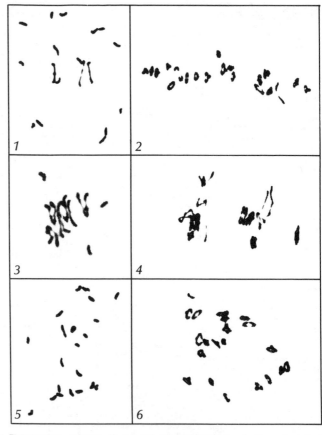

First metaphase of meiosis in Feulgen- and orcein-stained squashes of pollen mother cells of various derivatives of *T. vulgare*

(1) 21-Chromosome euhaploid with 4 bivalents (one widely stretched) and 13 univalents. (2) Euploid with 21 bivalents. (3) 20-Chromosome *HH* nulli-haploid with one 'pan-handle' and 4 chain trivalents, one bivalent and 3 univalents. (4) 40-Chromosome *HH* nullisomic with one ring of six, 2 chains of four and 13 bivalents. (5) 21-Chromosome haploid in which the *HH* chromosome is an isochromosome of the long arm ; one stretched bivalent, 18 normal univalents and the isochromosome paired interbrachially with two chiasmata. (6) 41-Chromosome *HH* monosomic with 20 bivalents and one univalent. (×770)

The utilization of related diploid species, for example, in the genus *Aegilops*, in wheat breeding is hampered because the chromosomes of the diploids do not pair with those of wheat, in intergeneric hybrids. To overcome this obstacle, Sears[10] was forced to use X-rays to induce the translocation of a disease-resistance gene of *Aeg. umbellulata* on to a wheat chromosome, and others[9][11,12] have been compelled to explore alien chromosome addition and substitution lines. However, this intergeneric allosyndetic pairing may well take place in the absence of the *HH* chromosome, just as it does between chromosomes of the different genomes; alien genes could then be introduced into wheat chromosomes by normal recombination. Critical hybrids of this constitution can be extracted from crosses of *HH* monosomics by the appropriate diploid.

Finally, nullisomic plants of this constitution may be used either within single varieties or in intervarietal hybrids in 'intergenome exchange' breeding. Thus a useful gene which showed a dose effect could be obtained duplicated or triplicated, represented on each chromosome of the homoeologous group. The chromosome structure within a variety could be re-patterned by breeding from *HH* nullisomics. Further, if an *HH* nullisomic were also an intervarietal hybrid, the release of variation, and the range of segregation in later generations, would be very much greater than from a euploid hybrid. Thus, the *HH* nullisomic may afford access to otherwise unavailable genetic variation.

However, the exploitation of *HH* nullisomics depends upon their fertility. There is no natural self-fertility, but a reasonable seed set has been obtained by pollinating *HH* nullisomics with Holdfast euploids, although the reciprocal cross has been unsuccessful. There were several large multivalents at meiosis in the one 41-chromosome derivative, so far examined, of an *HH* nullisomic pollinated by a euploid. The *HH* chromosome regained from the euploid pollen was always a univalent, and homoeologous pairing must have been prohibited by its presence. The multivalents were thus indicative of translocation heterozygosity, the outcome of homoeologous pairing in the nullisomic parent. Homozygotes for new structural conditions can be derived from such plants, although further back-crossing may first be necessary to reduce the extent of structural alteration.

Knowledge of the situation described already begins to make plain some of the problems of wheat cytogenetics and perhaps of polyploidy in general. Further investigation cannot fail to extend this advantage and may well contribute to the cytogenetic manipulation of wheat for practical breeding purposes.

The interest of Dr. G. D. H. Bell in the development of this work is gratefully acknowledged.

[1] Kihara, H., *Bot. Mag., Tokyo*, **32**, 17 (1919).
[2] Sax, K., *Genetics*, **7**, 513 (1922).
[3] McFadden, E. S., and Sears, E. R., *J. Hered.*, **37**, 81, 107 (1946).
[4] Sarkar, P., and Stebbins, G. L., *Amer. J. Bot.*, **43**, 297 (1956).
[5] Riley, R., Unrau, J., and Chapman, V., *J. Hered.* (in the press).
[6] Sears, E. R., *Res. Bul. Mo. Agric. Exp. Sta.*, **337** (1941).
[7] Sears, E. R., *Res. Bul. Mo. Agric. Exp. Sta.*, **572** (1954).
[8] Sears, E. R., *Amer. Nat.*, **87**, 245 (1953).
[9] Riley, R., and Chapman, V., *Heredity*, **12**, 301 (1958).
[10] Sears, E. R., Brookhaven Symp. in Biology, **9**, 1 (1956).
[11] Jenkins, B. C., Proc. Int. Genet. Symp., 1956, 298 (1953).
[12] Hyde, B. B., *Amer. J. Bot.*, **40**, 174 (1955).

Part IV

CROSSING-OVER

HISTORICAL PERSPECTIVES

One of the discoveries during the flurry of inheritance studies following the rediscovery of the Mendelian explanation of inheritance was the phenomenon of linkage that Bateson, Saunders, and Punnett (1906) first reported. Although the F_2 data obscured the ratios between parental and recombinant classes, Bateson and Punnett (1911) assumed that the same gametic series occurred in both male and female, different degrees of linkage being determined by a fixed gametic ratio, e.g., 3AB:1Ab:1aB:3ab (total = 8); 7:1:1:7 (total = 16); or 15:1:1:15 (total = 32), etc., for coupling and 1AB:3Ab: 3aB:1ab, etc., for repulsion. According to the *reduplication hypothesis* which they proposed to explain the different series, the members of linked pairs of factors undergo vegetative reduction early in development, and the cells containing the AB and ab factor combinations in coupling or Ab and aB in repulsion undergo more rapid division than cells with other combinations. The hypothesis was shown to be incorrect by the fact that similar linkage values were obtained from flowers in different parts of a plant.

Morgan (1912) showed that there was complete linkage in the male *Drosophila melanogaster*. The discovery that the genes could be placed in linkage groups and that within each group they could be placed in linear order assuming that the recombination frequency was related to physical distance between the genes in the chromosomes (Sturtevant 1913) led to the conclusion that exchanges did occur between the members of a chromosome pair. The linkage maps constructed for chromosomes I, II, and III of

Drosophila by Morgan and Bridges (1916–1923) were a logical out-growth. The methods used in arriving at the standard map were described in minute detail (Bridges and Morgan 1923).

Cytologists had noted previously that in the early stages of meiosis, the chromosomes were paired, but when each became visibly double, one pair of chromatids repelled the homologous pair of chromatids except at certain points at which there appeared to be a change of partners, a chiasma. One explanation for this configuration was that on one side of that point the separation between pairs of chromatids was reductional (that is, the disjunction was between pairs of sister chromatids), and on the other side, equational (that is a sister plus a non-sister chromatid passed to opposite poles). This came to be known as the "classical theory of chiasma formation" (Wenrich 1917 and others). That theory held that chiasmata were interpreted as points at which crossing-over might occur by breakage and reunion as the members of the pair separated. An alternate possibility, one of Janssens' (1909, 1924) proposals, was that the chiasma was the point at which an exchange had taken place between two of the four chromatids, "partial chiasmatypie."

Both Darlington and Belling realized that the configurations in multivalents would furnish convincing evidence as to which theory was correct. In a trivalent the critical configuration is one in which two chromosomes have two chiasmata, and one of them has an intercalary chiasma with the third chromosome. Belling (1929) studied the meiotic configurations of trivalents in hyacinths and concluded that the separation on both sides of the nodes was between homologues. As Belling stated, "Triploid configurations cannot be explained by the hypothesis of alternate separation of homologues and of sister strands, except with an additional hypothesis." The addition needed was the unlikely assumption that if separation on one side of this third chiasma is reductional, then in order for the separation to be equational on the other side, the six strands must fall apart and reassociate. Darlington concluded that certain configurations, if found in a quadrivalent, would demonstrate decisively that chiasmata were the result of previous exchange. Since the critical configurations were not observed in the first studies, Darlington (1929) concluded that random reassociation did take place. Later an intensive search in his laboratory was successful in finding many critical figures. Several of the less complicated ones are shown by Darlington (Paper 20). This report demonstrated cytological crossing-over and also demonstrated that the breaks preceded crossing-over.

Cytological observations combined with linkage tests by Stern (1931), Creighton and McClintock (Paper 21), and later by Brink and Cooper (1935) demonstrated that an actual exchange of chromosome segments had accompanied the genetically-observed recombinations.

Another approach to the question was to compare the chiasmat frequency with cross-over frequency. Brown and Zohary (1955) accomplished this using two terminal deficiencies and a paracentric inversion in *Lilium formosanum*. The data show nearly a 1:1 relationship. This relationship may not be universal since the authors cite data for other organisms in which there was poor agreement.

Darlington also stated that chiasmata were essential in holding the chromosomes together as pairs from then until metaphase. He also proposed 2-strand double cross-overs in the centromeric regions in the X and the Y chromosomes in *Drosophila* to account for the pairing and normal disjunction without any observable genetic recombination. Since chiasmata were not observed in the autosomal pairs, this would be pairing without chiasmata. Achiasmate pairing at meiosis has been reported in plants only in *Fritillaria* (Noda 1968) in micro- but not in megasporogenesis.

How cytological crossing-over occurs remains unexplained, but some information has accumulated as to *when* it probably does occur and *how* it does not occur.

That crossing-over occurs when the chromosomes are double stranded (the 4-strand stage) was suggested by genetic experiments in *Drosophila*, first by Bridges (Paper 46), later by Bridges and Anderson (1925) with triploids, and Anderson (1925) with an attached-X stock. Anderson's paper requires careful study to determine how he derived the values for the different cross-over combinations. The experiments all depend on a genetic system where more than one chromatid from the tetrad at meiosis is recovered in a resultant individual. If one of those chromatids is a cross-over and the other a noncross-over, crossing-over must have occurred when each member of the pair was duplicated. The attached-X data also showed that the end of the map with the yellow body color marker was not the centromere end, as previously assumed. Crossing-over at the 4-strand stage was shown to be true also for maize, as Rhoades (1933) demonstrated using trisomics for chromosome 5; and for *Neurospora* (Lindegren 1933) using ordered spores in individual asci.

Experiments in *Drosophila* similar to those of Anderson (1925) but on a larger scale showed that sister-strand crossing-over, if it

163

occurs, is relatively rare in that species (Beadle and Emerson 1935). Experiments by Schwartz (Paper 22) using a ring chromosome 6 were interpreted as indicating the occurrence of sister-strand crossing-over at meiosis in maize. This has been confirmed by Miles (1971) using a ring chromosome 10.

Breeding experiments with the various Bar alleles in *Drosophila* demonstrated conclusively that the occurrence of a more extreme Bar mutant, and also a wild type were the products of a rare cross-over event (Sturtevant 1925). From the genetic data, Sturtevant concluded that the best explanation was that the Bar phenotype was due to a duplication and the extreme types were the result of unequal crossing-over. Three different reports in 1936, Bridges (Paper 4), Muller, et al., and Dubinin and Volotov, furnished the cytological evidence that the original Bar mutant was itself a duplication and thus was the basis for its mutable behavior. Crossing-over at certain points in the duplication when pairing was misaligned ("unequal" crossing-over) could produce a normal chromosome with no duplication. Subsequent studies have revealed several other regions in *Drosophila* that behave similarly. One was reported by Lewis (1941).

Crossing-over may occur also in somatic cells. Bridges (1925) noted that in *Drosophila* flies carrying Minute-n, a dominant character for minute bristles and other effects, and a recessive in the other X-chromosome often had patches of varying size with the recessive character but without Minute. This was explained at first as a result of the loss of the entire X chromosome or a fragment carrying Minute. Stern, in 1936 (Paper 23), reported the results of experiments designed to determine how that elimination might take place. He used flies heterozygous for linked recessive markers that could be expressed in very small areas of the body surface and for dominant Minutes to increase the frequency of mosaic spots. When the recessives were in repulsion, he obtained flies with twin spots, but not when the markers were in coupling. The results were explainable by the hypothesis that somatic crossing-over had occurred between homologous chromosomes that were double stranded at the time of crossing-over. Stern's paper presents the results of numerous experiments that confirmed the hypothesis and also showed that the pattern of such crossing-over differed in several respects from that of meiotic crossing-over.

A process similar to somatic crossing-over occurs also in diploid nuclei that arise in certain fungi under certain conditions, referred to as *parasexuality* (Pontecorvo 1954). When this diploidization is followed by somatic reduction, a genetic analysis not dependent on meiosis is possible. This was accomplished for filamentous fungi. For a review, including studies of *Aspergillus nidulans*, see Pontecorvo and Käfer (1958). In the two maps for chromosome II, the order of the loci was the same, but the relative distances for certain regions in the meiotic map were longer, others shorter than in the somatic cross-over map.

Several models have been advanced describing the putative events leading to crossing-over. These have played a major role in stimulating discussion. Belling's rather simple model (Papers 24 and 25) stimulated considerable thinking at the time and in years hence. His model implies that crossing-over occurs at the time of chromosome duplication. At the pachytene stage, it was presumed that the paired homologous chromosomes formed a half-twist at the point of eventual crossing-over. The chromomeres then became duplicated and subsequently connected by new interchromomeric fibers that generated two parental and two recombinant strands.

Evidence available today suggests that crossing-over occurs after the time of chromosome duplication. Probably the best genetic evidence for this comes from gene segregation data in Ascomycetes; a tetra-type ascus could not occur if crossing-over preceded chromosome duplication. The same conclusion was reached through nondisjunction and attached-X studies in *Drosophila*.

In terms of DNA, the evidence suggests that crossing-over is not occurring at the premeiotic S-phase. In *Neottiella* (Rossen and Westergaard 1966) and in *Chlamydomonas* (Chiang and Sueoka 1967), the nuclei that are about to fuse and undergo meiosis have already synthesized their DNA. Thus the homologous chromosomes are not in the same nucleus when the bulk of the DNA is synthesized. Hotta, Ito, and Stern (1966) showed that some DNA is synthesized at pachynema in *Lilium* and that its base composition is different from DNA synthesized at zygonema. Later studies by Smyth and Stern (1973) have demonstrated that the pachytene-DNA is more repetitive than zygotene-DNA.

Other evidence suggesting that pachynema is the stage in which crossing-over occurs includes that from various cytogenetic investigations of chromosome aberrations that show a close relationship between pairing as seen in pachynema and cross-over frequencies. Howell and Stern (1971) have demonstrated also the existence of an endonuclease enzyme that is specific to pachynema and that the pachytene-DNA synthesis is a repair synthesis (a type not inhibited by hydroxyurea).

A model that encompasses previous information suggests that the chromosome bivalent is stabilized by the synaptonemal complex (as described in Part III) and in the process traps certain chromatin fibers. Presumably this chromatin includes the pachytene-DNA and is trapped because of its repetitive nature. Crossing-over then occurs between this DNA from homologous chromosomes by a breakage-and-reunion process that involves repair DNA synthesis. The cross-over process may proceed as Whitehouse outlined (Paper 26) involving DNA breakage, molecular hybridization of the broken DNA strands, and the necessary DNA synthesis. Complications introduced by certain genes being present in multiple copies in consecutive linear order (such as ribosomal genes in the nucleolus organizer) have been considered by Whitehouse (1967) in his "cycloid model" for the chromosome. He suggests that a master copy of the redundant genes stays in the chromosome during crossing-over and the slave copies are detached and unavailable for crossing-over.

Editors' Comments
on Papers 20 Through 26

Paper 20 includes a few of the less complicated configurations observed at metaphase I of meiosis in tetrasomic hyacinths which demonstrate that the change of partners (termed *chiasma*) had been preceded by a cross-over between two chromatids at or near that point. This proof is based entirely on cytological observations. Careful study of the models in Figs. 11 and 13 together with the text is needed to appreciate fully the meaning of the observations. Since the conclusion that each chiasma represented a cross-over was considered to be a universal principle, the Darlington school

*Ed. Note: his modified cross-over model.

167

proceeded to study chromosome pairing and to determine the number, locations, and behavior of chiasmata at meiosis in a wide range of species. The information they gathered was interpreted with respect to evolution; and they identified different cytogenetic breeding systems that have evolved in nature (Darlington 1960). For most of those species, there still is little or no genetical information. Perhaps some of the newer methods of cytogenetic analysis could be applied profitably to those species.

Creighton and McClintock (Paper 21) utilized linked genetic markers in maize together with two cytological markers in the same chromosome to demonstrate that when a genetic cross-over occurs, a physical exchange between the two chromosomes has occurred. Stern in the same year reported the results of similar experiments in *Drosophila*.

Paper 22 presents evidence in maize indicating the occurrence of sister-strand crossing-over at meiosis on the order of at least one per bivalent. A similar experiment using a heterozygote for a large ring for chromosome 10 led to the same conclusion (Miles 1970). This evidence is in contrast to the finding in *Drosophila* (Anderson 1925). The excerpts from Stern's report (Paper 23) include a few of the numerous and well-planned experiments Stern used to obtain information about aspects of chromosome behavior that lead to somatic segregation, i.e., the appearance of mosaic spots expressing the phenotypes for which the individual is heterozygous. The experiments led to the discovery of somatic crossing-over which accounted for the observed somatic segregation.

The excerpts from Paper 24 (1928) present Belling's first working hypothesis to explain exchange of chromosome segments and crossing-over. After studying pachynema and earlier stages of meiosis in several *Liliaceae*, he proposed in 1931 (Paper 25) a modification of that mechanism. His hypothesis generated much speculation as to the mechanism of crossing-over and often was the point of departure for later models that were proposed, even after the discovery of the double-helical structure of DNA.

The theory of crossing-over by molecular hybridization involving base pairing between complementary nucleotide strands from two homologous DNA molecules and subsequent repair-type DNA synthesis was proposed by Whitehouse in his paper (Paper 26). Several unusual aspects of crossing-over such as the occurrence of gene conversion could be accommodated with this model. It was the first well-developed bridge between crossing-over as understood by eukaryotic geneticists and DNA as understood by molecular geneticists. The "copy-choice" hypothesis proposed earlier relating crossing-over to the replication process and failed to explain certain facts about crossing-over.

REFERENCES

Anderson, E. G. 1925. Crossing over in a case of attached X chromosomes in *Drosophila melanogaster. Genetics* **10**:403–417.

Bateson, W., E. R. Saunders, and R. C. Punnett. 1906. Experimental studies in the physiology of heredity. *Reports to the Evol. Comm. Royal Soc. London* **3**:8–11.

Bateson, W., and R. C. Punnett. 1911. On gametic series involving reduplication of certain terms. *J. Genet.* **1**:293–302.

Beadle, G. W., and S. Emerson. 1935. Further studies of crossing over in attached-X chromosomes of *Drosophila melanogaster. Genetics* **20**:192–206.

Belling, J. 1929. Nodes and internodes of trivalents of hyacinths. *Univ. Calif. Publ. Bot.* **14**:379–388.

Bridges, C. B. 1925. Elimination of chromosomes due to a mutant (Minute-n) in *Drosophila melanogaster. Proc. Nat. Acad. Sci. U.S.A.* **11**:701–706.

_____ and E. G. Anderson. 1925. Crossing over in the X chromosomes of triploid females of *Drosophila melanogaster. Genetics* **10**:418–441.

Bridges, C. B. and T. H. Morgan. 1923. The third-chromosome group of mutant characters of *Drosophila melanogaster. Carnegie Inst. Washington Publ.* **327**:1–251.

Brink, R. A., and D. C. Cooper. 1935. A proof that crossing over involves an exchange of segments between homologous chromosomes. *Genetics* **20**:22–35.

Brown, S. W., and D. Zohary. 1955. The relationship of chiasmata and crossing over in *Lilium formosanum. Genetics* **40**:850–873.

Chiang, K.-S., and N. Sueoka. 1967. Replication of chromosomal and cytoplasmic DNA during mitosis and meiosis in the eucaryote, *Chlamydomonas reinhardi. Symp.* on chromosome mechanics at the molecular level. *J. Cell. Physiol.* **70**, Sup. 1:89–112.

Darlington, C. D. 1929. Meiosis in polyploids II. Aneuploid hyacinths. *J. Genet.* **21**:17–56.

_____. 1960. Chromosomes and the theory of heredity. *Nature* **187**:892–895.

Dubinin, N. P., and E. N. Volotov. 1936. Mutations arising at the Bar locus in *Drosophila melanogaster. Nature* **137**:869.

Hotta, Y., M. Ito, and H. Stern. 1966. Synthesis of DNA during meiosis. *Proc. Nat. Acad. Sci. U.S.A.* **56**:1184–1191.

Howell, S. H., and H. Stern. 1971. The appearance of DNA breakage and repair activities in the synchronous meiotic cycle of *Lilium. J. Mol. Biol.* **55**:357–378.

Janssens, F. A. 1909. La théorie de la chiasmatypie Nouvelle interprétation des cinéses de maturation. *LaCellule* **25**:389–411.

_____. 1924. La chiasmatypie dans les insectes. Spermatogenése dans (1) *Stethophyma grossum* (L.) (2) *Chortippus parallelus* (Zetterstedt). *LaCellule* **34**:135–359.

Lewis, E. B. 1941. Another case of unequal crossing over in *Drosophila melanogaster*. *Proc. Nat. Acad. Sci. U.S.A.* **27**:31–35.

Lindegren, C. C. 1933. The genetics of *Neurospora*. III. Pure bred stocks and crossing over in *N. crassa*. *Bull. Torrey Bot. Club* **60**:133–154.

Miles, J. H. 1971. Influence of modified K 10 chromosomes on preferential segregation and crossing over in *Zea mays*. *Diss. Abstr. Internat.* **32-1,** Part B:100B.

Noda, S. 1968. Achiasmate bivalent formation by parallel pairing in PMCs of *Fritillaria amabilis*. *Bot. Mag., Tokyo* **81**:344–345.

Morgan, T. H. 1912. Complete linkage in the second chromosome of the male of *Drosophila*. *Science* **36**:719–720.

Muller, H. J., A. A. Prokofjeva-Belgovskaja, and K. V. Kossikov. 1936. Unequal crossing-over in the Bar mutant as a result of duplication of a minute chromosome section. *C. R. (Doklady) Acad. Sci. URSS.* **10**:87–88.

Pontecorvo, G. 1954. Mitotic recombination in the genetic systems of filamentous fungi. *Proc. 9th Int. Congr. Genet., Caryologia* **6**, Sup. 1:192–200.

_____ and E. Käfer. 1958. Genetic analysis based on mitotic recombination. *Adv. Genet.* **9**:71–104.

Rhoades, M. M. 1933. An experimental and theoretical study of chromatid crossing over. *Genetics* **18**:535–555.

Rossen, J. M., and M. Westergaard. 1966. Studies on the mechanism of crossing over. II. Meiosis and the time of meiotic chromosome replication in the ascomycete *Neottiella rutilans* (Fr.) Dennis. *C. R. Trav. Lab. Carlsberg* **35**:233–260 (References:279–286).

Smyth, D. R., and H. Stern. 1973. Repeated DNA synthesized during pachytene in *Lilium henryi*. *Nature New Biol.* **245**:94–96.

Stern, C. 1931. Zytologisch-genetische Untersuchungen als Beweise für die Morgansche Theorie des Factorenaustauschs. *Biol. Zentralbl.* **51**:547–587.

Sturtevant, A. H. 1913. The linear arrangement of six sex-linked factors in *Drosophila*, as shown by their mode of association. *J. Exp. Zool.* **14**:43–59.

_____. 1925. The effects of unequal crossing over at the Bar locus in *Drosophila*. *Genetics* **10**:117–147.

Wenrich, D. H. 1917. Synapsis and chromosome organization in *Chorthippus (Stenobothrus) curtipennis* and *Trimerotropis suffusa* (Orthoptera). *J. Morph.* **29**:471–518.

Whitehouse, H. L. K. 1967. A cycloid model for the chromosome. *J. Cell Sci.* **2**:9–22.

20

Reprinted from *R. Soc. (London) Proc.* **107B**:50–59 (1930)

A Cytological Demonstration of " Genetic " Crossing-Over.

By C. D. DARLINGTON, John Innes Horticultural Institution, Merton.

(Communicated by Sir Daniel Hall, F.R.S.—Received July 1, 1930.)

[PLATE 5.]

1. *Introduction.*

A stage in meiosis where the paired chromosomes form loops, coming together at points distributed at random along their length, was first described by Rückert (1892) in *Pristiurus*. These points we now recognise in suitable material as exchanges of partner amongst the four pairing half-chromosomes. But although Rückert did not appreciate this detail of structure he grasped the possible significance of their relationship : " Die bei der Reifung der Geschlechtzellen eintretende Verklebung vorhergetrennter Chromosomen bezweckt einer Substanzaustausch derselben." He considered that here was the possibility of a chromosome " amphimixis," carrying a step farther Weismann's idea of nuclear " amphimixis."

Similarly Correns (1902) and Boveri (1904, contrary to his earlier view) foresaw the possibility of different parts of a chromosome behaving independently at meiosis. The discovery of linkage by Bateson, Saunders and Punnett in 1905 was a verification of these predictions, although not at first interpreted in this sense.

Again in 1909, Janssens pointed out the possibility of exchanges at such points of contact as those seen by Rückert. The hypothesis that exchanges of segments actually take place at these points provided the interpretation which Morgan (1911) and his school have used to show the linear arrangement of the hereditary materials. And their success is the highest testimony to the validity of Janssens' assumptions.

While logical construction on the cytological side has anticipated genetical observations, cytological observations affecting this question have lagged far behind. Before considering the present observations it will be necessary to review the theories relating to the two and give an outline of the chromosome behaviour which is known to determine segregation and is therefore supposed to be concerned with crossing-over. This description will define the use of terms.

171

2. *Chromosome Behaviour at Meiosis.*

Homologous chromosomes (each a single thread) associate in pairs laterally, particle by particle, at the prophase of meiosis. The double (pachytene) thread formed in this way splits longitudinally, giving a quadruple thread consisting of four half-chromosomes or "chromatids." Splits appear at diplotene which separate pairs of chromatids, but the associations of pairs are not continuous throughout : the chromatids change partners. A point at which this change occurs I shall refer to as a " chiasma," because the name describes the characteristic crossing of two of the four chromatids. I will, therefore, define a chiasma as *The occurrence of a single exchange of partner in a system of four chromatids associated in pairs.*

There is no direct evidence as to what causes the chiasma, but there are two opposed theories to account for it (fig. 1). The first is that it is the result of two splits meeting, the one equational, *i.e.*, separating identical chromatids, the other reductional, *i.e.*, separating the original somatic chromosomes as they came together. This is the view to which the majority of cytologists incline, though merely for the reason that it involves the fewest obvious cytological assumptions. It assumes no change in the identity of the chromatids because no such change is observed (Robertson, 1916 ; Wenrich, 1916 ; Wilson, E. B., 1925 ; Seiler, 1926 ; McClung, 1927 ; Bĕlař, 1928).

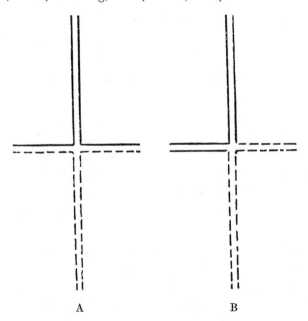

A   B

Fig. 1.—Genetical interpretations of chiasmata : A. Classical ; B. Chiasmatype.

The second view is that put forward by Janssens (1909, 1924). According to Janssens, chiasmata result from a change in the linear individuality of two of the four chromatids, so that if the two original chromosomes were of the structure ABCDEF and *abcdef*, and a chiasma is formed in the length between the paired points B*b* and C*c*, then the structure of the four continuous threads is to be taken as ABCDEF, AB*cdef*, *ab*CDEF and *abcdef*. In this case it is assumed that identical threads are always associated at diplotene. This change in the linear individuality of two of the four chromatids (partial chiasmatypy) is only one of the assumptions that Janssens made. He also assumed that all four threads might interchange segments at the same point (total chiasmatypy). This would result, not in a chiasma (in my sense) but in an interlacing. Janssens also believed that other types of less definite association occurred amongst paired chromosomes at diplotene and later* ("soudures dont le caractère semble plus transitoire"). Janssens attempted to show the validity of his assumptions directly and indirectly. His studies went far towards showing the physical structure of the chiasma after it was formed and they have been confirmed by the recent work of Belling (1925, 1926), Newton and Darlington (1927, 1929 and 1930) and Maeda (1930) in plants. His observations of discontinuity in the separating anaphase chromatids (confirmed by Bĕlař, 1928) and of their interlocking (*cf.* also Maeda, 1930) are indirect evidence of the origin of chiasmata that cannot be neglected.

The position now is that Janssens' assumptions are unproved. And since Janssens himself seemed to regard them as directly provable this lack of direct evidence has weighed against them. For example Robertson (not unreasonably) remarks (1915) : "The formation of cross-overs as a result of the opening out process of the four strands does away with the possibility of a 'compénétration graduelle de deux chromosomes au niveau d'un chiasma avec la soudure des filaments qui se touchent les premiers.'" And Seiler (1926) concludes that "Auch bei den bescheidensten Ansprüchen an den erklärenden Wert einer Hypothese diejenige von Janssens den crossing-over Befunden niemals gerecht werden kann, ganz abgesehen davon dass sie keine Tatsachenbasis hat."†

* Amongst these are what I should call "terminal chiasmata" (1929*b*).

† Other criticisms are less reasonable (*cf.* Darlington, 1930). For example, McClung (1927*a*) remarks that "Janssens quite overlooks the most important circumstances in the history of the meiotic chromosome, *i.e.*, the simultaneous existence of two planes of division at right angles to each other *proceeding from opposite ends of the chromosome.*" This begs the whole question, for it accepts Wenrich's *hypothesis* (1916) of what happens in *Phrynotettix* as applicable to *Mecostethus*, which Janssens (1924) and, less definitely, McClung (1927*b*) have shown not to be the case, for in *Mecostethus* which they have studied pachytene pairing is said to be incomplete.

But the agreement of Janssens' hypothesis with the genetical results of which it was to a great extent a prediction (*cf.* Belling, 1928 ; Darlington, 1930*b*) is at least as important an objection to the ordinary cytological hypothesis as the lack of particular observations is to Janssens' hypothesis.

The question therefore arises as to whether they could ever be demonstrated directly, and in my opinion this is in the last degree improbable. It is to be expected that an instantaneous change in relationship occurring between four indistinguishable and closely paired threads (each less than 100 $m\mu$ in diameter) is responsible for the changes that Janssens and the geneticists have imagined. Direct observation of such an occurrence is inconceivable. Surely then we ought to consider that Janssens' " chiasmatypy " and its alternative, the classical diplotene hypothesis are equally assumptions, equally incapable of direct demonstration and equally requiring demonstration by inference from material in which we can distinguish the results of the two hypotheses.*

I conclude that the assumption of " partial chiasmatypy," which demands crossing-over between two of the four chromatids at every chiasma, is the only possible working hypothesis for the correlation of the cytological and genetical observations. I have lately attempted to show (1930*b*) that there is no insuperable obstacle to this hypothesis in the genetical and cytological observations on *Drosophila* or *Primula*.

In considering the new cytological observations it must be remembered that the confusion in Janssens' thesis between the observations of chiasmata and the supposed evidence of their cause is unnecessary. My earlier hypothesis (1929*a* and *b*), that *chiasmata are the condition of chromosome pairing and of meiosis,* is distinct from the revised version of Janssens' hypothesis that I am now adopting (1930*b*) to the effect that *chiasmata are conditioned by crossing-over between two of the conjugant chromatids at pachytene.*† I therefore use the term chiasma, as above, in a morphological sense and without the genetic implications of Janssens (with which it is usually connected).

3. *The Crucial Test.*

Belling (1929) and Newton and Darlington (1929) have independently examined the evidence of configurations in triploids, both knowing that association occurs between pairs of chromosomes at pachytene. Belling has considered that the occurrence of certain configurations (similar to those

---

* See Appendix and Darlington (1930*b*).

† A third hypothesis suggesting *how* crossing-over might take place (Belling, 1928) is likewise independent of the other two. They stand or fall separately.

described later in tetrasomic forms) was conclusive evidence of crossing-over. We considered that the presence of the third thread at pachytene prevented a final conclusion that associations between all three threads could not take place at one point (perhaps successively). In the tetraploid no such doubt can exist. And since in tetraploids I found no configurations of these types (1929a) I reached a negative conclusion. The present observations are the result of a special search for the missing configuration in this critical material.

At pachytene in tetrasomic hyacinths the chromosomes come together in pairs as in disomic forms. But coming together at random they associate differently at different points; they therefore exchange partners (diagram model, Plate 5, fig. 12, and Darlington, 1929a, Plate 6, figs. 13–17). Although the intervening stages between pachytene and metaphase have not been studied (not being suitable for critical observation), the metaphase configurations (in disomic and tetrasomic forms) show the characteristic forms of chromosomes which have not undergone change since diplotene. Moreover the same types of configuration are found in tetraploid *Primula sinensis* at diplotene (Darlington, 1930b). That is, the chiasmata are distributed at random along the associated chromosomes. Since the chromosomes were assorted in pairs at random at pachytene the metaphase chiasmata are distributed at random amongst the different possible pairs of chromosomes.

Before studying the relations of these chiasmata to one another let us examine the possibilities under the two hypotheses, the classical and the chiasmatype. It is clear that configurations with simple exchanges are in accordance with both hypotheses. Similarly configurations where four chromatids engage in two chiasmata between chiasmata in which two of them engage with a third pair may be interpreted in both ways and are therefore inconclusive (Darlington, 1929a, p. 43). Such configurations have been illustrated. But there is a type of triangular configuration (such as has been found freely in triploid Tulipa and Hyacinthus, where the evidence of pachytene association is not unassailable, for the third free thread might be supposed to interfere with the paired threads) the occurrence of which would be a crucial test. The essential part of such a configuration is illustrated in the following diagram (fig. 2).

Analysis of Chiasmata.—The four chromatids of chiasma A have three in common with those of chiasma B (on either hypothesis). But the chromatids of one chiasma can only be derived from the four chromatids of two chromosomes. The fourth chromatid in A or B cannot be derived from a third chromosome, therefore the chiasmata A and B have all four chromatids in common.

If identical threads are associated at A and B, *i.e.*, on both sides of the chiasma C, the change of pairing at C must be accompanied by an exchange of identity between two of the chromatids at C.

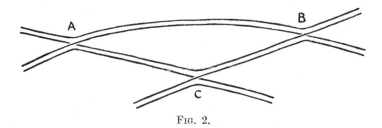

Fig. 2.

4. *The Demonstration.*

Such configurations occur in tetrasomic *Hyacinthus* and *Primula sinensis*. They are usually too complex to illustrate without models and the types taken (figs. 3–6) are the simplest (and probably rarest) found. The interpretation (figs. 7–10) is in accordance with the conclusions of Belling (1927) and myself (1929) that " nodes " where four chromatids come together are chiasmata. This conclusion cannot always be directly verified, but in the case chosen for illustration with models (fig. 6, diagram fig. 10, models figs. 11 and 13) the interpretation of critical chiasmata is not disputable. Fig. 3 provides the simplest example of the critical triangular configuration ; figs. 4, 5 and 6 provide compound examples.

The models (Plate 5, figs. 11–13) illustrate the interpretation of fig. 6 and show its incompatibility with the classical hypothesis. On this hypothesis chiasmata have to be represented as derived from chromatids of three or four original chromosomes. This assumption would be contrary to the observations at pachytene. In view of these considerations, the observations are only compatible with the assumption that *where a chiasma is formed amongst four chromatids an interchange of segments (genetic crossing-over) has occurred between two of them.*

Summary and Conclusion.

Chromosomes associate in pairs at the prophase of meiosis in polyploids. A chiasma is the occurrence of a visible change of partners amongst four half-chromosomes. It follows therefore that, where two chromatids, which form a chiasma with two other chromatids, also form chiasmata with a third pair of chromatids on either side of the first chiasma, then the chromatids must be assumed to be in the same combinations on the two sides of this chiasma.

In such cases genetic crossing-over (*i.e.*, an exchange of linear identity) must have taken place at the first chiasma.

Configurations of this kind occur in polyploid hyacinths and *Primula* and make it extremely probable that wherever an exchange of partners occurs

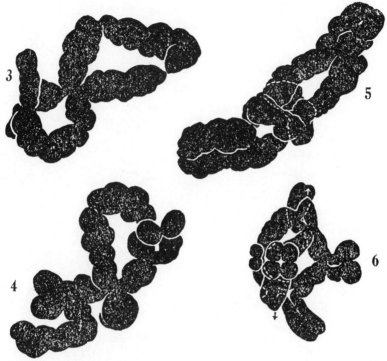

Figs. 3–6.—Quadrivalents from first division metaphases in the pollen mother-cells of the tetrasomic hyacinth "La Grandesse." ($2n = 28$.) × 5300 (drawn at 6400).

amongst four paired chromatids, crossing-over has occurred between two of them (Janssens' partial chiasmatypy). This conclusion is distinct from the hypothesis that the pairing of chromosomes depends on the formation of chiasmata between them.

Taking this as a working hypothesis it will be possible to study crossing-over from a cytological point of view. This is particularly important in considering its relation to structural change in the chromosomes. For example, it is possible to regard segmental interchange between non-homologous chromosomes as the result of crossing-over between small, relatively translocated, segments. This crossing-over would be between definite loci and with a definite frequency determined by the length of the segments in question. Such is the requirements on the basis of my hypothesis for explaining the origin of half-mutants (1929*a*). In a similar way all mutation in *Oenothera* not involving changes in chromosome-number can probably be explained in terms of crossing-over.

In these terms segmental interchange becomes a *secondary* structural change,

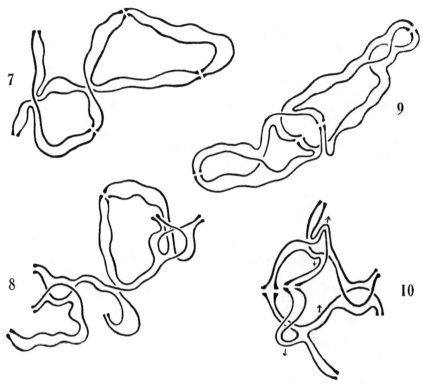

FIGS. 7–10.—Line diagrams to show chromatid structure of the quadrivalents. Arrows indicate points of attachment to the spindle in a quadrivalent seen in side view (figs. 6 and 10).

distinguished from sporadic *primary* structural changes by the regularity of its occurrence both in frequency and result. This question will be considered more fully at a later date.

APPENDIX.

There are other theories of crossing-over which need not be considered in detail. For example, Seiler's hypothesis of re-arrangement within multiple chromosomes does not seem to be generally applicable on either cytological or genetical grounds. Chromosome linkage in *Solenobia* like that in *Oenothera* seems to depend on special conditions of chromosome organisation. Nor need

Chodat's assumption (1925) of a new line of cleavage at anaphase be considered. It is contradicted by all the recent observations on material suitable for observing the relations of the chromatids at this and the earlier states (*cf.* Newton on *Tulipa*, Belling and Darlington on *Hyacinthus*, Maeda on *Vicia* and *Lathyrus*, Newton and Darlington on *Fritillaria*).

Similarly Janssens' total chiasmatypy (with which Belling's hypotheses, 1928, is to some extent involved) is no longer valid for the following reasons : (i) It is contrary to the genetical observations of crossing-over in the four-strand stage by Bridges and Anderson (1925) ; (ii) experiments in *Drosophila* have frequently shown over 50 per cent. of offspring without any crossing-over over considerable lengths of chromosome (*cf.* Jennings, 1923). In this 50 per cent. no total chiasmatypy can have occurred, and whether partial chiasmatypy occurs as well, or not, there must be chromosomes pairing without either type of exchange. This is a difficult assumption *a priori*, but I am inclined to reject it particularly because it is incompatible with the view that chiasmata are the condition of pairing ; (iii) I have never found any evidence of the type of interlocking the appearance of which suggested total chiasmatypy to Janssens ; (iv) on the contrary, evidence of various kinds is accumulating to indicate that the only association of chromosomes is that conditioned by chiasmata, although I appear to be alone in stating this categorically (Belling, 1926, 1927, 1928 ; Newton, 1927 ; Darlington, 1929*a*, 1929*b*, 1930 ; Maeda, 1930*a*, 1930*b*).

Note.—Certain genetical work on mosses appear to disagree with the observation of crossing-over in the four-strand stage in *Drosophila*. Wettstein (1924) and Allen (1926, 1930) have found " two-type " tetrads from sporophytes of *Funaria* and *Sphaerocarpos*, hydrid for supposedly linked factors. Even with Allen's suggestion that the linkage might be of the *Oenothera* type, a two-type tetrad would ordinarily mean crossing-over in the two-strand stage. The two-type tetrad could, however, be explained by the assumption that the linked factors are in different chromosomes which usually associate and disjoin to opposite poles but occasionally fail to associate and then segregate at random to give viable gametes. The cytological condition for this is found in *Rhoeo* (Darlington, 1929), *Campanula* (Gairdner and Darlington, 1930) and to a less extent in *Oenothera* (Kihara, 1926). In this case the observations have no bearing on crossing-over within the chromosome. This conclusion is justified by the genetical observations, which leave room for doubt as to the linkage being of the simple *Drosophila* type. It is not excluded by the cytological observations of Lorbeer (1927).

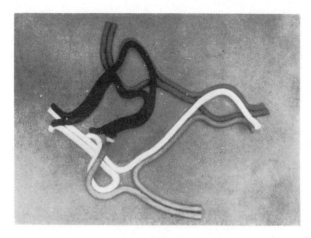

Fɪɢ. 11.—According to the assumption that no crossing over has taken place and that chiasmata are due to the opening out of successive equational and reductional splits. Certain chiasmata necessarily involve chromatids derived from more than two chromosomes.

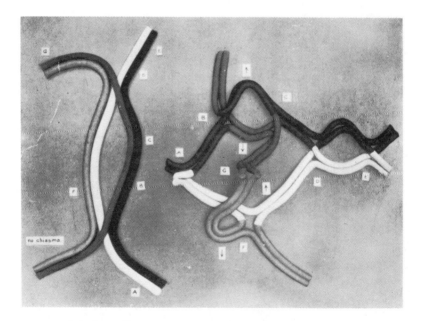

Fɪɢ. 12. — The pairing of chromosomes observed at pachytene in the variety " La Grandesse." Three chromosomes never associate at any one point.

Fɪɢ. 13.—The resultant configuration on the assumption that only identical threads are paired and that therefore every chiasma represents, and is the result of, a crossing-over between two chromatids. The chiasmata are lettered to show how they correspond with positions on the pachytene chromosomes (fig. 10).

Fɪɢs. 11–13.—Photographs of models to show possible genetical interpretations of chromatid relationships at figs. 6 and 10.

(Facing p. 58.)

REFERENCES.

Allen, C. E. (1926*a*). ' Proc. Nat. Acad. Sci.,' vol. **12**, p. 2.

Allen, C. E. (1930). ' Genetics,' vol. **15**, p. 150.

Bateson, W., Saunders, E. R., and Punnett, R. C. (1905). ' Rep. Evol. Comm. (Roy. Soc.) ' vol. **11**, p. 80.

Bělař, K. (1928). " Die cytoligischen Grundlagen der Vererbung." (Berlin.)

Belling, J. (1926). ' Biol. Bull.,' vol. **50**, p. 355.

Belling, J. (1927). ' Biol. Bull.,' vol. **52**, p. 480.

Belling, J. (1928). ' Univ. Calif. Pub. Bot.,' vol. **14**, p. 283.

Belling, J. (1928). ' Biol. Bull.,' vol. **54**, p. 465.

Belling, J. (1929). ' Univ. Calif. Pub. Bot.,' vol. **14**, p. 379.

Boveri, Th. (1904). " Ergebnisse über die Konstitution der chromatischer Substanz des Zellkerns." (Jena.)

Chodat, R. (1925). ' Bull. Soc. Bot. Genève,' 1925.

Correns, C. (1902). ' Bot. Z.,' vol. **60**.

Darlington, C. D. (1929*a*). ' J. Genet.,' vol. **21**, p. 17.

Darlington, C. D. (1929*b*). ' J. Genet.,' vol. **21**, p. 207.

Darlington, C. D. (1930*a*). ' J. Genet.,' vol. **22**, p. 65.

Darlington, C. D. (1930*b*). ' J. Genet.,' vol. **22** (*in press*).

Gairdner, A. E., and Darlington, C. D. ' Nature,' vol. **125**, p. 87.

Janssens, F. A. (1909). ' Cellule,' vol. **25**, p. 387.

Janssens, F. A. (1924). ' Cellule,' vol. **34**, p. 135.

Jennings, H. S. (1923). ' Genetics,' vol. **8**, p. 393.

Kihara, H. (1927). ' Jahrb. Wiss. Bot.,' vol. **66**, p. 429.

Lorbeer, G. (1927). ' Z. Indukt. Abstamm. Vererb. Lehre,' vol. **44**, p. 1.

McClung, C. (1914). ' J. Morphol.,' vol. **25**, p. 651.

Maeda, T. (1930*a*). ' Mem. Coll. Sci., Kyoto,' B, vol. **5**, p. 89.

Maeda, T. (1930*b*). ' Mem. Coll. Sci., Kyoto,' B, vol. **5**, p. 125.

Morgan, T. H. (1911). ' Science,' vol. **34**, p. 384.

Newton, W. C. F. (1927). ' J. Linn. Soc.,' vol. **47**, p. 339.

Newton, W. C. F., and Darlington, C. D. (1929). ' J. Genetic,' vol. **21**, p. 1.

Newton, W. C. F., and Darlington, C. D. (1930). ' J. Genetic,' vol. **22**, p.

Robertson, W. R. B. (1916). ' J. Morphol.,' vol. **27**, p. 179.

Renner, O. (1929). " Handb. Vererbungswissenschaft."

Rückert, J. (1892). ' Anat. Anz.,' vol. **7**, p. 107.

Seiler, J. (1926). ' Arch. Zellf.,' vol. **16**, p. 171.

Wenrich, D. H. (1916). ' Bull. Mus. Comp. Zool. Harvard,' vol. **60**, p. 57.

Wettstein, F. v. ' Z. Indukt. Amstamm. Vererb. Lehre,' vol. **33**, p. 1.

Wilson, E. B. (1925). " The Cell, New York."

21

Reprinted from *Natl. Acad. Sci. (U.S.A.) Proc.* **17**(8):492–497 (1931)

A CORRELATION OF CYTOLOGICAL AND GENETICAL CROSS-ING-OVER IN ZEA MAYS

By Harriet B. Creighton and Barbara McClintock

Botany Department, Cornell University

Communicated July 7, 1931

A requirement for the genetical study of crossing-over is the heterozygous condition of two allelomorphic factors in the same linkage group. The analysis of the behavior of homologous or partially homologous chromosomes, which are morphologically distinguishable at two points, should show evidence of cytological crossing-over. It is the aim of the present paper to show that cytological crossing-over occurs and that it is accompanied by genetical crossing-over.

In a certain strain of maize the second-smallest chromosome (chromosome 9) possesses a conspicuous knob at the end of the short arm. Its distribution through successive generations is similar to that of a gene. If a plant possessing knobs at the ends of both of its 2nd-smallest chromosomes is crossed to a plant with no knobs, cytological observations show that in the resulting F_1 individuals only one member of the homologous pair possesses a knob. When such an individual is back-crossed to one having no knob on either chromosome, half of the offspring are heterozygous for the knob and half possess no knob at all. The knob, therefore, is a constant feature of the chromosome possessing it. When present on one chromosome and not on its homologue, the knob renders the chromosome pair visibly heteromorphic.

In a previous report[1] it was shown that in a certain strain of maize an interchange had taken place between chromosome 8 and 9. The interchanged pieces were unequal in size; the long arm of chromosome 9 was increased in relative length, whereas the long arm of chromosome 8 was correspondingly shortened. When a gamete possessing these two interchanged chromosomes meets a gamete containing a normal chromosome set, meiosis in the resulting individual is characterized by a side-by-side synapsis of homologous parts (see diagram, figure 1 of preceding paper). Therefore, it should be possible to have crossing-over between the knob and the interchange point.

In the previous report it was also shown that in such an individual the only functioning gametes are those which possess either the two normal chromosomes (N, n) or the two interchanged chromosome (I, i), i.e., the full genom in one or the other arrangement. The functional gametes therefore possess either the shorter, normal, knobbed chromosome (n) or the longer, interchanged, knobbed chromosome (I). Hence, when such a plant is crossed to a plant possessing the normal chromosome complement, the presence of the normal chromosome in functioning gametes of the former will be indicated by the appearance of ten bivalents in the prophase of meiosis of the resulting individuals.   The presence of the interchanged

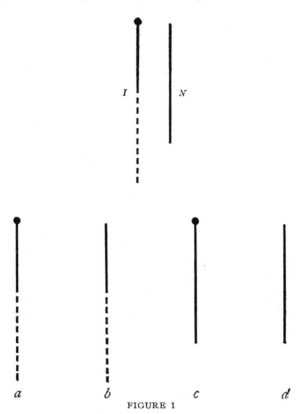

FIGURE 1

Above—Diagram of the chromosomes in which crossing-over was studied.

Below—Diagram of chromosome types found in gametes of a plant with the constitution shown above.

        a—Knobbed, interchanged chromosome.
        b—Knobless, interchanged chromosome.
        c—Knobbed, normal chromosome.
        d—Knobless, normal chromosome.
          a and *d* are non-crossover types.
          b and *c* are crossover types.

chromosome in other gametes will be indicated in other F_1 individuals by the appearance of eight bivalents plus a ring of four chromosomes in the late prophase of meiosis.

If a gamete possessing a normal chromosome number 9 with no knob, meets a gamete possessing an interchanged chromosome with a knob, it is clear that these two chromosomes which synapse along their homologous parts during prophase of meiosis in the resulting individual are visibly different at each of their two ends.   If no crossing-over occurs, the gametes formed by such an individual will contain either the knobbed, interchanged chromosome (*a*, Fig. 1) or the normal chromosome without a knob (*d*, Fig. 1).   Gametes containing either a knobbed, normal chromosome (*c*,

TABLE 1

$\dfrac{\text{KNOB-INTERCHANGED}}{\text{KNOBLESS-NORMAL}}$ ×		KNOBLESS-NORMAL, CULTURE 337 AND KNOBBED-NORMAL CULTURES A125 AND 340		
	PLANTS POSSESSING 2 NORMAL CHROMOSOMES		PLANTS POSSESSING AN INTERCHANGED CHROMOSOMES	
CULTURE	NON-CROSSOVERS	CROSSOVERS	NON-CROSSOVERS	CROSSOVERS
337	8	3	6	2
A125	39	31	36	23
340	5	3	5	3
Totals	52	37	47	28

Fig. 1) or a knobless, interchanged chromosome (*b*, Fig. 1) will be formed as a result of crossing-over.   If such an individual is crossed to a plant possessing two normal knobless chromosomes, the resulting individuals will be of four kinds.   The non-crossover gametes would give rise to individuals which show either (1) ten bivalents at prophase of meiosis and no knob on chromosome 9, indicating that a gamete with a chromosome of type *d* has functioned or (2) a ring of four chromosomes with a single conspicuous knob, indicating that a gamete of type *a* has functioned.   The crossover types will be recognizable as individuals which possess either (1) ten bivalents and a single knob associated with bivalent chromosome 9 or

TABLE 2

$\dfrac{\text{KNOB-}C\text{-}wx}{\text{KNOBLESS-}c\text{-}Wx}$ ×		KNOBLESS-c-wx					
C-wx		*c-Wx*		*C-Wx*		*c-wx*	
Knob	Knobless	Knob	Knobless	Knob	Knobless	Knob	Knobless
12	5	5	34	4	0	0	3

(2) a ring of four chromosomes with no knob, indicating that crossover gametes of types *c* and *b*, respectively, have functioned.   The results of such a cross are given in culture 337, table 1.   Similarly, if such a plant is crossed to a normal plant possessing knobs at the ends of both number 9 chromosomes and if crossing-over occurs, the resulting individuals should be of four kinds.   The non-crossover types would be represented by (1) plants homozygous for the knob and possessing the interchanged chromosome and (2) plants heterozygous for the knob and possessing two normal chromosomes.   The functioning of gametes which had been produced as the result of crossing-over between the knob and the interchange would give rise to (1) individuals heterozygous for the knob and possessing the

interchanged chromosome and (2) those homozygous for the knob and possessing two normal chromosomes. The results of such crosses are given in cultures A125 and 340, table 1. Although the data are few, they are consistent. The amount of crossing-over between the knob and the interchange, as measured from these data, is approximately 39%.

In the preceding paper it was shown that the knobbed chromosome carries the genes for colored aleurone (*C*), shrunken endosperm (*sh*) and waxy endosperm (*wx*). Furthermore, it was shown that the order of these genes, beginning at the interchange point is *wx-sh-c*. It is possible, also, that these genes all lie in the short arm of the knobbed chromosome. Therefore, a linkage between the knob and these genes is to be expected.

One chromosome number 9 in a plant possessing the normal complement had a knob and carried the genes *C* and *wx*. Its homologue was knobless and carried the genes *c* and *Wx*. The non-crossover gametes should contain a knobbed-*C-wx* or a knobless-*c-Wx* chromosome. Crossing-over in region 1 (between the knob and *C*) would give rise to knobless *C-wx* and knobbed-*c-Wx* chromosomes. Crossing-over in region 2 (between *C* and *wx*) would give rise to knobbed-*C-Wx* and knobless-*c-wx* chromosomes. The results of crossing such a plant to a knobless-*c-wx* type are given in table 2. It would be expected on the basis of interference that the knob and *C* would remain together when a crossover occurred between *C* and *wx;* hence, the individuals arising from colored starchy (*C-Wx*) kernels should possess a knob, whereas those coming from colorless, waxy (*c-wx*) kernels should be knobless. Although the data are few they are convincing. It is obvious that there is a fairly close association between the knob and *C*.

To obtain a correlation between cytological and genetic crossing-over it is necessary to have a plant heteromorphic for the knob, the genes *c* and *wx* and the interchange. Plant 338 (17) possessed in one chromosome the knob, the genes *C* and *wx* and the interchanged piece of chromosome 8. The other chromosome was normal, knobless and contained the genes *c* and *Wx*. This plant was crossed to an individual possessing two normal, knobless chromosomes with the genes *c-Wx* and *c-wx*, respectively. This cross is diagrammed as follows:

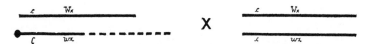

The results of the cross are given in table 3. In this case all the colored kernels gave rise to individuals possessing a knob, whereas all the colorless kernels gave rise to individuals showing no knob.

The amount of crossing-over between the knob and the interchange point is approximately 39% (Table 1), between *c* and the interchange approximately 33%, between *wx* and the interchange, 13% (preceding paper). With this information in mind it is possible to analyze the data given in table 3. The data are necessarily few since the ear contained but few kernels. The three individuals in class I are clearly non-crossover types. In class II the individuals have resulted from a crossover in region 2,

TABLE 3

KNOB-*C*-*wx*-INTERCHANGED			
KNOBLESS-*c*-*Wx*-NORMAL × KNOBLESS-*c*-*wx*-NORMAL / KNOBLESS-*c*-*wx*-NORMAL			
PLANT NUMBER	KNOBBED OR KNOBLESS	INTERCHANGED OR NORMAL	
Class I, *C*-*wx* kernels			
1	Knob	Interchanged	
2	Knob	Interchanged	
3	Knob	Interchanged	
Class II, *c*-*wx* kernels			
1	Knobless	Interchanged	
2	Knobless	Interchanged	
Class III, *C*-*Wx* kernels			*Pollen*
1	Knob	Normal	*WxWx*
2	Knob	Normal	. . . .
3	. . . .	Normal	*WxWx*
5	Knob	Normal	. . . .
6	Knob	. . . . . .	. . . .
7	Knob	Normal	. . . .
8	Knob	Normal	. . . .
Class IV, *c*-*Wx* kernels			
1	Knobless	Normal	*Wxwx*
2	Knobless	Normal	*Wxwx*
3	Knobless	Interchanged	*Wxwx*
4	Knobless	Normal	*Wxwx*
5	Knobless	Interchanged	*WxWx*
6	Knobless	Normal	*WxWx*
7	Knobless	Interchanged	*Wxwx*
8	Knobless	Interchanged	*WxWx*
9	Knobless	Normal	*WxWx*
10	Knobless	Normal	*WxWx*
11	Knobless	Normal	*Wxwx*
12	Knobless	Normal	*Wxwx*
13	Knobless	Normal	*WxWx*
14	Knobless	Normal	*WxWx*
15	Knobless	Normal	*Wx*—

i.e., between *c* and *wx*. In this case a crossover in region 2 has not been accompanied by a crossover in region 1 (between the knob and *C*) or region 3 (between *wx* and the interchange). All the individuals in class III had normal chromosomes. Unfortunately, pollen was obtained from only 1 of the 6 individuals examined for the presence of the knob. This one individual was clearly of the type expected to come from a gamete produced through crossing-over in region 2. Class IV is more difficult to analyze.

Plants 6, 9, 10, 13, and 14 are normal and $WxWx$; they therefore represent non-crossover types. An equal number of non-crossover types are expected among the normal $Wxwx$ class. Plants 1, 2, 4, 11 and 12 may be of this type. It is possible but improbable that they have arisen through the union of a c-Wx gamete with a gamete resulting from a double crossover in region 2 and 3. Plants 5 and 8 are single crossovers in region 3, whereas plants 3 and 7 probably represent single crossovers in region 2 or 3.

The foregoing evidence points to the fact that cytological crossing-over occurs and is accompanied by the expected types of genetic crossing-over.

Conclusions.—Pairing chromosomes, heteromorphic in two regions, have been shown to exchange parts at the same time they exchange genes assigned to these regions.

The authors wish to express appreciation to Dr. L. W. Sharp for aid in the revision of the manuscripts of this and the preceding paper. They are indebted to Dr. C. R. Burnham for furnishing unpublished data and for some of the material studied.

[1] McClintock, B., *Proc. Nat. Acad. Sci.*, **16**, 791–796 (1930).

ERRATA

Page 492, line 2 should read: "condition of two allelomorphic factor pairs ..."

Page 493, in figure 1, **I** is the larger interchange chromosome and **N** is the larger normal chromosome.

22

Reprinted from *Genetics* **38**(3):251–260 (1953)

EVIDENCE FOR SISTER-STRAND CROSSING OVER IN MAIZE[1]

DREW SCHWARTZ

Biology Division, Oak Ridge National Laboratory, Oak Ridge, Tennessee

Received November 9, 1952

CROSSING OVER, the exchange of chromatin between homologous chromosomes, is of utmost importance in genetic studies. Although little is known about the mechanism which is responsible for this exchange, a number of facts on this subject have been fairly well established: (1) Crossing over is produced by breakage and reunion of broken ends (even on BELLING's hypothesis, breakage and reunion have to be postulated to explain 3-strand and 4-strand doubles). (2) The position of the exchange is exactly between homologous regions of the chromosomes. (3) Crossing over occurs at prophase of the first meiotic division. (4) Only 2 of the 4 chromatids are involved in any one crossover. (5) The occurrence of one crossover decreases the probability of another occurring in its vicinity—the phenomenon of interference. It has generally been assumed that there is little or no sister-strand crossing over. This

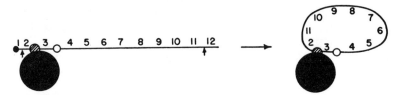

FIGURE 1.—Diagrammatic representation of the formation of the ring chromosome following breakage in the distal regions of chromosome 6. The nucleolus is represented by the large black circle and the centromere by the clear oval. The shaded area represents the nucleolar organizer. The arrows indicate the probable sites of X-ray-induced breakage. (Reprinted through courtesy of the American Naturalist.)

conclusion has been drawn from studies in Drosophila with attached-X chromosomes (BEADLE and EMERSON 1935) and with the Bar locus (STURTEVANT 1925, 1928; MULLER and WEINSTEIN 1933). However this is still an open question. The pioneering work of McCLINTOCK (1938, 1941b) on dicentric double-sized ring formation and the evidence to be presented in this paper suggest that crossing over does occur between sister chromatids. Such an event can be detected only with ring chromosomes with the possible exception of " unequal " crossing over in sister rod chromatids giving rise to duplicated segments.

In a previous paper (SCHWARTZ 1953) the behavior of a large ring in maize involving almost the whole of chromosome 6 was discussed. Gametophytes possessing nine chromosomes plus the ring are viable even though deficient for the terminal regions of chromosome 6 (fig. 1). Crossing over in the hetero-

[1] Work performed under USAEC Contract No. W-7405-eng-26.

zygote between the ring and its homologous rod may result in bridges in both anaphase I and anaphase II. Only crossovers in the long arm of the chromosome are being considered since those in the short arm between the centromere and the nucleolar organizer are very infrequent (McClintock 1941b). A

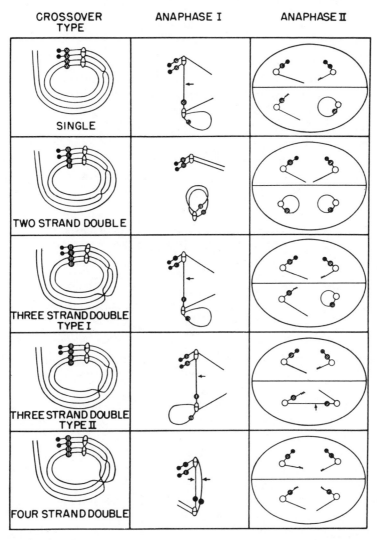

CROSSOVER TYPE	ANAPHASE I	ANAPHASE II
SINGLE		
TWO STRAND DOUBLE		
THREE STRAND DOUBLE TYPE I		
THREE STRAND DOUBLE TYPE II		
FOUR STRAND DOUBLE		

Figure 2.—Anaphase configurations resulting from crossing over between the ring and its homologous rod chromosome. (Reprinted through the courtesy of the American Naturalist.)

——— ⌇⌇⌇ = broken chromosome end   ———→ = position of breakage

single or a 3-strand double crossover (type I), in which the same ring chromatid is involved in both exchanges, gives a single bridge in AI only (fig. 2). Both 2-strand double crossovers and non-crossovers result in normal disjunction without bridge formation. A 3-strand double of the second type (II), in which the same rod chromatid is involved in both exchanges, gives a single bridge in both AI and AII. A 4-strand double gives a double bridge in AI. These bridges are not associated with fragments.

Two of the double crossover classes can easily be distinguished—type II 3-strand doubles and the 4-strand doubles. Equal frequencies of double bridges in AI and single bridges in AII would therefore indicate the absence of chromatid interference.

RESULTS AND INTERPRETATION

Four plants heterozygous for the ring and a rod chromosome were used in this study. The frequencies of the various anaphase configurations observed are listed in table 1. The anaphase II data are given in terms of daughter cell pairs which show a single or double bridge in one of the two cells. This frequency was calculated in the following manner. All cells in AII were counted regardless of whether they were found singly or in pairs. Of these cells 166 showed single bridges, 47 double bridges (from dicentric rings), and 737 no bridges. Type II-3-strand doubles and sister-strand crossovers in the ring give rise to bridges in only one of the two daughter cells, the other being normal. There-

TABLE 1

Meiotic anaphase configurations observed in plants heterozygous for a ring and a rod.

	Anaphase I				Anaphase II (daughter cell pairs)			
	Single bridge	Double bridge	No bridge	Total	Single bridge	Double bridge	No bridge	Total
Number	368	81	171	620	166	47	262	475
Percent	59	13	28	100	35	10	55	100

fore, 213 cells (166 + 47) must be subtracted from the 737 and the remaining 524 divided by 2 (i.e., 262 as given in table 1) to determine the number of daughter cell pairs which lacked bridges.

There are two important facts which should be pointed out in the data. First, the frequent occurrence of double bridges (dicentric rings) in AII. These would not be expected from single and double crossovers and only from 1/16th of the triple crossovers. Second, the frequency of single AII bridges.

It is apparent from table 1 that single bridges in AII are much more frequent than double bridges in AI, approximately a threefold difference. Can this difference be interpreted as being due to negative chromatid interference; i.e., the participation of one strand in a crossover enhances the probability that it will be involved in the second? This might appear to be the case, since anaphase configurations resulting from 3-strand doubles where one of the strands is involved in both crossovers, are more frequent than those arising from 4-strand doubles where each strand is involved in only one crossover. However, on this basis the frequency of 2-strand doubles should be even greater than 3-strand doubles. As the data indicate, this is not true. Only 28 percent of the AI cells showed no bridges and some of these are presumably due to non-crossovers.

If the 35 percent single bridges in AII are due entirely to type II 3-strand double exchanges, an equal frequency of type I might be expected. Thus, 70 percent of the AI cells would show 3-strand double crossing over and if one then added the frequency of double bridges, no bridges, and single bridges due to single crossovers, the total would come to much more than 100 percent. It seems clear therefore that negative interference cannot explain the data and that the frequency of single bridges in AII is not a true measure of the frequency of type II 3-strand doubles. The frequent occurrence of triple crossovers between the ring and the rod is unlikely and it would not account for the ratio of observed bridge configurations.

The high frequency of single bridges in AII can be accounted for by sister-strand crossing over (fig. 3). The term " sister-strand crossing over " as used

ANAPHASE I   **ANAPHASE II**

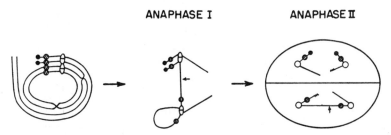

FIGURE 3.—Anaphase configurations resulting from a single non-sister crossover associated with a sister-strand crossover between the ring chromatids.

here refers simply to an exchange of chromatin between sister chromatids and does not imply any relationship to non-sister-strand crossing over in the time and manner of its occurrence, interference, etc.

There are four classes of anaphase configurations which result from double non-sister crossovers between the rod and ring (fig. 2). These are (1) single bridge in AI only, (2) single bridges in AI and AII, (3) double bridge in AI, and (4) no bridge in AI. Considering only non-sister crossing over, these classes would arise from 3-strand doubles of type I, 3-strand doubles of type

TABLE 2

Theoretical ratio of anaphase configurations resulting from double non-sister-strand crossovers associated with sister-strand crossing over. *

No. sister-strand crossovers	Single bridge AI only	Single bridge AI and AII	Double bridge AI	No bridge AI		Total
				No bridge AII	Double ring AII	
0	1	1	1	1	0	1
1	1	1	1	0.5	0.5	1
2	1	1	1	0.5	0.5	1
3	1	1	1	0.5	0.5	1
Many	1	1	1	0.5	0.5	1

*This table is based on the assumption of no chromatid interference and was calculated by following the consequences of the four non-sister double crossover types in all possible combinations with varying numbers of sister-strand crossovers.

II, 4-strand doubles, and 2-strand doubles, respectively. The proportions of these four classes are not changed by the occurrence of sister-strand crossing over (table 2). Regardless of the frequency of the latter, whether it be one or 10 per bivalent, the proportions of these classes would remain the same as those resulting from double non-sister crossovers alone. If the number of sister-strand crossovers per bivalent is high, an odd number occurring in any one region will appear as a crossover and an even number as a non-crossover. Thus, the frequency of single bridges in AII resulting from double non-sister crossovers associated with sister-strand crossing over should still be equal to the frequency of AI double bridges, i.e., 13 percent.

Another way by which single bridges in AII can arise is by a single non-sister crossover associated with sister-strand crossing over. An odd number of sister crossovers will give rise to a single bridge in AI and AII. An even number or no sister crossovers will result in a single bridge in AI only. It is not possible to estimate from the data the number of sister-strand crossovers which occur per bivalent except that it must be one or more to account for the high frequency of single AII bridges. One sister crossover together with a single non-sister crossover will give 50 percent bridges in AI only and 50 percent bridges in AI and AII if there is an equal probability that it will occur in the ring or in the rod, since only sister crossovers in the ring result in AII bridges. Likewise, a large number of sister-strand crossovers per bivalent will give 50 percent of each type if there is equal probability of an odd or an even number occurring in the rod and ring.

Single bridges were observed in 59 percent of the anaphase I configurations. These result from the single and half of the double non-sister crossovers. If there is no chromatid interference, 26 percent of the anaphase I cells would possess single bridges due to 3-strand double crossovers, since the frequency of 4-strand doubles was 13 percent. As stated previously, the frequency of single bridges in AI only and the frequency of single bridges in AI and AII resulting from double non-sister crossovers should each be equal to the frequency of AI double bridges. This leaves 33 percent AI single bridges due to single non-sister crossovers. As a result of sister-strand crossing over, half of these or 16.5 percent will also form single bridges in AII. Thus we can calculate that 29.5 percent single AII bridges would be expected on the basis of one or many sister-strand crossovers per bivalent, i.e., 16.5 percent from the single non-sister crossovers and 13 percent from the double non-sister crossovers. The observed value was 35 percent. This difference is not statistically significant.

In somatic mitoses the frequency of sister-strand crossing over in a ring chromosome can be determined by measuring the frequency of double-sized rings. The probability of two such crossovers occurring per ring and thus effectively cancelling each other is slight in view of the low frequency of double rings. This calculation cannot be made for meiosis where the ring and rod synapse, since only a portion of the sister-strand crossovers give rise to double rings. Dicentric rings arise from sister-strand crossing over in bivalents that had either no non-sister or double non-sister crossovers. The anaphase I cells showing no bridges, which result from double non-sister crossovers, should be equal in frequency to the cells showing AI double bridges—13 percent (table 2). Since the observed frequency of no bridge configurations in AI was 28 percent, this leaves 15 percent which did not have a non-sister crossover.

On the hypothesis of a large number of sister crossovers per bivalent, the frequency of double rings resulting from double non-sister crossovers would be one half the frequency of double bridges, or 6.5 percent (table 2). Also, half the bivalents not having a non-sister crossover, 7.5 percent, would give double rings. Thus, the expected frequency of double-sized rings would be 14 percent; the observed frequency was 10 percent.

The frequency of double rings in mitotic anaphases is quite low. McClintock (1938) counted only 8 percent double rings arising from a large ring which involved most of chromosome 2. The ring used in this study is somewhat smaller and gave 4.5 percent double rings, 57 out of 1275, as measured in meristematic root tissue. This striking difference between the meiotic and mitotic rates may offer a clue to the mechanism of sister-strand crossing over.

DISCUSSION

Matsuura (1940, 1948) has presented evidence that in Trillium the two chromatids in each chromosome are coiled in a relational spiral system at early meiotic metaphase I and in a parallel spiral system at late metaphase I. He has shown that the chromosome can undergo this change in coiling only by breakage and subsequent reunion of the chromatids at each half coil. He has used this mechanism to explain legitimate crossing over by postulating that the paired chromatids of each chromosome are non-sisters in two-thirds of the cases. There are many objections to such a hypohesis. The frequency of crossing over would be much higher than normally found. Also, crossing over should be reduced in the distal ends of a chromosome since the twisted chromatids will be able to separate without breakage in that region, as has been pointed out by Matsuura. However, in maize one of the most striking cases of non-correlation between cytological and crossover distances occurs in the very distal end of chromosome 9 where yg_2 and dt, which are cytologically very close, show 7 percent crossing over (Rhoades 1945).

Furthermore, this theory cannot explain the formation of anaphase bridges with fragments from heterozygous paracentric inversions. Matsuura has recognized this difficulty and in a later paper (1950) made the unlikely suggestion that there is no correlation between the occurrence of paracentric inversion loops at pachytene and bridges with fragments at anaphase.

Matsuura's mechanism for the parallelization of the spiral system could, however, account for sister-strand crossing over if we assume that the paired chromatids in each bivalent are sister strands. The number of such crossovers per bivalent would be high, roughly equal to the number of half coils. As has been pointed out, this is consistent with the data. This hypothesis would also explain the low mitotic rate of sister-strand crossing over since, in " somatic " mitosis, each of the two chromatids of a chromosome usually takes an independent spiral system from the beginning of its development (MATSUURA 1940).

Misdivision of the centromere is an alternative hypothesis to sister-strand crossing over which should be considered. Such a mechanism can be ruled out in a number of ways. The correlation of the frequency of dicentric ring formation with the size of the ring (McCLINTOCK 1938) is strong evidence against the production of double-sized rings by misdivision of the centromere. Double-sized rings arising from centromere misdivision would differ in the arrangement of the genes on the ring from those arising by sister-strand crossing over, and hence result in a different pattern of mosaicism (fig. 4). Endosperm mosaics resulting from the instability of a ring chromosome involving the short arm of chromosome 9 and carrying C (colored aleurone) and Wx (starchy endosperm) were analyzed (SCHWARTZ, unpublished). If the double-sized rings arise by centromere misdivision, twin sectors of colorless starchy and

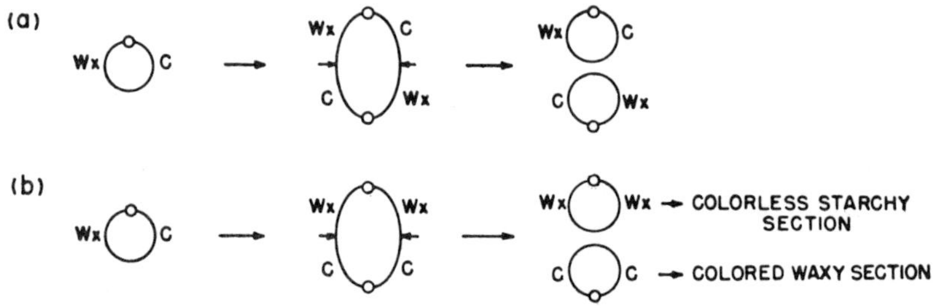

FIGURE 4.— Diagrammatic representation of endosperm mosaic pattern resulting from (a) dicentric ring arising by sister-strand crossing over, and (b) dicentric ring arising by misdivision of the centromere.

colored waxy tissue should be frequent. These would result from breakage of both bridge strands between C and Wx. No such twin sectors were found.

In order to explain the high frequency of single bridges in anaphase II by this mechanism, one would have to postulate that the undivided centromere of a dyad in which two of the arms are connected misdivides frequently while the centromere of a dyad with four free arms divides normally (fig. 5). It is difficult to visualize how such a modification in the morphology of the dyad would affect the plane of division of the centromere. Moreover, in the course of the breakage-fusion-bridge cycle described by McCLINTOCK (1941a), chromosome configurations are formed by the fusion of broken ends of sister chromatids which are similar to the dyads described above in that the centromere is undivided, and two of the arms are free and two are joined. Normal division of the centromere results in such a dicentric chromatid. Misdivision of the centromere would result in the formation of a centric ring and a centric rod. The breakage-fusion-bridge cycle, which results in a mosaic phenotype when in the proper genetic background, persists only in the gametophytic and endosperm tissue. The mosaicism is not carried over into the sporophyte since the broken ends heal in the embryo and the cycle is halted (McCLINTOCK 1941a). If rings are formed by centromere misdivision, the mosaic pattern should be evident in the sporophyte because of the unstable behavior of ring chromosomes.

Fusion of the broken ends of an AI double bridge might be thought of as an alternative hypothesis to explain the high frequency of single bridges in AII. However, McClintock has shown that following the breakage of an AI double bridge fusion did not occur in the telophase I nucleus. In our study no cases were found where each daughter cell of a pair contained an AII bridge.

The evidence which is most frequently cited against the occurrence of sister-strand crossing over is the frequency of homozygosis in attached-X females of Drosophila. If the distribution of chiasmata among the four strands at meiosis is at random, there are six possible combinations of these four strands and the frequency of homozygosis cannot exceed 16.7 percent. If only non-sister strand

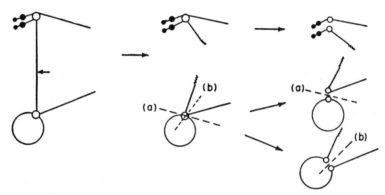

FIGURE 5.—Schematic diagram of alternative hypothesis involving misdivision of the centromere. (a) Represents normal plane of division of the centromere; (b) represents misdivision of the centromere.

crossing over occurs, the expected homozygosis value is 25 percent when one exchange occurs, 12.5 percent for two exchanges, 18.75 percent for three exchanges, etc. (SAX 1932). The finding of homozygosis frequencies significantly higher than 16.7 percent was interpreted as proving that little or no sister-strand crossing over occurs (BEADLE and EMERSON 1935). However, as these authors have pointed out, this evidence rules out only sister-strand crossing over which is equivalent to non-sister crossing over and shows chiasma interference. If the two crossover types are independent, as has been proposed in this paper, arising by different mechanisms and occurring at different times in the meiotic division, the maximum frequency of homozygosis expected from a combination of both sister- and non-sister-strand crossing over would remain at 25 percent.

From the study of the stable X^{c1} ring chromosome in Drosophila, MORGAN (1933) concluded that no sister-strand crossing over occurs. Her argument was as follows: Sister-strand crossing over in the ring would result in dicentric double-sized rings which would be eliminated in oogenesis. In females heterozygous for the ring and a rod X chromosome this elimination would result in a decrease in the frequency of gametes carrying a non-crossover ring chromosome. The fact that equal ratios of non-crossover ring and non-crossover rod gametes were produced is certainly strong evidence against the occurrence of sister-strand crossing over in the X^{c1} ring. However this does not warrant a generalization regarding such crossing over in other organisms. If dicentric ring formation is used as the criterion, the evidence is strongly in favor of sister-strand crossing over in maize. It is an established fact that dicentric double-sized rings do arise from ring chromosomes in maize (McCLINTOCK 1938, 1941b).

The behavior of ring chromosomes in Drosophila is quite different from that in maize and presents an extremely complicated and little understood picture. The X^{c2}, In(1)w^{vc} ring is highly unstable and gives rise to frequent gynandromorphs (BRAVER and BLOUNT 1950). The X^{c1} and X^{c2} rings are comparatively stable but the stability of the X^{c2} ring has been shown to be greatly influenced by environmental conditions (BROWN and HANNAH 1952). No stable ring chromosomes have been reported in maize.

In discussing the occurrence of those mutations at the Bar locus in Drosophila which were not associated with crossing over between forked and fused, STURTEVANT (1925, 1928) placed little emphasis on the possibility of explaining these mutations by unequal sister-strand crossing over. He was of the opinion that these exceptional cases arose as a result of contamination. In a similar study involving the compound A^b locus in maize, LAUGHNAN (1952) reported 8 percent exceptional cases where the separation of the a and β components of the locus was not associated with crossing over. It was possible to rule out contamination in these experiments. LAUGHNAN lists three possible explanations for these exceptional cases: (1) unequal sister-strand crossing over, (2) mutation of the β component to a null form, and (3) deficiency of the β component. The last two hypotheses are not very satisfactory. As LAUGHNAN pointed out, the exceptional cases are identical with those isolated from A^b by crossing over and are limited in the time of their occurrence to meiosis. If the non-crossover exceptions involve deficiencies of β, the deleted segments are very small including neither a nor sh_2 which are very closely linked with and situated on either side of β. While the A^b exceptional cases are not in themselves strong evidence for sister-strand exchange, they are at least suggestive and can be used as supporting evidence for such a form of crossing over.

SUMMARY

A large discrepancy in the expected ratio of anaphase configurations resulting from crossing over between a large ring involving chromosome 6 and its homologous rod chromosome is interpreted as due to sister-strand crossing over. The data indicate that at least one sister-strand crossover occurs per bivalent. An hypothesis is presented for the mechanism responsible for such crossing over based on Matsuura's observation on the parallelization of the spiral system in meiotic metaphase.

ACKNOWLEDGMENT

The author wishes to express his appreciation to Mrs. Rachel C. Cheniae for assistance in these studies.

LITERATURE CITED

Beadle, G. W., and S. Emerson, 1935 Further studies of crossing over in attached-X chromosomes of *Drosophila melanogaster*. Genetics **20**: 192–206.

Braver, G., and J. L. Blount, 1950 Somatic elimination of ring chromosomes in Drosophila melanogaster. Genetics **35**: 98. (Abstr.)

Brown, S. W., and A. Hannah, 1952 An induced maternal effect on the stability of the ring-X-chromosome of *Drosophila melanogaster*. Proc. Nat. Acad. Sci. **38**: 687–693.

Laughnan, J. R., 1952 The action of allelic forms of the gene *A* in maize. IV. On the compound nature of A^b and the occurrence and action of its A^d derivatives. Genetics **37**: 375–395.

Matsuura, H., 1940 Chromosome studies on *Trillium kamtschaticum* Pall. XII. The mechanism of crossing over. Cytologia **10**: 390–405.
　　1948 Chromosome studies on *Trillium kamtschaticum* Pall. and its allies. XXII. Critical evidence of the spiral theory of crossing over. Chromosoma **3**: 431–439.
　　1950 Chromosome studies on *Trillium kamtschaticum* Pall. and its allies. XIX. Chromatid breakage and reunion at chiasma. Cytologia **16**: 48–57.

McClintock, B., 1938 The production of homozygous deficient tissue with mutant characteristics by means of the aberrant behavior of ring-shaped chromosomes. Genetics **23**: 315–376.
　　1941a The stability of broken ends of chromosomes in *Zea mays*. Genetics **26**: 234–282.
　　1941b Spontaneous alterations in chromosome size and form in *Zea mays*. Cold Spring Harbor Symp. Quant. Biol. **9**: 72–80.

Morgan, L. V., 1933 A closed X chromosome in *Drosophila melanogaster*. Genetics **18**: 250–283.

Muller, H. J., and A. Weinstein, 1933 Evidence against the occurrence of crossing over between sister chromatids. Amer. Nat. **67**: 64–65. (Abstr.)

Rhoades, M. M., 1945 On the genetic control of mutability in maize. Proc. Nat. Acad. Sci. **31**: 91–95.

Sax, K., 1932 The cytological mechanism of crossing over. J. Arnold Arboretum **13**: 180–212.

Schwartz, D., 1953 The behavior of an X-ray induced ring chromosome in maize. Amer. Nat. **87**: 19–28.

Sturtevant, A. H., 1925 The effects of unequal crossing over at the bar locus in Drosophila. Genetics **10**: 117–147.
　　1928 A further study of the so-called mutation at the bar locus of Drosophila. Genetics **13**: 401–409.

23

Reprinted from pp. 625, 626–627, 628, 635–638, 644–648, 726–728 of *Genetics* **21**(6):625–730 (1936)

SOMATIC CROSSING OVER AND SEGREGATION IN DROSOPHILA MELANOGASTER*

CURT STERN

University of Rochester, Rochester, N. Y.

INTRODUCTION

IN 1925 BRIDGES found that females of *Drosophila melanogaster* containing the dominant factor Minute-n in one X chromosome and some recessive genes in the other X chromosome often exhibit a mosaic condition. While the main surface area of these flies showed the effect of the dominant Minute-n (I, 62.7) without the effect of the recessive genes, as was to be expected, smaller areas, in different regions of the body and of varying size, were not Minute-n, phenotypically, but displayed the effects of the recessive genes. BRIDGES' interpretation was this: the Minute-n factor has the property to eliminate occasionally the X chromosome in which it itself is located. The cells of mosaic spots are descended from one common ancestral cell in which such elimination had taken place. They possess, therefore, only one X chromosome and show the phenotype produced by its genes.

Minute-n is only one of a group of factors which are very similar in their phenotypical expression. The "Minutes" behave as dominants whose most striking phenotypic effect is a reduction in bristle size; in addition there is a strong retardation in development, tendency to rough eyes, etc. The homozygous Minute condition is lethal. Some Minutes have been shown to be deficiencies (SCHULTZ 1929). Many Minute factors have been found in different loci of all chromosomes. They are distinguished by adding different letters or numbers to the symbol M.

Following BRIDGES' discovery of mosaics with respect to sex-linked factors, the appearance of mosaic spots which exhibit autosomal characters was described (STERN 1927b). Such spots appear on flies which originally had a constitution heterozygous for genes determining the characters. These mosaics occurred in crosses in which autosomal Minute factors were present and the facts seemed to agree with the interpretation that the spots were due to an elimination of that arm or part of an autosomal chromosome which carried the Minute.

* A part of the cost of the accompanying tables and figures is paid by the Galton and Mendel Memorial Fund.

The present investigation was originally designed to attack the problem: How is a Minute factor able to eliminate the chromosome or that part of a chromosome in which it itself is located?

At the same time the solution of another problem was sought. The fact that small mosaic spots showed the phenotypic effect of certain genes contained in their cells whereas the remainder of the individual showed another phenotype was proof of the autonomous development of these characteristics. Among the very few genes which did not show phenotypical effects in spots was the recessive "bobbed" (I, 66.0) which produces short bristles: in $+^y Mn +^{bb}/y +^{Mn} bb$ females the $y +^M$ spots did not possess the bb-type bristle length, but a $+^{bb}$ length. Non-autonomous development of the bobbed character seemed improbable as typical gynandromorphs had shown clear demarcation lines for the bb and $+^{bb}$ areas (STERN 1927a). As bobbed is located at the extreme right end of the genetic X chromosome, next to the spindle fibre attachment, the following hypothesis was proposed: just as in the case of autosomal eliminations only part of the autosome disappears, so also in Mn mosaics merely a portion of the X chromosome is eliminated. The piece adjoining the spindle attachment and including the bobbed locus is assumed to be left in the cell, thus giving a constitution $+^{bb}/y +^{Mn} bb$, which, being heterozygous for bb, does not produce the effect of this gene (STERN 1928b). When PATTERSON (1930) using MULLER's Theta translocation showed that in his cases of X-radiated flies "not the whole X chromosome was eliminated" it was decided to use the same genetic technique to test the above hypothesis as to the partial elimination in the case of Mn. The Theta translocation was kindly put at my disposal by Prof. H. J. MULLER.

Both problems, the question as to the action of Minutes to bring about elimination and the question as to complete or partial elimination of the X chromosome (intimately bound up with the question as to autonomous or non-autonomous development of bobbed in small spots), proved to be based on an erroneous concept as to the origin of mosaic spots. While the investigation revealed this, it provided at least a partial solution of the problems by the discovery of somatic crossing over and segregation in Drosophila.

[*Editors' Note:* Material has been omitted at this point.]

METHODS

The methods were similar to those used by BRIDGES. Flies were made heterozygous for recessive genes whose phenotypic effects were of such a kind as to be exhibited by very small areas, preferably even single setae. The setae of Drosophila are divided into macrochaetae and microchaetae, the former generally called bristles, the latter, hairs. As far as the purpose of the present study is concerned the distinction is of no intrinsic importance. Genes mainly used were: (a) yellow body-color (y, 1, 0.0), producing an effect which can be distinguished in a single hair, making it yellowish-brownish as opposed to the black not-yellow condition (the general coloring effect of y on the hypodermis is often not very distinct in spots (STURTEVANT 1932)) and (b) singed-3 (sn^3, 1, 21.0), producing a thickened, curved or crooked condition of the setae, which generally can be distinguished in single hairs also. However, doubts occasionally remain as to whether a single hair on a heterozygous $+/sn^3$ fly is genotypically singed or whether it is normal but slightly more bent than usual. With spots of two or more hairs such doubts hardly ever occur.

[Editors' Note: Material has been omitted at this point.]

THE MECHANISM OF MOSAIC FORMATION
Various hypotheses

In the foregoing pages the effect of Minutes on the process of mosaic formation was discussed in general terms. The following part will contain an analysis of the process itself.

BRIDGES' work with females carrying *Mn* in one X and recessive genes in the other seemed to have established (1) that the cells of a spot do not

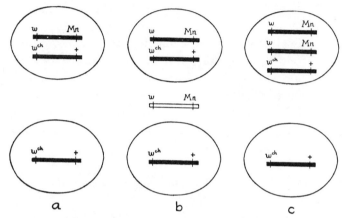

FIGURE 1 a–c. Three possibilities to account for elimination of an X chromosome.

contain the *Mn* chromosome and (2) that these cells are male in constitution, containing only one X with the recessive genes. Three main possibilities suggested themselves as mechanisms for the elimination of the *Mn* chromosome from the cells of the spot: (a) during a somatic division the *Mn* X chromosome does not divide and consequently passes into one daughter nucleus, leaving the other one in possession of only a division product of the not-*Mn* X chromosome; (b) the *Mn* X chromosome divides into two halves, but only one half passes into one of the daughter nuclei. The other half lags behind, is not included into a daughter nucleus, and degenerates in the cytoplasm; (c) the *Mn* X chromosome divides into two halves. These halves do not disjoin but pass together into one daughter nucleus.

The constitution of the daughter nuclei according to the three hypotheses is pictured in figure 1. No way of distinguishing between the mechanisms (a) and (b) was found, but a test between (a) and (b) on one side and (c) on the other seemed possible. The sister cell of the "elimination cell" in (a) and (b) would be of the same constitution as the original constitution of the fly, while in (c) it would have a different constitution, being triplo-X. By using suitable gene markers such a condition as (c) could be demonstrated if it occurred. Accordingly females were bred which had white in their Mn chromosome and cherry (w^{ch}) in the other X chromosome. The eye color in such females is a light cherry. Their eyes were searched for spots of normal, dark cherry color, indicating the elimination of the w Mn chromosome. When such spots were found it was determined whether their surroundings in the eye were of the w/wch coloration as expected according to (a) and (b) or whether they contained a very light cherry spot, indicating the $w/w/w^{ch}$ constitution expected according to (c). On 1702 w Mn/wch females which were inspected 20 eye spots were found. They all were cherry colored and unaccompanied by a twin spot of $w/w/w^{ch}$ coloration or of the irregular facet arrangement characteristic for 3X+2A eyes. Sixteen of these spots covered an area of at least 3 or 4 but not more than about 50 facets; 4 spots included more than 150 and less than 300 facets. These facet numbers have not been determined accurately but have been estimated from sketches. The total number of facets in one eye is between 700 and 800 in females. Similarly, in 168 $w^{ch}Mn/w$ females, two white eye spots were observed not accompanied by a twin spot of the darker colored triplo-X constitution $w^{ch}Mn/w^{ch}Mn/w$. These results rule against the hypothesis (c) provided that cells with two Mn factors and one normal allele are viable and can give rise to cell patches large enough to be recognized as mosaic spots. Nothing was known originally about this point.

Somatic segregation

y/sn^3 flies; preliminary discussion

The solution was brought about by chance. When it had been found that the "Minute condition" caused by an "uncovered" Blond-deficiency was necessary for the elimination of the deficiency X chromosome, an experiment was made in order to determine whether a deficiency which in itself does not produce a Minute effect would, together with an independent Minute factor, bring about the elimination of the deficient chromosome. The deficiency chosen was the well known Notch-8 (N^8) in the X chromosome and the Minute was the autosomal Mw. The following cross was made: N^8/y Hw dl-49 ♀ by sn^3; $Mw/+$ ♂, and the frequencies of head-thorax spots in N^8 and not-N^8 sister females were compared (table 4). The presence of Hw and of the dl-49 inversion is irrelevant. From 280 N^8/sn^3;$+^M$ or Mw females 9 mosaic spots were obtained while 381 control y Hw/dl-49/sn^3;$+^M$ or Mw females yielded 15 spots. Thus the frequency of spots in N^8 was 3.6 per cent, in controls 3.9 per cent. There was no interaction of the Notch deficiency with the autosomal Minute-w.

The nine spots in the N^8/sn^3 flies were recognized by singed setae occurring presumably as a consequence of elimination of the N^8 chromosome. In cases where the sn^3 containing X chromosome would have been eliminated, no visible spot would have been produced, for the N^8 chromosome contained no recessive genetic marker which would have expressed itself in a spot (provided that a cell containing only the Notch-deficient X chromosome is able to reproduce sufficiently to give rise to a large enough cell-patch).

TABLE 4

$N^8/y\ Hw\ dl$-49 by sn^3; $Mw/+$. Head and thorax spots only.

	N^8		$+^\lambda$	
	$+^M$	Mw	$+^M$	Mw
y spots	—	—	1	1
sn^3 spots	4	5	0	2
$\left.\begin{array}{l} y- \\ sn^3 \end{array}\right\}$ twin spots	—	—	2	9
Total spots	4	5	3	12

The situation was different in the $y\ Hw$ dl-49/sn^3 control flies. If there was no preference which of the two X chromosomes would be eliminated, then two different types of spots might be expected to occur in about equal numbers: (1) y spots in case of elimination of the sn^3 X chromosome and (2) sn^3 spots in case of elimination of the $y\ Hw$ dl-49 X chromosome. The 15 spots found consisted of (1) two y spots and (2) two sn^3 spots while (3) the remaining 11 spots showed an unexpected structure: *they were twin-spots formed by a yellow not-singed area adjacent to a singed not-yellow area.*

The obvious explanation is that, during a somatic division of one of the cells of these 11 $y\ Hw$ dl-49/sn^3 females, a segregation had taken place whereby one daughter cell obtained the yellow gene carried originally by one of the X chromosomes while the other cell obtained the singed gene carried originally by the other X. A process similar to gametic segregation of genes lying in opposite members of a pair of chromosomes had occurred in a somatic cell. The further division and normal somatic differentiation of the two daughter cells finally gave rise to mosaic twin areas.

These findings of somatic segregation suggested that the so-called chromosome elimination in certain cells leading to the appearance of mosaic spots was in all or most cases the consequence of somatic segregation. This theory is substantiated by three facts. (1) Further experiments demonstrated the general occurrence of twin spots in flies of suitable constitutions. (2) In appropriate experiments it could be shown that nearly all mosaic spots exhibit the results of somatic crossing over. This makes simple elimination hypotheses improbable. (3) The theory solves the difficulties encountered by the assumption that the sex-linked Minutes eliminate only their own chromosomes.

We shall first discuss point (3). The more important statements (1) and (2) will be dealt with in later sections.

Minute-n and Blond-Minute "elimination"
as somatic segregation

Somatic segregation of the X chromosomes in a female carrying Mn in one X and a recessive gene in the other will lead to two daughter cells, one containing only Mn, the other only the recessive. Mn is known to be lethal to a male or a homozygous Mn female zygote. If we assume Mn in such a condition to be lethal to a somatic cell also we shall expect the one daughter cell to die while the other one, containing only the recessive, will give rise to the observed spot. (It might be argued that a cell containing only Mn is viable but phenotypically not different from the non-segregated surrounding tissue. This, however, is excluded by having a recessive gene together with Mn in the one X chromosome but not in the other X. Segregation without lethal effect of the Mn segregation product should exhibit Mn spots which also show the recessive gene effect. In the experiments discussed on p. 635 the Mn chromosome contained a white or cherry gene, but no spots showing the respective eye colors were found. Other experiments of similar nature are described in later sections of this paper.)

[*Editors' Note:* Material has been omitted at this point.]

Somatic segregation and crossing over
Experiments involving $y\ sn^3/+$ flies

In the experiments summarized in table 5, y and sn^3 were in opposite chromosomes. When both mutants were in the same chromosome, some new results were obtained (table 8): Three kinds of spots appeared with

TABLE 8

Spots in flies of the basic constitution $y\ sn^3/+$.

EXP.	CONST.	IND.	SPOTS	$y\ sn^3$			y			sn^3			OTHER SPOTS
				1	2	>2	1	2	>2	1	2	>2	
(1)	$y\ sn^3bb/car\ bb$	83	34	10	4	6	3	3	5	2	—	1	—
(2)	$y\ sn^3bb/car\ bb$	495	72	25	15	10	12	2	5	2	—	1	—
(3)	$y\ sn^3bb/+$ †	508	35	10	11	4	5	4	—	1	—	—	—
(4)	$y\ sn^3/+$ †	321	21	8	3	4	1	1	2	—	—	—	2*
				53	33	24	21	10	12	5	—	2	
Totals		1407	162		110			43			7		2

† Partly $M\dot{w}/+$.
* $+^y+^{sn}$; ♂-colored.

different frequencies, namely 110 $y\ sn^3$, 43 y and 7 sn^3 spots. The finding of $y\ sn^3$ spots was expected, for segregation in $y\ sn^3/+$ females (disregarding the presence of bb and car) should give rise to $y\ sn^3$ and $+$ cells, which would be visible as $y\ sn^3$ spots. The occurrence of y and of sn^3 spots needs an additional interpretation. Somatic crossing over between y and sn^3 would separate these two genes from each other and thus afford an explanation. If the crossover process occurred during a two strand stage, the resulting strands would be $y+^{sn}$ and $+^y sn^3$, and if segregation ensued a y and a sn^3 twin spot would be produced. No twin spots were found, making the two strand crossing over assumption invalid. If, however, somatic crossing over occurred at a four strand stage between two of the four strands, and segregation two strands by two strands followed, then the facts can be explained (fig. 2). It is seen that following single crossing over, different types of chromatid segregation, namely **x** and **y,** give rise to either y or sn^3 single spots. (Throughout this paper the term "chromatid" is used in reference to the strands which *constitute* a multivalent chromosome group during prophase and metaphase, as well as in reference to those chromosomes of anaphase and telophase which *originated* from a multivalent.)

If crossing over occurs at the four strand stage the subsequent segregation process will be expected to lead to either one of two results. (1) A separation of the four chromatids into two daughter cells will occur, followed

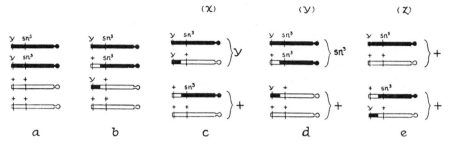

FIGURE 2. y sn³/+. Crossing over between y and sn³ at a four strand stage. a. Non-crossover chromatids. b. Two crossover and two non-crossover chromatids. c-e. Three different types of chromatid segregation.

later by normal mitosis or (2) the initiated segregation will bring about a true reduction of chromatids leading to a second segregating division with resulting cells containing only single X chromatids. In females of the constitution y sn³/++, such a reduction process, after crossing over between two of four strands, should lead to four cells with the genes y sn³, y+^{sn}, +^{y}sn³ and ++. The areas resulting from later cell-divisions of the segregation products would exhibit the phenotypes y sn³, y, and sn³. The visible result would be a triple spot. Not a single spot of this nature was observed; the hypothesis of a complete somatic reduction process is thus refuted.

We can test the assumption of somatic crossing over at a four strand stage by applying it to the earlier experiments in which the constitution of the flies was y/sn³. Single crossing over between y and sn³ at a two strand stage would yield y sn³ and ++ strands; segregation would result in y sn³ spots. No such spots have been found (table 5). Crossing over at a four strand stage, however, would, according to different types of chromatid segregation, produce y or sn³ single spots (fig. 3). Such spots did occur and while many of them could be regarded as vestiges of potential twin spots, it is possible to assume a certain number of them to have been products of crossing over between y and sn³ in the manner suggested.

Again it is obvious that no reduction of chromatids occurred which

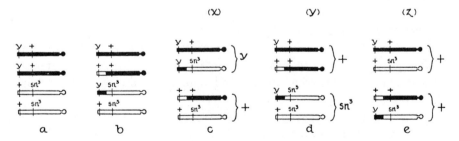

FIGURE 3. y/sn³. Crossing over between y and sn³ at a four strand stage.

would have resulted in triple spots y, sn^3, and $y\ sn^3$. No triple spots were found and the phenotype $y\ sn^3$ did not occur even in single spots.

If one characterizes the two types of chromatid segregation **x** and **y** by the constitution of the right chromatid ends, then **x**-segregation would be

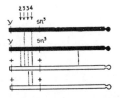

Figure 4 (left):

strands involved		(X)	(Y)	(Z)
2		sn^3	y	$+$
		$+$	$+$	$+$
3		sn^3	$+$	$+$
		$+$	$+$	y
3		sn^3	$+$	y
		$+$	$+$	$+$
4		sn^3	$+$	$+$
		$+$	y	$+$

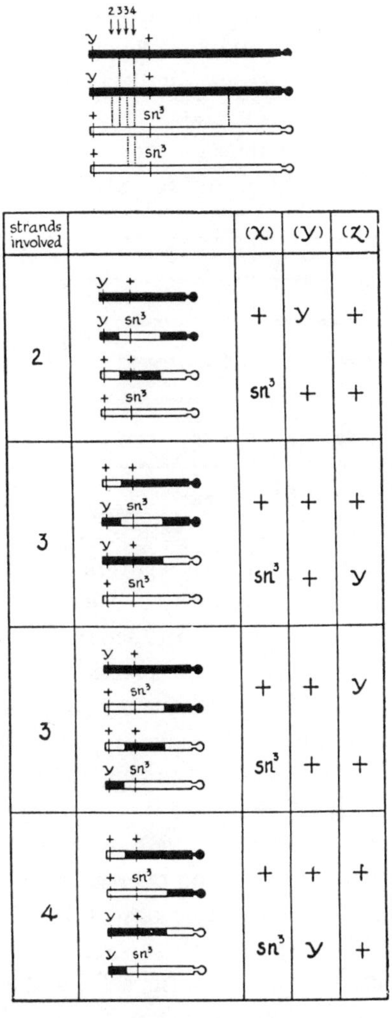

Figure 5 (right):

strands involved		(X)	(Y)	(Z)
2		$+$	y	$+$
		sn^3	$+$	$+$
3		$+$	$+$	$+$
		sn^3	$+$	y
3		$+$	$+$	y
		sn^3	$+$	$+$
4		$+$	$+$	$+$
		sn^3	y	$+$

FIGURES 4 (LEFT) AND 5 (RIGHT). $y\ sn^3/+$ and y/sn^3. The four different types of double cross-overs, involving 2, 3, and 4 chromatids and the results of the three different types of segregation.

equational and **y**-segregation reductional. As the right end of the X chromosome is known to contain the fibre attachment point, processes **x** and **z** would result from separation of sister attachment points while **y** would occur only when sister points stay together. It is possible to account for all observed spots with the assumption of equational segregation for the right end, if one considers the possibility of occurrence of somatic double crossovers simultaneously to the left and to the right of sn^3 (figs. 4, 5). Four different kinds of double crossover processes within the four strand "tetrad" are possible, involving two, three, and all four strands. After single crossing over **x**-segregation results in y single spots in flies of both constitutions $y\ sn^3/++$ and $y+/+sn^3$, after double crossing over it results in sn^3 spots. If **y** segregation occurs, it gives rise to y spots in one-half of all cases; similarly **z** segregation produces visible y spots in two out of the four cases.

While a decision between the hypotheses of single crossing over and different kinds of segregation or single and double crossing over and equa-

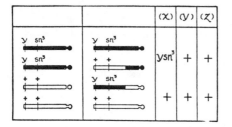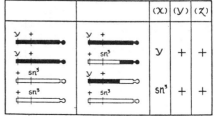

FIGURES 6 (LEFT) AND 7 (RIGHT). $y\ sn^3/+$ and y/sn^3. Results of crossing over to the right of sn^3.

tional **x**- and **z**-segregation only cannot be derived from the present data, the second hypothesis seems to be in better agreement with the known facts in regard to the separation of daughter chromosomes in mitosis. We shall return to this question in the next chapter.

What is the frequency of the different crossover types in the $y\ sn^3/+$ experiments? We shall regard (and shall justify this hypothesis later) the $y\ sn^3$ spots as due to single crossing over between sn^3 and the right end of the X chromosome (fig. 6; see also fig. 7 for the y/sn^3 flies). The sn^3 spots are considered to be due to double crossing over. In determining the frequency of crossing over we cannot follow the usual procedure in germinal crossing over and base the calculation on the total of observed non-crossovers and crossovers, since the absence of crossing over in the great majority of somatic cells makes the number of non-crossovers rather meaningless. However, we can determine the relative frequencies of observed crossovers to the right and to the left of sn^3 and the frequency of double crossovers. The frequencies of $y\ sn^3$, y, and sn^3 spots were 110, 43, and 7. As was shown earlier, some y spots have to be regarded as products of **z**-segregation after double crossing over. As they constitute one-third of all visible double crossover products, their number is taken as 3.5. After subtracting this number from the total number of y spots, we find the proportion of spots produced after crossing over to the right of sn^3 ($y\ sn^3$ spots) to spots from crossing over to the left of sn^3 (y spots) to be 110:39.5. Only the **x**-type of segregation in single crossing over yields visible spots, while the equally frequent **z**-type remains undetected. In double crossing over both **x**- and **z**-segregations result in visible spots, a total of 75 per cent of all cases. In order to have comparable figures we have to insert two-thirds of the number of visible double crossover spots, namely $(7+3.5)\cdot\frac{2}{3}=7$. Thus the proportion, single crossovers to the right of sn^3:singles to the left:doubles is 110:39.5:7. Comparing these values with the corresponding frequencies in meiosis, we find for the single crossovers 36:20.5. (MORGAN, BRIDGES, and STURTEVANT 1925, modified; no satisfactory data are available for the double crossovers.) In mitosis there seems to occur a higher proportion of all crossovers to the right of sn^3 than in meiosis. Even this frequency of somatic crossovers to the left of sn^3 was unusually high, as further experiments show.

[*Editors' Note:* Material has been omitted at this point.]

SUMMARY

(1) Mosaic areas on the body of *Drosophila melanogaster* appear on flies which are heterozygous for genes whose homozygous effect can be recognized in a small spot.

(2) The frequency of spots is increased by the presence of Minute factors.

(3) Spots of sex-linked characters occur with higher frequency when either sex-linked or autosomal Minutes are present, but sex-linked Minutes are more powerful than autosomal ones. Autosomal spots are more frequent in the case of presence of autosomal Minutes than of sex-linked

Minutes. Different Minutes show different degrees of ability to induce spot formation.

(4) The mechanism of mosaic formation is not based on simple elimination of chromosomes but on processes of somatic crossing over involving two strands of a four strand group. Segregation of the four strands occurs in an equational, typically mitotic mode in respect to the fibre points. It leads to homozygosis of originally heterozygous genes. No reduction of number of chromatids takes place in normal cases. These conclusions are derived from an analysis of types and frequencies of twin and single spots and of the number of X chromosomes present as judged by the secondary sexual characters of favorable spots. Interpretations based on experiments with certain combinations of genes have been verified by tests of validity in other combinations of the same genes.

(5) The increase of frequency of sex-linked spots is not directly dependent on the localized, material constitution of the chromosomes involved in somatic crossing over but rather on the general "phenotypic Minute reaction" in development.

(6) The relative frequencies of somatic crossovers in different regions of the X chromosomes are different from those of germinal crossovers. Somatic crossing over is more frequent near the fibre point. The presence of Minute-n accentuates this shift.

(7) The X chromosome duplication "Theta" frequently undergoes somatic crossing over with the X chromosome—more frequently in the homologous right than in the homologous left regions. Germinal crossing over involving Theta is very rare.

(8) Somatic crossing over involving Theta followed by equational segregation leads to twin segregates of the constitution 3X chromosomes-1X chromosome.

(9) The apparently exceptional behavior of the bobbed character, which does not become visible in spots, is understandable under the assumption that no somatic crossing over occurs to the right of the bobbed locus.

(10) Somatic crossing over involving the sex chromosome occurs in superfemales and in males.

(11) Somatic autosomal crossing over takes place in both sexes, though more frequently in females. A peculiar specificity of the Minute effect leads to crossovers in that arm of the third chromosome in which the Minute itself is located. Most crossovers are concentrated near the fibre point region.

(12) Somatic crossing over between X chromosomes heterozygous for the bb^{Df} inversion occurs within the inversion. It leads to a chromatid

which possesses no fibre point and is thus eliminated, and to a complementary chromatid with two fiber points. This chromatid becomes fragmented and each fragment is included in a daughter nucleus.

(13) When Theta is present in cells heterozygous for the bb^{Df} inversion the most frequent type of somatic crossover involves Theta and the not-inverted, not-bb^{Df} chromosome. (A discrepancy is pointed out between this interpretation and the observed facts.)

(14) The presence of an extra Y chromosome in flies discussed under (12) increases the frequency of somatic crossing over within the inversion to the right of sn^3 as well as the frequency of fragmentation of the two-fibre-point chromatid, also to the right of sn^3.

(15) In flies discussed under (13) the presence of an extra Y chromosome increases the frequency of crossovers involving Theta.

(16) An exceptional series of cultures with XXY females gave results which can be interpreted as caused by somatic crossing over between X and Y chromosomes, leading to XXX and X segregates.

(17) Somatic crossing over in flies heterozygous for a ring-shaped X chromosome leads to a two-fibre-point "tandem" chromatid. Segregation occurs preferentially so that the two non-crossover chromatids go to one pole and the tandem chromatid to the other. In the following division the tandem chromatid becomes fragmented.

(18) In different experiments certain segregated constitutions are not sufficiently viable to give rise to mosaic areas. Others, though not viable as zygotic constitutions, permit the formation of hypodermal spots.

(19) Under different genetic conditions different patterns of mosaics are formed. The proportion of small to larger spots can vary. In Minute-n flies crossovers to the left of Mn occur later in development than crossovers to the right. Various genetic constitutions have differential effects on frequency and size of spots in various body regions. Different types of spots in flies of the same constitution are differently distributed over the head, thorax and abdomen or over the different abdominal segments.

[*Editors' Note:* Material has been omitted at this point. The references from these excerpts appear on the following page.]

REFERENCES

Bridges, C. B. 1925. Elimination of chromosomes due to a mutant (Minute-n) in *Drosophila melanogaster*. *Proc. Nat. Acad. Sci. U.S.A.* **11**:701–706.

Morgan, T. H., C. B. Bridges, and A. H. Sturtevant. 1925. Genetics of *Drosophila*. *Bibliogr. Genet.* **2**:1–262.

Patterson, J. T. 1930. Proof that the entire chromosome is not eliminated in the production of somatic variations by X-rays in *Drosophila*. *Genetics* **15**:141–149.

Schultz. J. 1929. The Minute reaction in the development of *Drosophila melanogaster*. *Genetics* **14**:366–419.

Stern, C. 1927a. Ein genetischer und zytologischer Beweis für Vererbung im Y-Chromosom von *Drosophila melanogaster*. *Z. Indukt. Abstamm.-Vererbungsl.* **44**:187–231.

_____. 1927b. Über Chromosomenelimination bei der Taufliege. *Naturwissenschaften* **15**:740–746.

_____. 1928b. Elimination von Autosomenteilen bei *Drosophila melanogaster*. *Z. Indukt. Abstamm.-Vererbungsl.* Sup. **2**:1403–1404.

Sturtevant, A. H. 1932. The use of mosaics in the study of the developmental effects of genes. *Proc. Sixth Int. Congr. Genet.* **1**:304–307.

24

Reprinted by permission of the University of California Press from pp. 283–284, 288–290 of *Univ. Calif. Publ. Bot.* **14**(8):283–291 (1928)

A WORKING HYPOTHESIS FOR SEGMENTAL INTERCHANGE BETWEEN HOMOLOGOUS CHROMOSOMES IN FLOWERING PLANTS*

BY

JOHN BELLING

In scientific investigations a working hypothesis is needed as a guide to the discovery of important facts, and also serves to prevent distraction of the attention by unimportant facts. Even a wrong hypothesis may be better than none at all; e.g., the mutation hypothesis of Oenothera. If, however, a working hypothesis is in great part correct (as is, in the writer's opinion, Janssens' working hypothesis as to segmental interchange between homologous chromosomes), then it may lead, as Janssens' has led, to fruitful work and to discoveries of the first importance. Of course from time to time a fruitful working hypothesis should be scrutinized and corrected to bring it up to date.

Perhaps affairs are now ripe for a renovation of the hypothesis with which the name of Janssens must always be connected. This hypothesis has had a curious fate. It has been accepted more or less by most geneticists, in the lack of any other working hypothesis; and the discoveries made with its aid have justified its use. On the other hand, Janssens' hypothesis has been almost unanimously disregarded or rejected by other cytologists. This hypothesis, as it appears to the writer after reading Janssens' last paper (1924), may be stated briefly as follows. At some stage after they have split lengthwise, the homologous chromosomes of a bivalent twist around one another corkscrew fashion. In some unknown way the two corkscrews which form a bivalent come into contact at a series of points. At these points one of the sister strands of each homologue breaks; and then, instead of uniting again in the original manner, it unites with one end of the broken strand of the other homologue. In this way the cross or X, called a chiasma, is made by two of the halves, while the other two are unbroken. There are some minor, more or less vague, additions to the

* This paper is the result of cytological work done under the Carnegie Institution of Washington.

hypothesis which need not be recapitulated here.   One weak point, which has already been noted by critics (Morgan, 1925) is that, even if corkscrew twisting of the homologues was proved to occur regularly (which is not the case, in the writer's opinion), it would not lead to contact between the two homologues at a definite series of points unless the corkscrews were also flattened out more or less in one plane. A second weak point is that, even if such a series of points of contact were produced, there is no known reason why the twin halves of the homologues should break at these points; and if they did break, there is no reason why only two should break.   Janssens was doubtless aware of these weak points, for he does not stress the explanation in his last paper (1924).

[*Editors' Note:* Material has been omitted at this point.]

WORKING HYPOTHESIS FOR SEGMENTAL INTERCHANGE OF CHROMOSOMES AND CROSSING OVER OF GENES

It is natural to suppose, as Morgan (1925) has pointed out, that breaks in a chain of genes (not breaks in the thread in which the chain is) occur, if at all, when it is drawn out into a tenuous length at leptophase. We know from genetic work that the homologous chromosomes are already split into twin halves at the time of crossing over (Anderson, 1925; Bridges and Anderson, 1925). Mutual attraction of the genes is assumed, for otherwise they could hardly hold together in a thread. If the breaks in each of the twin halves of each homologous chromosome at the leptophase take place independently, as seems natural to assume; and if the breaks are at random, except for one break hindering another close to it (genetic interference); and if the chance of two breaks, in any *particular* two of the four twin halves, coinciding at pachyphase, at any particular point, is $1/x$; then the chance for such coincidence in any two of the four twin halves is $6/x$. One-third of these cases of coincidence will be between sister strands of the same chromosome, hence the chance for coincidence between breaks at a particular point in twin halves of different homologous chromosomes is $4/x$. The chance for coincidence between four such breaks, one in each of the four twin halves, at pachyphase, is only $1/x^2$, at any particular point.

In any coincidence of two breaks in strands of different homologues, for which the chance at any particular point is $4/x$, we may suppose that in half the cases the ends of the same threads are reunited, and in the other half of the cases ends of different threads unite, forming a chiasma. Hence the chance of a chiasma, or point of segmental interchange, between homologues, at any one point in the bivalent is $2/x$. Since there is no such chiasma between the other two strands, $1/x$ represents the fraction of segmental interchange at any one point in the resulting chromosomes. (Allowance is to be made for the occurrence of a second, or third, chiasma.) Hence $1/x = c/100$, where c is the observed percentage number of cross-overs between two particular adjacent genes.

This working hypothesis includes the following assumptions. That breaks in the chain of genes (not in the chromosome thread) occur at leptophase, with the homologues already split. That these breaks are at random in each of the two sister strands, except for there being a

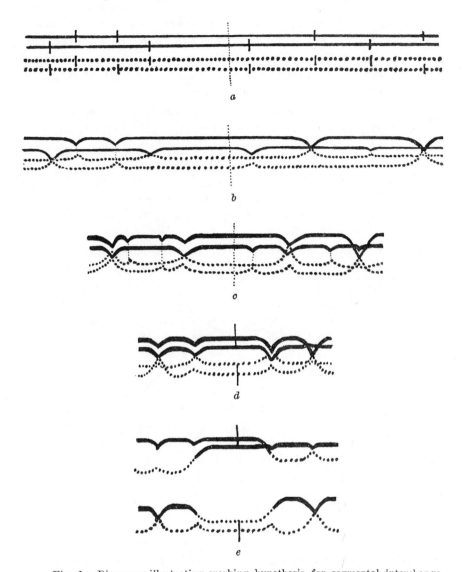

Fig. 1. Diagrams illustrating working hypothesis for segmental interchange between homologous chromosomes. *a*, represents coincident breaks in the chains of genes at leptophase and zygophase. It should be supposed relatively longer. Non-coincident breaks are omitted, as are also coincident breaks between sister strands. Four coinciding pairs of breaks are supposed to reunite as before breaking. *b*, represents union of the strands at pachyphase. It is not impossible that the other four pairs of coinciding breaks form temporary nodes. *c*, represents diplophase and early diaphase. The four chiasmas and the temporary nodes are visible because of the opening out. They are not seen at pachyphase because the threads are in contact laterally. *d*, represents late diaphase and metaphase. The temporary nodes have disappeared. Successive internodes become more or less at right angles, which is not shown. *e*, represents early anaphase. Note that one difference between this diagram and those of Janssens is that he posits a previous twist at each node.

certain minimum and maximum limit between consecutive breaks. That when two breaks in different strands happen to coincide at pachyphase (being between the same two adjacent genes, or allelomorphs), as the bivalent shortens the two broken left-hand ends unite at random with the two broken right-hand ends; so that either the original ends reunite, or a cross (chiasma, or point of segmental interchange or of crossing over) is formed. This working hypothesis seems to fit the facts already known, and may perhaps be used until superseded by a better.

The diagrams shown in figure 1 may be compared with those of Janssens (1924). In the leptophase and zygophase (fig. 1*a*) there are assumed to be eight coincidences of breaks between strands of different homologues, which is near the maximum number seen in Lilium. These hypothetical breaks are shown by crossbars. Coincident breaks between the sister strands of each homologue are not shown. Other breaks than coincident ones are perforce omitted, as nothing has yet been calculated as to their number. In the pachyphase (fig. 1*b*) the union of the coincident breaks takes place; in half the cases (shown by fine dotted lines) through restoration of the original junctions, and in half the cases by the formation of chiasmas. In figure 1*c* the bivalent is shortened, and the chiasmas have become nodes, at diplophase and early diaphase. At late diaphase and metaphase (fig. 1*d*) only the four chiasmas remain as nodes. In the anaphase (fig. 1*e*) it is to be noted that the two strands which pass to each pole are always united at the point of attachment of the spindle fiber; and that, if this point were in a different position, different pairs of strands would pass to each pole.

REFERENCES

Anderson, E. G. 1925. Crossing over in a case of attached X chromosomes in *Drosophila melanogaster. Genetics* **10**:403–417.

Bridges, C. B., and E. G. Anderson. 1925. Crossing over in the X chromosomes of triploid females of *Drosophila melanogaster. Genetics* **10**:418–441.

Janssens, F. A. 1924. La chiasmatypie dans les insectes. *LaCellule* **34**:135–359.

Morgan, T. H. 1925. The bearing of genetics on the cytological evidence for crossing-over. *LaCellule* **36**:113–123.

25

Reprinted by permission of the University of California Press from pp. 153, 159–161 of *Univ. Calif. Publ. Bot.* **16**(5):153–170 (1931)

CHROMOMERES OF LILIACEOUS PLANTS*

BY

JOHN BELLING

[*Editors' Note:* In the original, material precedes this excerpt.]

A MODIFIED WORKING HYPOTHESIS FOR CROSSING-OVER

The writer's previous working hypothesis for crossing-over (Belling, 1928a) was framed after he had studied the diplotene and diaphase, but before he had succeeded in staining the chromomeres at leptotene, zygotene, and pachytene. Hence more stress was naturally laid on the threads than on the chromomeres. Work during the last three years has shown that the chromomeres are the important parts of the threads, the fine connecting fibers being apparently a product of the chromomeres. For such fibers may be seen often to have grown transversely between homologous chromomeres, where they must have been formed *de novo*. So it is likely that they will grow afresh longi-

---

* This paper presents the results of cytological work done under the auspices of the Carnegie Institution of Washington.

tudinally between chromomeres adjacent in the rows, where connecting fibers may be absent. When the chromomeres are divided by the secondary split into two chromioles, in each case one of these chromioles can remain connected by the old longitudinal fiber; but the sister chromiole requires a new longitudinal fiber (fig. 3). This may arise, presumably, either by the longitudinal division of the previous fiber, or by the formation of a new fiber (fig. 5). That new longitudinal fibers may sometimes be formed is known, because in every case of crossing-over (and crossing-over is a *vera causa*), new longitudinal fibers *must* arise at the point of crossing-over, and could not arise by the splitting of old fibers. There are other cases, too, in which longitudinal fibers arise *de novo*. When this possibility is kept in mind, a modified hypothesis for crossing-over comes into view.

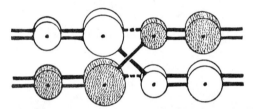

Fig. 5. Diagram illustrating the writer's modified hypothesis of crossing-over. At late pachytene, the two homologous chromosomes, or strings of chromomeres (one white, the other shaded), show a half-twist. After the secondary division of the chromomeres, the new chromomeres become connected by new fibers. These new fibers at the half-twist take the shortest way (broken lines). This constitutes crossing-over.

At the pachytene stage in Lilium, the writer not seldom remarked an X formed by a half-twist which had occurred probably at the zygotene stage. Similar half-twists, between two synapsed chromomere pairs have been previously figured (Gelei, 1921). When the chromomeres synapse individually, it is too much to expect that the two rows will lie always in one plane. There are as yet no proofs of any special lateral *orientation* of the genes in a string. That there is no end-to-end orientation of the genes follows from the facts that a segment of a chromosome can be inverted (Sturtevant, 1926); also that two chromosomes can be combined by homologous ends (L. V. Morgan, 1922); and that parts of non-homologous chromosomes can be joined to form a new chromosome (Sturtevant and Dobzhansky, 1930; Belling, 1927). Hence half-twists, in the presumed absence of any lateral orientation of the genes, may be regarded as occurring regularly in most pachytene coils. (The average distance between half-twists may depend mainly on physical causes, including the mutual attraction of homologous genes, or chromomeres.) At each such half-twist, which may be in one direction as often as in the other, the two threads form an X (figs. 5 and 9). This is before the secondary division has occurred, and while the constituent chromomeres of the pairs and their longitudinal connecting threads are single. Figure 5 shows that at late pachytene, when the chromomeres have split but before the chromioles have separated, the new longitudinal threads will probably take the shortest way and be attracted to the chromomeres closest to them in a straight line. This is all that is needed as a cause of crossing-over, and it postulates little beyond what has already been seen.

In figure 5 the new connecting threads are not distinguished from the old, except at the twist. However, it is probable that their position is determined by chance. This will make no difference in the longitudinal threads; but, when there is a half-twist, it will probably be a matter of chance (for each separate chromiole out of the four pairs of chromioles concerned) whether there is a longitudinal or oblique attachment. Then it would follow that any of the four chromatids concerned has an equal chance (at the half-twist) of crossing over or of not crossing over.

Any working hypothesis to account for chiasmas, that does not include crossing-over may, in the writer's opinion, be safely rejected; for crossing-over (which is a known fact) must, on any probable hypothesis, be itself a cause of chiasma formation. On the present hypothesis, the two cross-over chromatids are the ones that pass straight on; and the chromatids which form the X are non-cross-overs.

[*Editors' Note:* Material has been omitted at this point. The references from these excerpts appear on the following page.]

REFERENCES

Belling, J. 1928a. A working hypothesis for segmental interchange between homologous chromosomes in flowering plants. *Univ. Calif. Publ. Bot.* **14**:283–291.

Gelei, J. 1921. Weitere Studien über die Oogenese des *Dendrocoelum lacteum. Arch. Zellforsch.* **16**:88–169.

Morgan, L. V. 1922. Non-criss-cross inheritance in *Drosophila melanogaster. Biol. Bull.* **42**:267–274.

Sturtevant, A. H. 1926. A crossover reducer in *Drosophila melanogaster* due to inversion of a section of the third chromosome. *Biol. Zentralbl.* **46**:697–702.

———— and T. Dobzhansky. 1930. Reciprocal translocations in *Drosophila* and their bearing on *Oenothera* cytology and genetics. *Proc. Nat. Acad. Sci. U.S.A.* **16**:533–536.

Reprinted from *Nature* **199**(4898):1034–1040 (1963)

A THEORY OF CROSSING-OVER BY MEANS OF HYBRID DEOXYRIBONUCLEIC ACID

By Dr H. L. K. WHITEHOUSE

Botany School, University of Cambridge

THE term 'crossing-over' was introduced into genetical literature by Morgan in 1912 to describe the phenomenon he had discovered whereby new combinations of linked characters arose. Several theories to explain the process have been put forward in the intervening half-century. Ideas suffered a radical change in 1955 when the classical theory of reciprocal recombination was overthrown. This followed acceptance of the notion that, within small segments of the hereditary material, the products of recombination are not always complementary. Lindegren[1] had first suggested this on the basis of his work with *Saccharomyces cerevisiae*; but the idea did not gain wide support until Mitchell[2] demonstrated the phenomenon in *Neurospora crassa*. One of the first examples of non-reciprocal recombination which she found was remarkable in that conversion (as the non-reciprocal process is often called) had taken place in opposite directions at allelic sites.

In attempting to formulate a general theory of the mechanism of recombination, a difficulty has been the diversity in the results obtained with different organisms. In *Aspergillus nidulans*, where in ordinary linkage studies there appears (unlike *Neurospora*) to be no interference between cross-overs, Pritchard[3] found it necessary to postulate the frequent occurrence of more than one cross-over in short segments, which he called regions of 'effective' pairing, while non-reciprocal recombination has been somewhat rarely recorded[4]. In *Ascobolus immersus*, on the other hand, Lissouba and Rizet[5] and Lissouba *et al.*[6] have shown that non-reciprocal phenomena occur regularly at loci concerned with spore pigmentation, and moreover that a polarity is evident in the data. Within a group of apparently linearly sited alleles, if one is found to show conversion more frequently than another placed to the left of it, then the right-hand member of any other pair of them is found to show a similar difference from the left-hand member. Such polarity has not been demonstrated in *Neurospora*.

Variability is also shown within species. In *Neurospora*, Lindegren and Lindegren[7] found a significant excess of two-strand double cross-overs across the centromere of one of the linkage groups, while others (for example, Stadler[8]) have found no such chromatid interference. In *Ascobolus*, Lissouba *et al.*[6] have found that within a series of multiple alleles different pairs may show diversity in the relative frequencies of occurrence of reciprocal and non-reciprocal recombination.

Copy-choice Hypothesis

The hypothesis that has most frequently been used to explain non-reciprocal recombination is the theory of 'switch synthesis' or 'copy-choice', which relates recombination to replication of the hereditary material. This hypothesis requires that homologous chromosomes shall be paired, at least at certain points along their length, at the time of replication and that the new strands may be replicas of one parental strand in one region and of the other one in another region (Fig. 1). This idea provides a neat explanation of how within a tetrad one site may show a 2 : 2 ratio and a neighbouring one a 3 : 1 ratio.

The theory has been extended by Lissouba *et al.*[6] to take account of their observations on polarity, by postulating that the direction in which replication occurs is from left to right, that is, towards sites with the higher conversion frequency, and, furthermore, that once a copying 'error' has been started it continues to the end of the group of sites. They have coined the term 'polaron' for such a group of sites within which crossing-over appears to be absent.

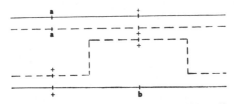

Fig. 1. Diagram to illustrate the copy-choice hypothesis. Old chromatids are shown by unbroken lines, and newly synthesized chromatids by broken lines: *a* and *b* represent the sites of allelic mutations, and the + signs their normal counterparts

The copy-choice theory, and its extension in the polaron concept, are open to a number of serious criticisms:

(1) The theory of the polaron, as proposed by Lissouba *et al.*[8], demands that few or none of the copying errors should continue through the intervening so-called 'linkage structure' to the next polaron, yet Stadler[9] has shown that, according to Lissouba's own data, a majority do so.

(2) Watson and Crick[10] postulated that the method of replication of deoxyribonucleic acid (DNA) was such that each new DNA molecule contained one of the nucleotide chains from the parent molecule and one new one. The work of Meselson and Stahl[11] showed that DNA replication in *Escherichia coli* followed the Watson and Crick method, and these observations were extended to a chromosomal organism, *Chlamydomonas reinhardi*, by Sueoka[12]. Taylor[13] showed that the chromosomes in root-tips of various flowering plants replicated by the same method, so far as their DNA was concerned. These observations suggest that the terms 'old' and 'new' must be applied, not to the two daughter-chromatids from one chromosome, but rather to half of each DNA molecule within them. On the other hand, the idea of copy-choice implies the contrary, namely, that replication is at the chromatid and not the sub-chromatid level.

(3) Meselson and Weigle[14] and Kellenberger *et al.*[15] showed that in lamba phage of *Escherichia coli*, genetic recombination can occur in the absence of replication of DNA.

(4) The necessity, on the copy-choice theory, to associate conversion with replication of DNA presents difficulties, since replication appears to occur at the end of the resting-stage prior to meiosis, before homologous chromosomes are thought to have associated[16]. Furthermore, if reciprocal recombination (crossing-over) also occurs at the time of DNA replication, then it is presumably necessary to postulate that chiasmata must already exist at the earliest stages of meiosis, though not observed. Alternatively, if crossing-over is thought to occur at the pachytene stage, as generally supposed, then a mechanism is required to associate the places of occurrence of reciprocal and non-reciprocal recombination in each meiosis. For it is evident from tetrad analyses in *Neurospora crassa*, such as those by Case and Giles[17], that about half the conversions were associated with crossing-over of 'marker' genes on either side of the pair of alleles. Yet the marker genes were so closely linked that they would have been expected to remain in their parental combinations in more than 90 per cent of instances if the two phenomena had been independent of one another. Earlier work[18,19] in which prototrophs at a nutritional locus were selected among random ascospores from crosses between alleles, had pointed to the same conclusion. Thus, if non-reciprocal and reciprocal recombination are thought to occur at different times, it is necessary to associate them, for example, by postulating that they occur in the same effective pairing segments.

In view of these difficulties and contrary evidence, and above all on account of the fundamental dilemma presented by the occurrence of DNA replication at the sub-chromatid level, the copy-choice theory must in my opinion be discarded as an explanation of genetic recombination between chromatids at meiosis.

Search for an Alternative Theory

In looking for an alternative to the copy-choice theory, one feature of conversion which appears highly significant is the association between it and crossing-over. Not only are the two phenomena often associated in position, as indicated here, but they also appear to be associated as regards the strands involved. Data from the analysis of tetrads in *Neurospora crassa* are given in Table 1. The early results obtained by Mitchell[2] and by Case and Giles[17] had indicated that, when reciprocal and non-reciprocal recombination occur in proximity, two strands might be participating more often than three, although the difference was scarcely significant. However, additional data obtained in this laboratory by Threlkeld[21] and by Cooke[22] make it appear certain that this excess is real. This conclusion is confirmed by the extensive data, not yet published, of Stadler and Towe[23]. Indeed, on the basis of the total data, three chromatids appear to participate so rarely (6 instances out of 42) as to suggest that a third strand may be involved only when the cross-over concerned is not in fact in the immediate vicinity of the non-reciprocal event.

The data of Kitani *et al.*[24] on the segregation of marker genes on either side of the locus for grey *versus* normal black spores in *Sordaria fimicola* point to the same conclusion. Out of 11 asci showing a 6 : 2 ratio of spore colours, and also showing recombination of marker genes on either side, the frequencies of those involving 2 and 3 strands were 9 and 2, respectively. Similarly, out of 52 asci showing 5 : 3 ratios and marker recombination, the 2- and 3-strand frequencies were 48 and 4, respectively. The markers spanned about 4 units of the linkage map.

In searching for an alternative to copy-choice as an explanation of non-reciprocal recombination, it was also evident that some replication of DNA must be postulated, in addition to the doubling which occurs just before meiosis. The occurrence within a tetrad of 3 : 1 ratios at a mutational site demands that such additional replication shall have taken place. Since there is clearly some connexion, from the evidence given here, between crossing-over and conversion, it would be reasonable to suggest that short segments of the DNA of a chromatid may undergo an extra replication at about the same time as crossing-over occurs. This would presumably be at the zygotene or pachytene stage. Was it possible that such further replication could occur in the process of crossing-over itself ?

Theory of Crossing-Over by Molecular Hybridization

It has been customary to regard crossing-over as a process of breakage of two chromatids at corresponding points, followed by rejoining the other way round. As applied to two DNA molecules, this would mean breakage of internucleotidic links in each pair of nucleotide chains

Table 1. THE FREQUENCIES WITH WHICH TWO AND THREE CHROMATIDS TAKE PART WHEN RECIPROCAL AND NON-RECIPROCAL RECOMBINATION OCCUR IN PROXIMITY IN *Neurospora crassa*

Author	No. of tetrads		Distance apart in map units of marker genes
	Two chromatids	Three chromatids	
Mitchell (ref. 2)	2	0	6
Case and Giles (ref. 17)	7	2	10
Suyama *et al.* (ref. 20)	1	0	17
Threlkeld (ref. 21)	6	1	15
Cooke (ref. 22)	2	0	17
Stadler and Towe (ref. 23)	18	3	15
Total	36	6	—

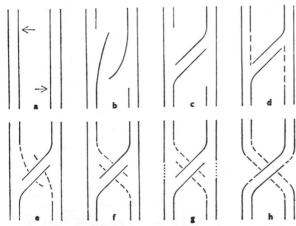

Fig. 2. Hypothetical stages in the process of crossing-over between two DNA molecules, according to the first model, in which the initial points of breakage (arrows) are non-homologous. Old nucleotide chains are shown by unbroken lines, newly synthesized chains by broken lines, and chains which are breaking down by dotted lines. No attempt has been made to show the spiral coiling

in each molecule, followed by re-establishment of the links with new partners. Such behaviour is out of keeping with the known peculiarities of the molecule, of which the most remarkable is the complementary base-pairing between the two nucleotide chains. Was it possible to explain crossing-over by means of this base pairing ?

If both chromatids were broken at precisely corresponding places, crossing-over by base-pairing would appear to be impossible (but see following). On the other hand, if the breaks were at not quite identical places, and invo.ved one nucleotide chain from one molecule and the complementary chain from the other, then the two longer broken ends would have terminal portions which overlapped and hence were complementary to one another. Base-pairing would thus be possible between them, giving rise to one of the cross-over DNA molecules. The reciprocal cross-over molecule could not, however, be formed in the same way, because the other two broken ends would necessarily both be deficient for the segment which the longer broken ends had in common. It was evident that to construct the reciprocal cross-over molecule some replication of DNA was required. Such replication was just what was needed to explain non-reciprocal recombination, and so the possibility was opened of explaining both kinds of recombination by a common mechanism.

On the hypothesis now proposed, broken ends of nucleotide chains of DNA do not join terminally, but instead all reconstruction is by lateral association of chains through complementary base-pairing. Two ways of bringing about crossing-over then appear possible.

First Model

The steps in crossing-over would be as follows:

(1) Complementary nucleotide chains, one from each DNA molecule, break at nearly but not exactly homologous points (Fig. 2a).

(2) The parts of the two broken chains between the two breakage points uncoil from their complementary chains (Fig. 2b), and then coil round each other so that complementary base-pairing occurs between their terminal portions to form a cross-over molecule (Fig. 2c).

(3) New nucleotide chains are synthesized alongside the old chains that have not crossed over (Fig. 2d).

(4) The new chains uncoil from their complements, peel back (Fig. 2e) and unite by complementary base-

pairing to form the other cross-over molecule (Fig. 2f).

(5) The old nucleotide chains that have not crossed over break down (Fig. 2g) and complementary nucleotides fill the gaps in the two cross-over molecules (Fig. 2h).

The detailed consequences of crossing-over, if it occurred by this means, are shown in Fig. 3, where the individual nucleotides are indicated by the letters of the alphabet, capitals (A, B, C, D . . .) for those contributed by one parent, and lower case (a, b, c, d . . .) for those from the other. (There are, of course, only 4 different nucleotides, but the whole alphabet is required in order to define a sufficient length of the unique sequence of them.) The situation in Fig. 3 (i) corresponds to that in Fig. 2e, where the first cross-over molecule has formed and the second one is about to do so. The end-products of the cross-over are shown in Fig. 3 (ii). Both cross-over molecules necessarily have a segment of the DNA in which one nucleotide chain is derived from each parent. The question of whether such DNA is heterozygous or homozygous is discussed later. The expression 'hybrid DNA' as used here is intended merely to imply that it is of dual parentage. The ends of the first-formed hybrid segment (G–T in the diagram) are determined by the positions of the initial breakages, while those of the second one (H–Q in the diagram) are determined by the points where the formation of the two new chains ceased. Thus, as in the example illustrated, the hybrid regions in the two molecules need not correspond and, moreover, might be expected to vary in length and relative position from one cross-over to another. An important consequence of the theory is evident, namely, that a cross-over does not occur between specific nucleotides of the two DNA molecules,

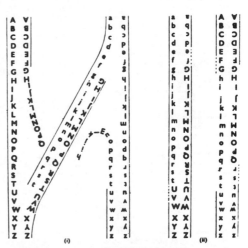

Fig. 3. Stages in the process of crossing-over, according to the first model. The unique sequence of nucleotides is indicated by the letters of the alphabet (although there are, of course, only four different nucleotides), and the parentage by inverting the letters. An unbroken line alongside the lettering indicates old nucleotide chains, the absence of a line implies a new chain synthesized during crossing-over, and broken lines show chains which might be either old or new. No attempt has been made to show the spiral coiling. (i) Stage of crossing-over shown in Fig. 2 (e); (ii) end-products of the cross-over

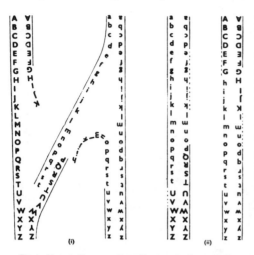

Fig. 4. Stages in the process of crossing-over, according to the first model, corresponding to those in Fig. 3, to show how loss of primary segment G–O and secondary segment L–Q will cause non-reciprocal recombination (conversion) at sites L–O

but instead occupies a significant length of the nucleotide chains, and therefore has a number of mutational sites lying within it.

The account of crossing-over just given does not explain non-reciprocal recombination, but the occurrence of additional replication of DNA gives to the mechanism the potentiality of doing so. A slight modification to the steps outlined here will suffice to bring it about. This alteration is illustrated in Fig. 4, where stages are shown corresponding to those in Fig. 3. The essential difference is that certain nucleotides are missing from the contribution to the cross-over by one parent. Such loss could no doubt readily occur as a result of breakage of one or other or both of the segments of the chain when necessarily unpaired prior to formation of the cross-over molecules. In Fig. 4 (i), as compared with Fig. 3 (i), losses are shown in both segments contributed by one parent, G–O from the primary chain and L–Q from the secondary one, giving a deficiency for nucleotides L–O. After formation of the cross-over molecules, the gaps caused by this deletion can only be filled by nucleotides complementary to those in the other chain, with the result that both cross-over chromatids will have this segment derived from the same parent (Fig. 4 (ii)). This implies that non-reciprocal recombination has taken place over this region. As can be seen from the diagram, the converted segment (l–o) lies between the two regions of hybrid DNA (P–T in the primary cross-over molecule and H–K in the secondary one). The direction in which conversion takes place will depend on which parental molecule suffers the deletion of a segment from the chain that is about to cross over.

Second Model

It was suggested earlier that if the initial breaks were in corresponding positions instead of being staggered, there would be no opportunity for complementary base-pairing, and hence for crossing-over. However, if, following such breakage, the nucleotide chains separated in one direction from the points of breakage, as in Fig. 5 (b or n), chain synthesis could occur alongside the unbroken chains (Fig. 5, c or o), and cross-over molecules could arise by base-pairing between the complementary old and new broken chains (Fig. 5, d and e, or p and q). Breakdown

of the parental previously unbroken chains would complete the process of crossing-over (Fig. 5, f and g, or r and s).

In the first model, one cross-over molecule was constructed from the two old chains and the other from the two new ones, and so their formation was likely to be sequential. In the present scheme, the formation of the cross-over molecules could be simultaneous, as each has one old and one new chain. As with the first model, loss of segments from the two free ends of the chains belonging to one parent, such as to leave this parent's contribution to the cross-over deficient for one or more nucleotides, will automatically lead to a corresponding converted segment in the middle of the cross-over.

On this second model, crossing-over could occur in the manner outlined, provided the nucleotide chains unwound in one direction only from the point of breakage, that is, either upwards (Figs. 5, b–g) or downwards (Fig. 5, n–s). If they separated in both directions, a remarkable situation would develop, illustrated in Fig. 5, h–m. Chain synthesis could occur alongside the unbroken chains (Fig. 5, i), but the new chains could not become attached at either end because of the old ones. If they separated from their unbroken complements, they could pair with the broken chains, provided they changed sides as in Fig. 5, j, k. The consequence of this, if the process followed a similar course to that outlined here for crossing-over, would be to produce hybrid DNA in both molecules, but no cross-over (Fig. 5 l, m).

The detailed effects of this process of nucleotide exchange are shown in Fig. 6, where diagram (i) corresponds to the stage reached in Fig. 5 k. The secondary chains have changed sides, and the pairing with their complementary broken primary chains is almost complete. In Fig. 6 (ii), the end-products of the process are shown, the molecules being hybrid from E to R and from I to V, respectively. The process that has taken place is effectively one of exchange of a segment between one nucleotide chain of each of two DNA molecules, but with the provision that the exchanged parts need not be exactly the same as regards their length and position in the molecules.

The effects will now be considered of loss of terminal segments from one or more of the four primary free ends in the configuration under discussion. If the lost part of a chain is long enough to give a deficiency in that parent's contribution to the process of exchange, conversion will result. In Fig. 6 (iii) it has been supposed that primary segment G–M of Fig. 6 (i) had been lost, thereby shortening this arm to a point beyond the end of the secondary chain (I–V) of the same parentage. In consequence, sites G and H show conversion to the other parent. In this diagram it has also been imagined that primary segment n–t has been lost, and since the secondary chain of this parentage ends at site r, conversion to the other parent will occur at sites s and t. It is evident that losses of the kind indicated, that is, from a primary chain and extending beyond the end of the secondary chain of the same parentage, will produce conversion without crossing-over. In general, they may cause conversion in either direction in either or both of two nearly adjacent regions. In any particular instance, the interval between these regions will be determined by the distance between the appropriate ends of the secondary chains (I to r in the example given).

Fig. 6 (iii) also illustrates an even more remarkable possibility. It is supposed that primary segment k–m has been lost. In conjunction with the loss of the homologous primary segment derived from the other parent (that is, K–M), this will give rise to reciprocal conversion, that is, a two-strand double cross-over in a short interval. Since this end-result merely requires loss of homologous segments, however short, from the primary chains, it can occur in the absence of conversion, which, as indicated

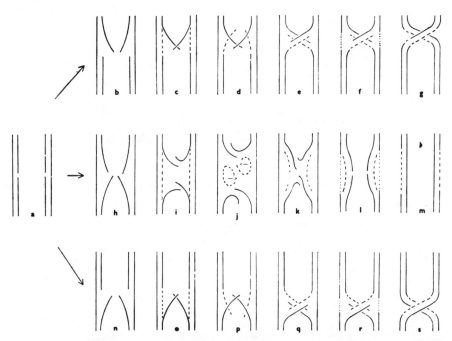

Fig. 5. Hypothetical stages in the processes of crossing-over and of nucleotide-chain exchange, according to the second model in which the initial points of breakage are homologous. Old nucleotide chains are shown by unbroken lines, newly synthesized chains by broken lines, and chains which are breaking down by dotted lines. No attempt has been made to show the spiral coiling. Diagram (*a*) shows the initial breakage; diagrams (*b*)–(*g*) show stages of crossing-over when the upper chains separate from their complementary chains; (*h*)–(*m*) the stages of nucleotide chain exchange when both upper and lower chains separate; and (*n*)–(*s*) the stages of crossing-over when only the lower ones separate

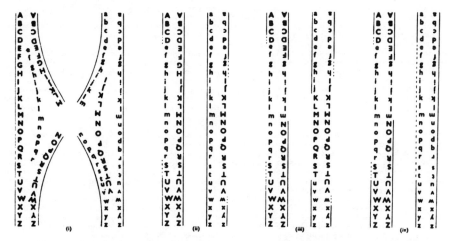

Fig. 6. Diagrams to illustrate the process of nucleotide chain exchange. The significance of the lettering and lines is the same as in Figs. 3 and 4. (i) Stage of nucleotide exchange shown in Fig. 5 (*k*); (ii) end-products of the exchange process; (iii) end-products if primary segments *G–M*, *k–m* and *n–t* have been lost; sites *G* and *H* show conversion in one direction, and *S* and *T* in the opposite direction; sites *K–M* show a reciprocal conversion, that is, they are spanned by a two-strand double cross-over; (iv) end-products if primary segment *G–M* and the entire secondary chain (*I–V*) have been lost; sites *G–M* show conversion

227

here, requires the lost segments to be long enough to extend beyond the end of the secondary segments.

Another variation which might be expected to occur, arising from this configuration with the four primary free ends, is the result of a failure of one of the secondary chains to be formed, or having formed, to pair with its complementary broken primary chain from the other molecule. Such failure would lead to hybrid DNA in one molecule, but to parental DNA in the other. Loss of a segment, however small, from either of the primary broken ends belonging to the parent which failed to produce a secondary chain will then cause conversion. This is illustrated in Fig. 6 (iv), where it has been supposed that one of the secondary chains (*I–V* in Fig. 6 (i)) failed to develop. It has further been supposed that primary segment *G–M* has been lost, with the result that these sites show conversion.

If the initial breakage-points were at homologous positions in similar instead of complementary nucleotide chains of the two DNA molecules, no exchange of material would be possible, unless segments from one molecule broke off and became directly incorporated, without replication, in the other. On the other hand, if the breakage-points were non-homologous, as in the first model of crossing-over, and if the chains unwound in both directions from the points of breakage, then conversion without crossing-over would be possible. It would be necessary for a secondary chain to form against one of the longer broken arms, and then for this chain to become detached and pair with the broken chain of the other molecule, and to span the gap between its two broken ends. This would give hybrid DNA in one molecule, and if the tip of one of the broken chains had been lost, conversion would occur. The reciprocal process could also take place, thereby giving rise to conversion in opposite directions in neighbouring but not quite adjacent regions. On the other hand, reciprocal conversion (a two-strand double cross-over) could be generated only by making the improbable assumption that both the longer arms lost segments of particular lengths while they were exchanging secondary chains.

To sum up, the first model (with non-homologous points of breakage) appears able to account for crossing-over with and without conversion, but in order to explain conversion without crossing-over, it is necessary to postulate that similar instead of complementary chains were broken in the two DNA molecules. On the other hand, the second model (with homologous breakage-points) will account for all three possibilities as the result of complementary strand breakage. This model also predicts the occurrence of two-strand double cross-overs in short intervals, which are unlikely on the first model.

Hybrid DNA

A fundamental feature of the theory of crossing-over that has been proposed here is the occurrence of hybrid DNA in the region of cross-overs or potential cross-overs. Heterozygous DNA would imply segregation of genetic differences at the first mitoses after meiosis, and such an occurrence, although known, is quite rare. It is presumed, therefore, that the hybrid DNA, although often potentially heterozygous, usually becomes homozygous in the process of formation of the hybrid molecules. This could be achieved by the removal from one chain at each heterozygous site or region of those nucleotides which did not conform to those in the other chain, followed by the insertion of nucleotides complementary to those in the first chain. At any particular site of heterozygosity it is to be expected that the two ways of making it homozygous (such as changing *Aa* to *AA* or to *aa*) will not necessarily occur with equal frequency. Indeed, the relative frequencies of the two alternatives might be expected to depend on the nature of the heterozygosity itself, that is, on whether the mutation is a nucleotide substitution or a

minute structural change such as a deficiency or a duplication.

This process of replacement of the nucleotide sequence peculiar to one parent by that of the other may lead to non-reciprocal recombination, in the same way as does loss of segments from the primary broken chains. Whether or not conversion is manifest will depend on the direction of the change from the heterozygous to the homozygous condition, and on what has happened at this site in the other molecule. If both molecules contain hybrid DNA, there are four possibilities: if nucleotide replacement occurs in the same direction in both molecules, conversion will also take place in that direction (to *A* or to *a*). If the replacements are in opposite directions, either the parental condition will be restored or a reciprocal conversion (two-strand double cross-over) will have been produced spanning the site.

There would thus appear to be two ways in which conversion can take place: first, by a failure of one parent to contribute to the formation of potentially heterozygous DNA, and secondly, by nucleotide replacement in such DNA so as to make it homozygous. The frequency of occurrence of conversion by the first method will depend on the frequency of loss of the appropriate segments, and by the second method on the frequencies of nucleotide substitution in the two directions. The relative importance of the two methods can be established only from experimental data. It may be that the nucleotide substitution, whereby the DNA is rendered homozygous, is the method by which conversion is usually brought about, even though the alternative, namely, loss of particular segments of the primary broken chains, would automatically cause conversion.

Discussion

Detailed examination of the data on aberrant tetrads in *Neurospora crassa*, *Aspergillus nidulans*, *Sordaria fimicola* and *Ascobolus immersus*, in the light of the present theory, has not yet been attempted. It should assist discrimination between the two models of crossing-over, and provide information about the behaviour of potentially heterozygous DNA. Thus, the observation by Kitani et al.[24] that, in *Sordaria* heterozygous at the grey spore site, asci with 6 : 2 ratios and others with 5 : 3 ratios for spore colour may occur in the same cross, is of particular interest and implies on the present theory that when the DNA is hybrid at this site, its heterozygosity is sometimes maintained through meiosis. It is evident that, on either model, the theory will explain the association, both in position and in strands, between conversion and crossing-over, and also the occurrence of either phenomenon in the absence of the other. It will also explain how two conversions can take place in proximity but in opposite directions, as first found by Mitchell[2] in *Neurospora*.

The observation that reciprocal and non-reciprocal recombination are affected differently by heat treatment[19] or ultra-violet light[25] is not unexpected on the present theory, since the two kinds of recombination, although related, are caused in different ways. On the postulated mechanism of crossing-over, although there is no net change in DNA content, inhibition of DNA synthesis at the appropriate time will prevent crossing-over. This suggests that the theory might be more readily tested by looking for suppression of crossing-over or chiasma formation, as a result of such inhibition, than by attempting to detect the DNA synthesis directly.

Explanation of the polarity of the *Ascobolus* conversion data of Lissouba et al.[5] would be facilitated if marker genes were available on either side of the sites of the spore pigment mutations. However, Hastings[26] has pointed out that the polarity could be explained by making the following two postulates:

(1) In any particular region of the chromosome, when crossing-over or potential crossing-over is about to occur,

either the process of homologous pairing of chromatids or the formation of hybrid DNA takes place progressively along the chromosome in a fixed direction, say, from right to left.

(2) With certain mutations, the formation of hybrid DNA often ceases immediately beyond (to the left) of their sites, presumably as a consequence of the heterozygosity, which is perhaps structural.

Conversion would thus occur less frequently at sites with heterozygosity to the right of them than when this region was homozygous, since the heterozygous site would shield a region to its left (which would often include the other site) by preventing hybridization of DNA, either directly or through inhibition of pairing of chromatids.

The second model described here, where two free ends give rise to crossing-over, and four to conversion without crossing-over, suggests an explanation for the phenomenon of interference in the position of neighbouring cross-overs. As is evident from Fig. 5, both the crossing-over and the conversion have a common starting-point in the breakage at corresponding positions of complementary nucleotide chains from the two DNA molecules. It follows from this that the initial points of breakage might be randomly distributed, but the cross-overs show interference, if the effect of a cross-over was to divert neighbouring potential cross-overs into the other course leading to conversion.

Hastings[24] has suggested that the coiling of complementary nucleotide chains round one another, which is an essential part of the process of constructing the crossover molecules, might itself lead to the uncoiling of the same chains at the next points of breakage and thereby favour diverting these cross-overs into the conversion path. If transmission of spiral torsion were the correct explanation of interference, the lack of it between crossovers in opposite arms of a chromosome would imply that the DNA was not continuous from one arm to the other, or at least that the spiral torsion was not transmitted (cf. Darlington[27]). Conversely, the occurrence of interference throughout a chromosome arm would suggest that the DNA was continuous, in agreement with Taylor's observations[12]. That interference should exist within a chromosome arm, even when the cross-overs involve different chromatids, would presumably be a consequence of the pairing behaviour of chromosomes, such that only two chromatids participate in crossing-over in any particular region.

It is of interest that the occurrence of crossing-over and conversion in proximity and involving three strands altogether (Table 1) appears to happen more often than would two cross-overs in the same interval, implying that there is less interference (or perhaps none) between the places of occurrence of the two phenomena (cf. Stadler[23]). Moreover, in *Aspergillus*, where interference is absent, Strickland[4] found conversion to be comparatively rare, and his data are consistent with the view that conversion takes place only in conjunction with crossing-over. On the theory of interference suggested above, this would mean that, in *Aspergillus*, coiling of nucleotide chains round each other at one point fails to lead to their uncoiling at a neighbouring point.

The second model for crossing-over also offers an explanation for the excess of two-strand double cross-overs which Lindegren and Lindegren[7] found between genes placed on either side of the centromere of a *Neurospora* chromosome. Their results have sometimes been attributed to overlapping of the spindles at the second division of meiosis and not to crossing-over. However, analysis of their data (Whitehouse[20]) has shown that the crossover phenomenon is real, and it has, moreover, been confirmed in this laboratory (Rifaat[30], Scott-Emuakpor[31]). A curious feature of these two-strand double cross-overs spanning the centromere is that they are frequently associated with an additional cross-over in the proximal

region of either or both chromosome arms. On the present hypothesis of interference, if crossing-over is rare near the centromere, as is usual, conversion and its associated variant, reciprocal conversion, would be correspondingly frequent. If two mutant sites on either side of the centromere both fell in regions of reciprocal conversion in the same chromatids, provided there was no intervening cross-over, the result would appear as a two-strand double cross-over between the mutants, together with further cross-overs on either side, in agreement with observation.

Summary and Conclusions

It is suggested that crossing-over occurs by basepairing between complementary nucleotide chains from two homologous DNA molecules. Two possibilities are discussed, depending on whether the initial breakage-points are homologous or not. Either model will account for non-reciprocal recombination (conversion), for its frequent association with crossing-over, and for such an apparent anomaly as double conversions in proximity but in opposite directions. An explanation for polarity in conversion frequencies is also offered. The second model (with homologous breakage) will explain, in addition, the phenomenon of interference between cross-overs, both as regards their position and the strands involved.

If the theory were accepted, the inference would be that the chromosome is single-stranded, in the sense that each gene is represented only once, and that the DNA is arranged longitudinally within it. The gulf that has existed hitherto between the study of heredity at the molecular and the chromosomal levels would then have been bridged.

I thank Mr. P. J. Hastings for invaluable discussion during the preparation of this paper, and for reading the manuscript. I also thank Mr. F. Cooke, Mr. M. B. Scott-Emuakpor and Dr. D. R. Stadler for permission to refer to their unpublished data, and Mr. G. J. Clark for assistance in the preparation of the diagrams.

[1] Lindegren, C. C., *J. Genet.*, **51**, 625 (1952).
[2] Mitchell, M. B., *Proc. U.S. Nat. Acad. Sci.*, **41**, 215, 935 (1955).
[3] Pritchard, R. H., *Heredity*, **9**, 343 (1955).
[4] Strickland, W. N., *Proc. Roy. Soc.*, B, **148**, 533 (1958).
[5] Lissouba, P., and Rizet, G., *C.R. Acad. Sci.*, **250**, 3408 (1960).
[6] Lissouba, P., Mousseau, J., Rizet, G., and Rossignol, J. L., *Adv. Genet.*, **11**, 343 (1962).
[7] Lindegren, C. C., and Lindegren, G., *Genetics*, **27**, 1 (1942).
[8] Stadler, D. R., *Genetics*, **41**, 623 (1956).
[9] Stadler, D. R., *Heredity*, **18**, 233 (1963).
[10] Watson, J. D., and Crick, F. H. C., *Nature*, **171**, 737, 964 (1953).
[11] Meselson, M., and Stahl, F. W., *Proc. U.S. Nat. Acad. Sci.*, **44**, 671 (1958).
[12] Sueoka, N., *Proc. U.S. Nat. Acad. Sci.*, **46**, 83 (1960).
[13] Taylor, J. H., *Proc. Tenth Intern. Cong. Genet.*, 1, 63 (Univ. Toronto Press, 1958).
[14] Meselson, M., and Weigle, J., *Proc. U.S. Nat. Acad. Sci.*, **47**, 857 (1961).
[15] Kellenberger, G., Zichichi, M. L., and Weigle, J., *Proc. U.S. Nat. Acad. Sci.*, **47**, 869 (1961).
[16] Taylor, J. H., *Amer. Nat.*, **91**, 209 (1957).
[17] Case, M. E., and Giles, N. H., *Proc. U.S. Nat. Acad. Sci.*, **44**, 378 (1958); *Cold Spring Harb. Symp. Quant. Biol.*, **23**, 119 (1959).
[18] St. Lawrence, P., *Proc. U.S. Nat. Acad. Sci.*, **42**, 189 (1956). Freese, E., *Genetics*, **45**, 671 (1957).
[19] Mitchell, H. K., in *The Chemical Basis of Heredity*, edit. by McElroy, W. D., and Glass, B., 94 (Johns Hopkins Press, Baltimore, 1957).
[20] Suyama, Y., Munkres, K. D., and Woodward, V. W., *Genetics*, **36**, 293 (1959).
[21] Threlkeld, S. F. H., Ph.D. thesis, Univ. Cambridge (1961).
[22] Cooke, F. (personal communication).
[23] Stadler, D. R., and Towe, A. M., *Genetics* (in the press).
[24] Kitani, Y., Olive, L. S., and El Ani, A. S., *Amer. J. Bot.*, **49**, 697 (1962).
[25] Roman, H., and Jacob, F., *Cold Spring Harb. Symp. Quant. Biol.*, **23**, 155 (1959).
[26] Hastings, P. J. (personal communication).
[27] Darlington, C. D., *Recent Advances in Cytology*, second ed. (London, 1937).
[28] Stadler, D. R., *Proc. U.S. Nat. Acad. Sci.*, **45**, 1625 (1959).
[29] Whitehouse, H. L. K., *New Phytol.*, **41**, 23 (1942); and Ph.D. thesis, Univ. Cambridge (1947).
[30] Rifaat, O. M., Ph.D. thesis, Univ. Cambridge (1956). Cf. Whitehouse, H. L. K., *Proc. Tenth Intern. Cong. Genet.*, **2**, 312 (Univ. Toronto Press, 1958).
[31] Scott-Emuakpor, M. B. (personal communication).

Part V

CHROMOSOME FUNCTION

HISTORICAL PERSPECTIVES

Sex Chromosomes

Sex determination was one of the first areas in which cyto-geneticists studied chromosome function. In nature usually the fertilized eggs of a particular species develop into about equal numbers of male and female individuals. Counts of chromosome numbers in certain Orthoptera and Hemiptera showed that in males there was an odd or accessory chromosome. McClung (1902) was the first to suggest the association with sex determination. Wilson (Paper 27) discussed the relationship between the accessory chromosome and sex determination. The first evidence as to the mechanism involved was from breeding experiments with the currant moth (*Abraxas*) in which a pair of genetic factors was linked closely with sex (Doncaster and Raynor 1906).

In all the preceding cytological studies, the male was the heterogametic sex. The genetic linkage with sex in the currant moth indicated that the female was heterogametic. It was not until 1913 that Seiler's (1913) cytological observations in Lepidoptera showed that the female could be the sex with a heteromorphic pair of chromosomes associated with sex determination.

The first report in plants of a correlation between a chromosome difference and sex determination was in the Liverwort, *Sphaerocarpos Donnellii* (Allen, Paper 28). This report was followed by many such studies in the Bryophytes (as stated by Cleland 1956, over 100 papers in less than 30 years). In 1923 there were four published papers that dealt with sex chromosomes in Angiosperms, Santos for *Elodea*, Kihara and Ono for *Rumex*, Blackburn for *Melandrium*, Winge for *Melandrium, Humulus*, and *Vallisneria*.

The persistence of heterochromatic regions in somatic cell interphases as *prochromosomes* may facilitate sex identification, similar to the Barr body story in mammals. In one insect species, the spruce budworm (*Archips fumiferana*), only the Y-chromosome is heterochromatic, making it possible to distinguish male and female by cytological examination of somatic cell nuclei (Smith 1945).

Goldschmidt conducted a long series of experiments using crosses within and between different races of *Lymantria*, the Gypsy moth, from Japan and Europe. Basically, there was a segregation of 1 male:1 female within a particular race; that is, the mechanism was Mendelian and followed the segregation expected for one pair of chromosomes. But when crosses were made between races, one of the classes might have a mixture of the characters of both sexes (termed *intersexual*). Often in reciprocal crosses, one cross would have the normal ratio of 1 male:1 female, but in the reciprocal cross the class that cytologically should have been male actually was intersexual. The degree of intersexuality varied depending on which races were crossed. Goldschmidt was able to classify the races into different strengths, from strong to weak, and from this could predict the degree and type of intersexuality. His interpretation was that intersexuality was a genetic characteristic that affects the stage in development at which the secondary sex characters begin to express themsleves. What he termed *strong* followed the distribution through the cytoplasm or the Y-chromosome (Goldschmidt 1921). This led to the suggestion that a certain threshold in physiological development was necessary for a particular character to express itself. In Goldschmidt's studies, the chromosome numbers were constant.

Chromosome Balance: Effects of Changes in Chromosome Dosage

The early genetic studies dealt with characters that showed the typical Mendelian behavior in reciprocal crosses and subsequent generations. In *Datura* Blakeslee (1910) found a mutant type that was transmitted almost entirely through the female parent. Subsequent studies, as reported by Blakeslee, Belling, and Farnham (Paper 29) showed that these mutants had an additional chromosome and that the extra one was different in different mutants. They concluded that the phenotypic changes resulted from the new chromosome balance brought about by the extra chromosome. Since each chromosome carries a different set of genes, each of the different primary trisomics had an extra dose of different genes, which resulted in a phenotype distinguishable from the normal diploid and from each other. When the extra chromosome was an interchange chromosome (a tertiary trisomic), the plant showed some of the characters seen in each of the two respective primary trisomic types. This concept has had far-reaching impacts such as in studies of sex determination (discussed below), in genetic and breeding studies of polyploid species, and in direct application to studies of abnormal syndromes in man.

That a specific chromosome, when added to the normal chromosome complement in man, produces the abnormal Down's syndrome (Mongolism) was first reported by Lejeune and co-workers in 1959 (Paper 30). This report led to a rapid expansion of cytological studies in man and other animal species that are still in progress.

Chromosome Balance: Sex Determination

Among the offspring of triploid *Drosophila*, Bridges (1922 and Paper 31) found a wide variation in expression of male and female secondary sex characters ranging from superfemale through normal female, intersex, normal male, to supermale. There was also a variation in the number of X-chromosomes and the number of sets of autosomes. Flies with 3 sets of X-chromosomes and 2 of autosomes (3X:2A) were superfemales, ones with one X-chromosome and 3 sets of autosomes (1X:3A) were supermale. A normal female had a ratio of 1X:1A (e.g., 2X:2A, 3X:3A). A normal male had a ratio of 1X:2A. Bridges concluded that the net effect of the X was female determining: the net effect of the autosomes was male. Hence sex expression was a matter of chromosome balance.

Later studies using flies that were aneuploid for fragments of the X demonstrated that femaleness was not determined by a single factor. Each aneuploid caused a shift in the female direction (Pipkin 1940).

In the honeybee and in certain other Hymenoptera, males in the wild were haploid. This was contrary to the finding in *Drosophila* that haploid areas in mosaic flies were female. Whiting (1933) and Bostian (1934) found that inbreeding could produce diploid males. The explanation is that there is a series of alleles for sex determination; females are heterozygotes, haploids are males, but homozygotes are also males.

233

Expression of Genes

1. Lyon hypothesis: Various modes of regulating gene activity have evolved among species with sex chromosomes. A very interesting one may apply to the sex chromosomes of many or all mammals. Barr and Bertram (1949) reported a cytologically visible difference between neurone cells from male versus female cats. They observed that a dark-staining body commonly present in the female cells was absent in male cells. This body has been termed the *Barr body, sex-chromatin body*, and more recently the *X-body* since the Y-chromosome can now be visualized in interphase cells by fluorescence staining and is called the *Y-body*. X-bodies are common in mammals and are rare in most other organisms with a few exceptions.

Russell and Bangham (1959, 1960), during a series of mutation-rate studies in the mouse, discovered variegated-type position effects that were expressed in XX females but not in the XO females or in XY males. The fact that XO females developed normally indicated that the second X was not necessary for the expression of variegation (Welshons and Russell 1959). Russell (1961) explained the V-type position effect observed in X-autosome translocation heterozygotes as being a heterochromatic response when more than one X-chromosome was present. The additional observation that several sex-linked mutants for coat color (including one reported by Lyon 1960) when heterozygous, showed a mosaic phenotype led Lyon (Paper 33) to propose the hypothesis that females heterozygous for sex-linked genes would be mosaics because one of the X chromosomes at random, maternal or paternal, becomes inactivated early in development and remains inactive throughout development. It is the inactive X that is heterochromatic at interphase, forming the X-body. This hypothesis has been very useful in the interpretation of gene expression in mammals, including man.

2. Puffs: Beermann (Paper 35) and Breuer and Pavan (1955) studying *Chironomus* and *Rhynchosciara*, respectively, provided much of the basic information needed to make scientists cognitive of the value of Dipteran polytene chromosomes for gene activity studies. One of their useful findings was that the bands appear in the same relative positions in various tissues and provide a constant banding pattern. One difference observed in homologous chromosomes between different tissues of the same individual was in the appearance of *puffs*, swollen areas of a diffuse nature arising from a single band region. The process of forming puffs has been studied extensively and is known to be characteristic of a specific stage of development.

3. Lampbrush Loops: Lampbrush-chromosome loops are related to polytene chromosome puffs in that they represent chromosome parts that are uncontracted and actively involved in RNA synthesis. Most loops synthesize RNA along their entire axis. However, Gall and Callan (1961) reported that in one particular loop of *Triturus*, RNA synthesis occurred only at the thin-insertion end of the loop. The RNA, once synthesized, was found on succeeding days at progressively further points from that end of the loop, which was constant in size during the experiment. Their interpretation was that the chromatin of the lampbrush loop was spun-out from the chromomere at the thin-insertion end, transcribed RNA, and moved like a conveyor belt until it again became part of the chromomere at the thick-insertion end. Miller and his co-workers (1970) observed the normal lampbrush loop in the process of transcription with the electron microscope. RNA synthesis appeared to be occurring over the entire loop and the ribonucleoprotein (RNP) fibrils were very closely spaced, indicating a high frequency of initiation of transcription. The RNP fibrils were extremely long, suggesting that a large part of the loop was involved in synthesizing a single RNA molecule.

4. Nucleolus Organizers: The nucleolus organizer is now known to be the chromosomal site of the genes that synthesize the precursor for 18S and 28S ribosomal RNA. McClintock (Paper 36) brilliantly analyzed various types of cytogenetic stocks to reach the conclusion that there is a specific chromosomal element in maize responsible for nucleolar formation. She viewed the element as an organizer of nucleolar material synthesized elsewhere. We now know that the maize nucleolus organizer is the site of rDNA (Phillips, Kleese, and Wang 1971 and Phillips *et al*. 1974) and thus is a synthesizer (of precursor rRNA). The region is probably a synthesizer and an organizer, perhaps an organizer of other nucleolar components such as 5S rRNA and proteins. McClintock's observations and perceptive interpretations provided the basis for nucleolus organizer studies in other species as well.

5. Position Effects: The fact that the position of a gene could affect the phenotype was discovered first in *Drosophila* by Sturtevant (1925) in studies of mutations at the Bar (B) locus. The expressions of the bar character for the *BB*/normal versus *B/B* arrangement were not identical. Many other genes in *Drosophila* have shown what is termed *variegated position effect* (Schultz, Paper 32). The genes are suppressed when brought into close juxtaposition with heterochromatin which itself has been broken, for exam-

ple, in a translocation where one break is in the centromeric heterochromatin and the other in the euchromatin close to the gene in
question.

The first reported case of a position effect in higher plants is in
Oenothera for sepal color (Catcheside 1939). Control of gene action mediated by shifts of genetic elements such as in the maize
mutable gene systems first described by McClintock (1950, reports
of early observations are given in the 1942–49 Yearbooks of the
Carnegie Institution of Washington) may be another example of
position effect in higher plants. The original source of controlling
elements in maize was a stock undergoing the chromosome type
of breakage-fusion-bridge cycle. This cycle resulted in numerous
structural modifications of various chromosomes of the genome
and gave rise to several instances of mutable loci. One of the first
systems described involved a locus in chromosome 9 at which
there were chromosome breakage events that were correlated
with the occurrence of the mutations. The locus was called *Ds* for
"dissociator," and for it to cause breakage events another element
was required called *Ac* for "activator." Both elements were capable of moving around in the genome resulting in new linkage relationships and sometimes causing other genes to become mutable.
Another case in maize of altered gene expression is that of para-
mutation (Brink 1956). Brink found that certain alleles may be
changed in their subsequent phenotypic expression due to their
having been associated with a particular allele in heterozygotes.
No transposable elements, however, have been discovered in the
paramutagenic system. The altered expression of the paramutable
allele continues in subsequent generations, but it is a reversible
process. Although considerable information of a descriptive nature is now available, the molecular mechanism is not understood.

Visualization of Gene Action

The possibility of observing a gene, particularly one in the
process of synthesizing its product, was far beyond cytogeneticists' expectations for many years. Miller and Beatty (Paper 34),
however, creatively utilized an excellent system and were able to
visualize genes in action. The amphibian oocyte amplifies its level
of rDNA and forms about a thousand nucleoli that line the periphery of the nucleus. Miller and Beatty developed a method to recover the DNA from the extrachromosomal nucleoli of the amphibian
oocyte and prepare it properly for electron microscopy. Many aspects of ribosomal gene action that had been elucidated previous-

ly by biochemical means were visualized. They observed (1) a repeated arrangement of rRNA genes, (2) a high level of simultaneous undirectional RNA transcription, and (3) the previously unsuspected existence of intergene segments.

Editors' Comments
on Papers 27 Through 37

Wilson (Paper 27) pointed out that the constancy of a difference in the chromosomes of the two sexes indicated a connection with the determination of sex. He showed that one chromosome had no mate in one sex of one group of species and that one chromosome had a smaller chromosome as a mate in another group. This sex produced in each case two kinds of gametes. The other sex had no such chromosome difference and therefore produced only one kind of gamete.

Paper 28 by Allen was the first report of a correlation between a chromosome difference and sex determination in a plant, *Sphaerocarpos*, a Bryophyte. This, along with his genetic studies of this species, stimulated many such studies in Bryophytes and in dioecious higher plants.

Blakeslee and co-workers (Paper 29) report in *Datura* ($n = 12$) the occurrence of 12 different morphological mutants, each with $2n + 1 = 25$ chromosomes, the extra chromosome being a duplication of one of the basic 12. This was the first report that distinctive phenotypes may be the result of changes in gene balance caused by the presence of an extra chromosome. This was also the first report of breeding tests in plants using trisomic ratios to identify a particular Mendelian character with a particular chromosome.

Paper 30 had a tremendous impact on human genetics and firmly initiated a new era in cytogenetics—a period when cytogenetics would be applied as a clinical tool to diagnose several different human diseases or syndromes. This is continuing with much enthusiastic activity. The authors reported the presence of 47 chromosomes in the cultured fibroblast cells of 5 male and 4 female Down's Syndrome (Mongolian) infants, the extra chromosome being one of the small acrocentrics. This was the first demonstration that living human beings, albeit ones with mental retardation and heart and other malformations, might possess an extra chromosome. The idea that an extra chromosome could result in an abnormal phenotype due to gene balance effects and thus be the cause of the syndrome was acceptable to the scientific community because of related findings in other species, notably *Datura* and *Drosophila*, first reported more than 40 years before.

By studying the progeny of triploid females in *Drosophila*, Bridges (Paper 31) showed that sex expression depended on the balance between the number of X-chromosomes and the number of sets of autosomes. This marked a significant advance in our knowledge about sex determination and again the importance of gene balance in relation to phenotypic expression.

Schultz (Paper 32) reports the results of his experiments with 13 different variegations in *Drosophila melanogaster*, all of which were associated with chromosomal rearrangements. By characterizing these both genetically and cytologically, he discovered an apparent relation between the nature of the rearrangements and the production of the variegation and proposed a scheme of chromosome replication as a possible explanation. This phenomenon is known now as "variegated position effect."

Paper 33 was the first to hypothesize that one member of a chromosome pair may become inactive at some stage during development. This theory was advanced to explain the mosaic phenotypes of mice that were heterozygous for sex-linked coat-color mutants. The hypothesis was an outgrowth of phenotypic observations on coat color combined with previously reported cytological observations on the inactivation of the X-chromosome. The hypothesis has had far-reaching effects toward understanding the mechanism of dosage compensation in mammalian species, especially man. Lyon makes specific reference to man as well as to other species including the tortoise-shell cat.

Paper 34 is a classic not only because it furnished biologists with their first opportunity to actually see genes producing their product, in this case ribosomal RNA precursor molecules, but also because it united several areas of genetics, most notably molecular genetics and cytogenetics. The cytological evidence confirmed many conclusions reached previously by molecular techniques with regard to the transcription process. It also revealed an unexpected feature of rDNA—the presence of spacer regions between the structural genes. Observations similar to those reported here have since been made in several other species, including man and bacteria, and for genes transcribing messenger RNA as well as ribosomal RNA.

Puffs are transient, reversible, localized, diffuse swellings in the giant polytene chromosomes of Diptera. Beermann (Paper 35) demonstrated that a group of bands in homologous polytene chromosomes may occur also as a puff, depending on cell type and developmental stage. The translated excerpts from his 1952 paper illustrate his analysis of the constancy, structure, and function of puffs. Breuer and Pavan simultaneously and independently discovered the same relationships in *Rhynchosciara* (see Breuer and Pavan 1955). Puffs represent areas of the chromosome exhibiting differential gene activity thereby allowing cytological analysis of gene expression. The evidence bears on the relationship between differentiation and gene activity.

As in the case of puff formation in Dipteran polytene chromosomes, nucleoli represent cytological manifestations of chromosomal activity and are important in the developmental process. The excerpts from McClintock's paper (Paper 36) include observations that led to a hypothesis on nucleolar formation. She demonstrated that the nucleolus is formed as the result of the activity of a specific cytologically distinguishable genetic element in the chromosome. She hypothesized that "the function of the nucleolar-organizing element of chromosome 6 apparently is to organize the nucleolar substance present in each chromosome into a definite body, the nucleolus." The information in this paper has served as the basis for understanding the genetic control of nucleolus formation in many organisms.

Sometimes the importance of novel phenomena is not realized for many years after the initial report. Paper 37, describing what was later termed "neocentric activity" caused by abnormal maize chromosome 10, may well be of this nature. Several races of maize from Latin America and southwestern United States have a chromosome 10 that differs from normal chromosome 10 in having a different chromosome pattern in a portion of the long arm and the addition of a large heterochromatic segment. The future may show that the induction of neocentric activity will be useful in understanding what is responsible for forming centromeric substance(s), the molecular nature of the substance(s), their interaction with chromatin and microtubules, and the movement of the substance(s) along the chromosome.

REFERENCES

Barr, M. L., and E. G. Bertram. 1949. A morphological distinction between neurones of the male and female, and the behaviour of the nucleolar satellite during accelerated nucleoprotein synthesis. *Nature* **163**:676–677.

Blackburn, K. B. 1923. Sex chromosomes in plants. *Nature* **112**:687–688.

Blakeslee, A. F. 1910. The botanic garden as a field museum of agriculture. *Science* **31**:685–688.

Bostian, C. H. 1934. Biparental males and biparental ratios in *Habrobracon. Biol. Bull.* **66**:166–181.

Breuer, M. E., and C. Pavan. 1955. Behavior of polytene chromosomes of *Rhynchosciara angelae* at different stages of larval development. *Chromosoma* **7**:371–386.

Bridges, C. B. 1922. The origin of variations in sexual and sex limited characters. *Am. Nat.* **56**:51–63.

Brink, R. A. 1956. A genetic change associated with the R locus in maize which is directed and potentially reversible. *Genetics* **41**:872–889.

Catcheside, D. G. 1939. A position effect in *Oenothera. J. Genet.* **38**:345–352.

Cleland, R. E. 1956. Analysis of trends in biological literature—Plant Sciences. *Biol. Abstr.* **30**:2459–2462.

Doncaster, L., and G. H. Raynor. 1906. On breeding experiments with Lepidoptera. *Proc. Zool. Soc., London* **1**:125–133.

Gall, J. G., and H. G. Callan. 1961. Patterns of synthesis in lampbrush chromosome loops (Abstract). *Genetics* **46**:867.

Goldschmidt, R. 1921. The determination of sex. Nature **107**:780–784.

Kihara, H., and T. Ono. 1923. Cytological studies on *Rumex* L. I. Chromosomes of *Rumex acetosa* L. *Bot. Mag., Tokyo* **37**:84–90 (Japanese) (Eng. summary 35–36).

Lyon, M. F. 1960. A further mutation of the mottled type in the house mouse. *J. Hered.* **51**:116–121.

McClintock, B. 1950. The origin and behavior of mutable loci in maize. *Proc. Nat. Acad. Sci. U.S.A.* **36**:344–355.

McClung, C. E. 1902. The accessory chromosome—sex determinant? *Biol. Bull.*, 3:43–84.

Miller, O. L., Jr., B. R. Beatty, B. A. Hamkalo, and C. A. Thomas, Jr. 1970. Electron microscopic visualization of transcription. *Cold Spring Harb. Symp. Quant. Biol.* **35**:505–512.

Phillips, R. L., R. A. Kleese, and S. S. Wang. 1971. The nucleolus organizer region of maize (*Zea mays* L.): Chromosomal site of DNA complementary to ribosomal RNA. *Chromosoma* **36**:79–88.

Phillips, R. L., D. F. Weber, R. A. Kleese, and S. S. Wang. 1974. The nucleolus organizer region of maize (*Zea mays* L.): Tests for ribosomal gene compensation or magnification. *Genetics* **77**:285–297.

Pipkin, S. B. 1940. Multiple sex genes in the X-chromosome of *Drosophila melanogaster*. *Univ. Texas Publ.* *4032*:126–156.

Russell, L. B. 1961. Genetics of mammalian sex chromosomes. *Science* **133**:1795–1803.

——— and J. W. Bangham. 1959. Variegated-type position effects in the mouse (Abstract). *Genetics* **44**:532.

———. 1960. Further analysis of variegated-type position effects from X-autosome translocations in the mouse (Abstract). *Genetics* **45**:1008–1009.

Santos, J. K. 1923. Differentiation among chromosomes in *Elodea*. *Bot. Gaz.* **75**:42–59.

Seiler, J. 1913. Das Verhalten der Geschelchtschromosomen bei Lepidopteren. *Zool. Anz.* **41**:246–251.

Smith, S. G. 1945. Heteropycnosis as a means of diagnosing sex. *J. Hered.* **36**:195–196.

Sturtevant, A. H. 1925. The effects of unequal crossing over at the Bar locus in *Drosophila*. *Genetics* **10**:117–147.

Whiting, P. W. 1933. Selective fertilization and sex-determination in *Hymenoptera*. *Science* **78**:537–538.

Welshons, W. J., and L. B. Russell. 1959. The Y-chromosome as the bearer of male determining factors in the mouse. *Proc. Nat. Acad. Sci. U.S.A.* **45**:560–566.

Winge, Ø. 1923. On sex chromosomes, sex-determination, and preponderance of females in some dioecious plants. *C. R. Trav. Lab. Carlsberg* **15**(5):1–26.

27

Reprinted from *Science* **22**(564):500–502 (1905)

THE CHROMOSOMES IN RELATION TO THE DETERMINATION OF SEX IN INSECTS

Edmund B. Wilson

Zoological Laboratory, Columbia University

MATERIAL procured during the past summer demonstrates with great clearness that the sexes of Hemiptera show constant and characteristic differences in the chromosome groups, which are of such a nature as to leave no doubt that a definite connection of some kind between the chromosomes and the determination of sex exists in these animals. These differences are of two types. In one of these, the cells of the female possess one more chromosome than those of the male; in the other, both sexes possess the same number of chromosomes, but one of the chromosomes in the male is much smaller than the corresponding one in the female (which is in agreement with the observations of Stevens on the beetle *Tenebrio*). These types may conveniently be designated as *A* and *B*, respectively. The essential facts have been determined in three genera of each type, namely, (type *A*) *Protenor belfragei, Anasa tristis* and *Alydus pilosulus,* and (type *B*) *Lygæus turcicus, Euschistus fissilis* and *Cænus delius.* The chromosome groups have been examined in the dividing oogonia and ovarian follicle cells of the female and in the dividing spermatogonia and investing cells of the testis in case of the male.

Type *A* includes those forms in which (as has been known since Henking's paper of 1890 on *Pyrrochoris*) the spermatozoa are of two classes, one of which contains one more chromosome (the so-called 'accessory' or heterotropic chromosome) than the other. In this type the somatic number of chromosomes in the female is an even one, while the somatic number in the male is one less (hence an odd number) the actual numbers being in *Protenor* and *Alydus* ♀ 14, ♂ 13, and in *Anasa* ♀ 22, ♂ 21. A study of the chromosome groups in the two sexes brings out the following additional facts. In the cells of the female all the chromosomes may be arranged two by two to form pairs, each consisting of two chromosomes of equal size, as is most obvious in the beautiful chromosome groups of *Protenor,* where the size differences of the chromosomes are very marked. In the male all the chromosomes may be thus symmetrically paired with the exception of one which is without a mate. This chromosome is the 'accessory' or heterotropic one; and it is a consequence of its unpaired character that it passes into only one half of the spermatozoa.

In type *B* all of the spermatozoa contain the same number of chromosomes (half the somatic number in both sexes), but they are, nevertheless, of two classes, one of which contains a large and one a small 'idiochromosome.' Both sexes have the same somatic number of chromosomes (fourteen in the three examples mentioned above), but differ as follows: In the cells of the female (oogonia and follicle-cells) all of the chromosomes may, as in type *A,* be arranged two by two in equal pairs, and a small idiochromosome is not present. In the cells of the male all but two may be thus equally paired. These two are the unequal idiochromosomes, and during the maturation process they are so distributed that the small one passes into one half of the spermatozoa, the large one into the other half.

These facts admit, I believe, of but one interpretation. Since all of the chromosomes in the female (oogonia) may be symmetrically paired, there can be no doubt that synapsis in this sex gives rise to the reduced number of symmetrical bivalents, and that consequently all of the eggs receive the same number of chromosomes. This number (eleven in *Anasa,* seven in *Protenor* or *Alydus*) is the same as that present in those spermatozoa that contain the 'accessory' chromosome. It is evident that both forms of spermatozoa are functional, and that in type *A* females are produced from eggs fertilized by spermatozoa that contain the 'accessory' chromosome, while males are produced from eggs fertilized by spermatozoa that lack this chromosome (the reverse of the conjecture made by McClung). Thus if n be the somatic number in the female $n/2$ is the number in all of the matured eggs, $n/2$ the number in one half of the spermatozoa (namely, those that contain the 'accessory'), and $n/2 - 1$ the number in the other half. Accordingly:

In fertilization

$$\text{Egg } \frac{n}{2} + \text{spermatozoon } \frac{n}{2} = n \text{ (female)}.$$

$$\text{Egg } \frac{n}{2} + \text{spermatozoon } \frac{n}{2} - 1 = n - 1 \text{ (male)}.$$

The validity of this interpretation is completely established by the case of *Protenor,* where, as was first shown by Montgomery, the 'accessory' is at every period unmistakably recognizable by its great size. The spermatogonial divisions invariably show but one such large chromosome, while an equal pair of exactly similar chromosomes appear in the oogonial divisions. One of these in the female must have been derived in fertilization from the egg-nucleus, the other (obviously the 'accessory') from the sperm-nucleus. It is evident, therefore, that all of the matured eggs must before fertilization contain a chromosome that is the maternal mate of the 'accessory' of the male, and that females are produced from eggs fertilized by spermatozoa that contain a similar group (*i. e.,* those containing the 'accessory'). The presence of but one large chromosome (the 'accessory') in the somatic nuclei of the male can only mean that males arise from eggs fertilized by spermatozoa that lack such a chromosome, and that the single 'accessory' of the male is derived in fertilization from the egg nucleus. In type *B* all of the eggs must contain a chromosome corresponding to the large idio-

chromosome of the male. Upon fertilization by a spermatozoon containing the large idiochromosome a female is produced, while fertilization by a spermatozoon containing the small one produces a male.

The two types distinguished above may readily be reduced to one; for if the small idiochromosome of type *B* be supposed to disappear, the phenomena become identical with those in type *A*. There can be little doubt that such has been the actual origin of the latter type, and that the 'accessory' chromosome was originally a large idiochromosome, its smaller mate having vanished. The unpaired character of the 'accessory' chromosome thus finds a complete explanation, and its behavior loses its apparently anomalous character.

The foregoing facts irresistibly lead to the conclusion that a causal connection of some kind exists between the chromosomes and the determination of sex; and at first thought they naturally suggest the conclusion that the idiochromosomes and heterotropic chromosomes are actually sex determinants, as was conjectured by McClung in case of the 'accessory' chromosome. Analysis will show, however, that great, if not insuperable, difficulties are encountered by any form of the assumption that these chromosomes are specifically male or female sex determinants. It is more probable, for reasons that will be set forth hereafter, that the difference between eggs and spermatozoa is primarily due to differences of degree or intensity, rather than of kind, in the activity of the chromosome groups in the two sexes; and we may here find a clue to a general theory of sex determination that will accord with the facts observed in hemiptera. A significant fact that bears on this question is that in both types the two sexes differ in respect to the behavior of the idiochromosomes or 'accessory' chromosomes during the synaptic and growth periods, these chromosomes assuming in the male the form of condensed chromosome nucleoli, while in the female they remain, like the other chromosomes, in a diffused condition. This indicates that during these periods these chromosomes play a more active part in the metabolism of the cell in the female than in the male. The primary factor in the differentiation of the germ cells may, therefore, be a matter of metabolism, perhaps one of growth.

28

Reprinted from *Science* **46**(1193):466–467 (1917)

A CHROMOSOME DIFFERENCE CORRELATED WITH
SEX DIFFERENCES IN SPHAEROCARPOS

Charles E. Allen

University of Wisconsin

THE chromosome group found in the cells of the female gametophyte of *Sphœrocarpos Donnellii* contains one large element which considerably exceeds both in length and in thickness any of the older chromosomes. The chromosome group of the male gametophyte contains no element similarly distinguished by its size; on the other hand, the male possesses a very small chromosome which seems not to correspond in size to any element in the female.

The other chromosomes in the cells of either sex have the form of slender rods; there are noticeable differences in length between those of each group. The bending and not infrequent overlapping of the ends of the chromosomes place difficulties in the way of an exact determination of their number; but, subject to modification by further study, it may be said with reasonable assurance that the chromosome number for each sex is eight. As to seven of the eight, the chromosomes of the male seem to resemble those of the female; but the eighth chromosome of the female is probably corresponding to it in the male is the the large one already referred to, and the one very small chromosome.

Of the two spindles formed in each spore mother cell at the time of the homœotypic division, one shows a large body which is sometimes plainly two-parted; no element appears on the other spindle that approximates in size this large chromosome. It has been reported that in at least one species of *Sphœrocarpos* two of the spores of each tetrad develop into male plants and the other two into females. Observations which I have made, although as yet in limited number, indicate that the same rule holds for *S. Donnellii.* The cytological results here reported seem to show that in consequence of the chromosome distribution in the reduction divisions two of the four spores derived from a single mother cell receive each a large chromosome (and seven of smaller size), and these spores develop into female plants; and that each of the other two spores receives a small chromosome instead of the large one, and, on germination, gives rise to a male plant.

The resemblance between this history and that of the chromosomes of certain insects, such as *Lygœus* and *Euschistus,* which possess a large X- and a small Y-chromosome, is obvious. It is too early to conclude that the particular chromosomes with respect to which the male and female gametophytes of *Sphœro-carpos* differ are the bearers of definite sex-determining factors; but it seems not unlikely at least that the greater size and vigor of growth of the female gametophyte are associated with the greater amount of chromatin that its cells contain.

29

Reprinted from *Science* 52(1347):388–390 (1920)

CHROMOSOMAL DUPLICATION AND
MENDELIAN PHENOMENA IN DATURA MUTANTS

Albert F. Blakeslee,
John Belling, and M. E. Farnham
Carnegie Station for Experimental Evolution

THERE are 12 separate and distinct mutants of the Jimson weed (*Datura Stramonium*) which have recurred with more or less frequency in our cultures of this species during the past six years. The majority of these 12 mutants have been already briefly described or figured elsewhere.[1]

The twelve have certain characteristics which distinguish them from the normal stock from which they arose. They are of feebler growth than normals and have a relatively high degree of pollen sterility, while pollen from normals is relatively good with less than 5 per cent. obviously imperfect grains when examined in unstained condition. The breeding behavior of the twelve is peculiar in that the mutant character is transmitted almost entirely through the female sex. Usually about one quarter or less of the offspring only from a given mutant reproduce the parental mutant type. The pollen entirely fails to transmit the mutant character, or transmits it only to a small percentage of its offspring. This is concluded from the fact that normal female plants crossed with mutant pollen produce no mutant offspring or only a small per-centage, and from the fact that the pollen of any of the 12 mutants seems to be no more potent in reproducing the mutants than pollen from normals.

Another type of mutant, provisionally called "New Species" because of the difficulty or impossibility of crossing it with normals has relatively good pollen and breeds true.

A study has been begun by the present authors of the relationship which exists in

[1] Blakeslee, A. F., and Avery, B. T., "Mutations in the Jimson Weed," *Jour. of Heredity*, X., 111–120, Figs. 5–15, March, 1919.

Datura between the cytological condition and the related phenomena of mutation and Mendelian inheritance. The cytological findings are based on counts of over 350 groups of chromosomes. We can confirm the report of others as to the presence of 12 pairs of chromosomes in the somatic cells of normal jimsons. The somatic number is accordingly twenty-four in contrast to the gametic number twelve. Chromosomal counts from the first division of pollen mother cells show that the gametic number in all the 12 mutants is apparently 12 and 13 giving a calculated somatic number of 25 instead of the 24 found in normals. Whereas in normals all the gametes have 12 chromosomes, in our dozen mutants presumably half the gametes have 12 and half have 13 chromosomes. Apparently in the 13-chromosome gamete the extra chromosome is brought in by a duplication of one of the regular twelve.

The suggestion lies near at hand that each of our 12 mutants is associated with, if not actually determined by, the duplication of a different individual chromosome to make up the calculated total of 25 characteristic of their somatic cells.

If each of our dozen mutants is characterized by the presence of an additional chromosome in a definite one of the 12 chromosome sets, it should be possible by breeding tests to identify the mutant which has as its extra chromosome the one which carries the gene for any particular Mendelian character. This we apparently have been able to do for two of the twelve sets.

The mutant *Poinsettia* (1) which appears to be caused by a duplication of one of the chromosomes carrying determiners for purple or white flower color will serve as an example.

Poinsettia plants have 2 chromosomes in all the sets except in the one carrying the gene for flower pigmentation, which has three. Considering only the latter, we may have *Poinsettia* mutants, as regards their purple pigment, either triplex PPP, duplex PPp, simplex Ppp or nulliplex ppp.

A duplex purple *Poinsettia* with the formula PPp should, if the chromosomes assort at random, be expected to form egg cells of the following types: $2P + p + pp + 2Pp$. The pollen grains should have the same constitution; but, since the *Poinsettia* character fails to be carried by the pollen to any significant extent, the effective male gametes are $2P + p$. Combining male and female gametes in selfing we expect the following zygotes: $4PP + 4Pp + pp + 2PPP + 5PPp + 2Ppp$. The zygotes with 2 chromosomes in the set are normals, the zygotes with 3 chromosomes are *Poinsettia* mutants. We should have therefore among the normals 8 purples to 1 white, and among the Poinsettias 9 purples to no whites. The expectation of an equal number of normals and mutants is practically never realized, probably because of differential mortality in early stages favoring the normals.

A simplex purple heterozygote with the formula Ppp should have the following female gametic formula: $P + 2p + 2Pp + pp$. Its effective male gametes should be $P + 2p$. Selfing a simplex purple heterozygote therefore should give offspring showing a ratio of purples to whites in normals of 5:4 and in the *Poinsettias* of 7:2. Several *Poinsettia* plants of these two heterozygous purple types have been selfed and found to give color ratios in their offspring in close agreement with the calculated values above. When *Poinsettia* mutants are made heterozygous for the other known Mendelian factors, segregation occurs in normal manner giving the customary 3:1 ratio for the characters involved, in both normals and *Poinsettias*.

Two of the 12 mutants have each a single varietal type, which may be due to factors modifying the expression of the more typical complex. In addition two new mutant forms have arisen each of which in appearance seems to be a combination of two of the typical 12 recurrent mutants. It has not been possible as yet to count their chromosomes nor to study their breeding behavior.

We have discussed the duplication of a single chromosome from only one of the 12 sets, producing mutants with 25 somatic chromosomes, with 3 chromosomes in one set and 2 chromosomes in the other 11. We have obtained in addition the duplication of a single chromosome from each of the 12 sets producing a mutant triploid for all the 12 homologous sets.

The duplication may bring about a doubling of all the chromosomes, producing Gigas-like tetraploid mutants—the "New Species" type already mentioned. Such tetraploid plants have presumably 48 chromosomes in somatic cells and 24 in the gametes. From a study of the color ratios in over eight thousand offspring from tetraploid plants, it is possible to assert with some confidence that independent assortment of the chromosomes in the homologous sets of such tetraploid mutants is the rule. Selfed duplex purple heterozygotes throw 35 purples to 1 white, while the back-cross gives a ratio of 5:1. Simplex purple heterozygotes on the other hand give 3:1 ratios when selfed and 1:1 ratios when back-crossed.

Evidence is at hand which indicates that we may have plants with other of the theoretically possible combinations of chromosomes than those mentioned in the present paper.

The significance of the findings in *Datura* in relation to the peculiarities in inheritance in *Gigas* and other mutant types in *Œnothera* will be pointed out later. It is hoped that it may be possible to publish in the near future a series of more detailed papers on the phenomena of chromosomal duplication in the *Daturas*. The present preliminary publication will suffice to emphasize the distinction which must be kept in mind between chromosomal mutations and mutations affecting only single genes.

30

STUDY OF SOMATIC CHROMOSOMES OF NINE MONGOLIAN CHILDREN

Jérôme Lejeune, Martha Gautier, and Raymond Turpin

Presented by Léon Binet

This article was translated expressly for this Benchmark volume by C. R. Burnham, University of Minnesota, from "Étude de chromosomes somatiques de neuf enfants mongoliens," C. R. Acad. Sci. (Paris) **248**:1721–1722 (1959) *with the permission of the publisher, Gauthier-Villars.*

The culture of fibroblasts of 9 mongolian children reveals the presence of 47 chromosomes, the supernumerary chromosome being a small telocentric. The hypothesis of chromosome determination of mongolism is considered.

Among nine mongolian children the study of mitosis of fibroblasts in recent culture(1) permitted us to verify the regular presence of 47 chromosomes. The observations made in these nine cases (five boys and four girls) are given in Table 1.

Table 1. Number of cells examined in each case.

| | Number of chromosomes | Diploid Cells | | | | | | Tetraploid cells | | | |
| | | Doubtful cells | | | Perfect cells | | | Perfect cells | | | |
		46	47	48	46	47	48	—	94	—	—
Boys	Mg 1	6	10	2	—	11	—	—	1	—	30
	Mg 2	—	2	1	—	9	—	—	—	—	12
	Mg 3	—	1	1	—	7	—	—	2	—	11
	Mg 4	—	3	—	—	1	—	—	—	—	4
	Mg 5*	—	—	—	—	8	—	—	—	—	8
Girls	Mg A	1	6	1	—	5	—	—	—	—	13
	Mg B	1	2	—	—	8	—	—	—	—	11
	Mg C	1	2	1	—	4	—	—	—	—	8
	Mg D	1	1	2	—	4	—	—	—	—	8
		10	27	8		57			3		105

*This child is from a pregnancy involving twins. His normal twin possesses 46 chromosomes, 5 of which are small telocentrics.

The number of cells included in each case may seem relatively small. This is because the table includes only the cells that required a minimum of interpretation.

The apparent variation of chromosome number in the "doubtful" cells, that is, those cells in which each chromosome could not be

Jérôme Lejeune, Martha Gautier, and Raymond Turpin

counted with absolute certainty, has been observed by many authors(2). This phenomenon does not seem to correspond to a cytological reality, but simply reflects the difficulties of a delicate technique.

It therefore seems logical to us to prefer a small number of absolutely certain counts ("perfect" cells in the table) to an accumulation of doubtful observations of which the statistical variance depends only on the lack of precision of observations.

The analysis of the chromosome garniture of "perfect" cells reveals among the mongolian boys the presence of six small telocentrics (instead of five for the normal man) and among the mongolian girls five telocentrics (instead of four for the normal woman).

The "perfect" cells of the nonmongolian individuals never show these characteristics(1). It seems legitimate to conclude that there exists among mongolians a small telocentric supernumerary chromosome, accounting for the abnormal number of 47.

Discussion

To explain these observations, the hypothesis of nondisjunction at the time of meiosis of a pair of small telocentrics could be envisioned. Since we know that in *Drosophila* nondisjunction is strongly influenced by maternal age, such a mechanism would account for the increase in frequency of mongolism as a function of advanced age of the mother.

It is, however, impossible to affirm that the small supernumerary telocentrics are certainly normal chromosomes, and one is not able at this time to dispel the possibility that it is due to a fragment resulting from another type of aberration.

REFERENCES

1. J. M. Lejeune, M. Gautier, et R. Turpin. 1959. Les chromosomes humains en culture detissus. *C. R. Acad. Sci. (Paris)* **248**:602–603.
2. P. A. Jacobs and J. A. Strong. 1959. A case of human intersexuality having a possible XXY sex-determining mechanism. *Nature* **183**:302–303.

31

Reprinted from *Am. Nat.* **59**(661):127–137 (1925)

SEX IN RELATION TO CHROMOSOMES AND GENES

DR. CALVIN B. BRIDGES

CARNEGIE INSTITUTION OF WASHINGTON AND COLUMBIA UNIVERSITY

DURING the three years since the report at the Toronto meeting (Bridges '22) considerable new information has accumulated with regard to the series of different sex-types that has arisen in the breeding work with Drosophila (Table I). Each of these different sex-types is the result of a particular combination of chromosomes. They occur principally among the offspring of females that are triploids, that is, that have three X-chromosomes and three of each kind of autosome. The possession of an extra X and at the same time of an extra set of autosomes does not change this individual in its sexual characters from the normal type of female. However, in gametogenesis the 3N group is an unstable one. Each egg receives a full set of chromosomes and a full set goes to the polar body. The members of the extra set are distributed between the egg and the polar body in all possible combinations. Thus, a common type of egg of a 3N female has an extra set of autosomes. If we represent a set of autosomes by A, this egg can be formulated as X + A + A or X,2A. When such an egg is fertilized by an ordinary sperm, that can be formulated X + A, the zygote is 2X,3A. This type of zygote develops into an intersex, that is, into an individual that is neither male nor female but an intermediate, or rather a mixture of male and female parts, very similar in type to the intersexes that Goldschmidt has worked with so extensively in *Lymantria dispar* (Goldschmidt '20).

TABLE I

RELATION OF SEX TO CHROMOSOMES IN DROSOPHILA MELANOGASTER

Sex Type		X (100)	A (80)	Sex Index	Interval	X = −6 A = +2
Superfemale		3	2	1.88	50%	−14
Female	4N	4	4	1.25	......	−20
	3N	3	3	1.25	......	−12
	2N	2	2	1.25	......	−8
	N[1]	1	1	1.25	50%	−4
Intersex	♀ type	2	3	.83	......	−6
	♂ type	2	3–IV	.83−	33%	−6
Male		1	2	.63	50%	−2
Supermale		1	3	.42	......	0

The interpretation of these intersexes in terms of genes carried in the chromosomes was made possible as the result of the very extensive studies of the manner in which ordinary characters are determined, and especially from the study of the contrasted character changes

[1] The haploid type has not been discovered.

brought about on the one hand by the loss, and on the other hand by the gain, of one of the small round chromosomes, the fourth chromosome. From such studies the view had been reached that each character of an individual is the index of the point of balance in effectiveness of a large but unknown number of genes, some of which have a tendency to change development in one direction and others in the opposite (Bridges '22). This conception of "genic balance" was applied to the sex characters of the intersexes as follows: In chromosome constitution the intersexes differ from females only in that they have an extra set of autosomes. This proves that the autosomes are concerned with the determination of sex. Moreover, they are male-determining in their action, since the addition of a set of autosomes causes the female to assume male characteristics. That is to say, in the autosomes there are genes that tend to produce the characters that we call male, and these are more effective, either through greater numbers or through greater potency, than the total of autosomal genes tending to produce the alternative characters that we call female. On the other hand, the X has a net female tendency, as shown by the fact that the addition of an X to a male group changes the individual into a female. The net male tendency of a set of autosomes is less than the net female tendency of an X. This is seen in the fact that in the individual in which there are two of each, namely 2X,2A, the female genes outweigh the male and the result is a female. If we represent the net effectiveness of the female tendency genes in the X by 100, then we should represent the net male effectiveness of a set of autosomes by some lower number; let us say 80. In a 2X,2A individual the ratio of female effectiveness to male effectiveness is 200:160, or 1.25 to 1; and on this formulation the sex index of 1.25 corresponds to the normal female. In the X,2A individual the ratio of female to male effectiveness is 100:160; or the sex index of a normal male is 0.63. In the 2X,3A intersex the ratio is 200:240, and the sex index is 0.83, which is intermediate between the indices for female and male. In the 3N female the ratio is 300 : 240, and the sex index is 1.25, exactly the same as in the normal female. This identity of sex indices for the 3N and 2N forms corresponds to the observation that there seems to be no strictly sexual differences between them. The larger size, coarser texture of eye, etc., of the 3N can be directly attributed to the changed volume of the nucleus, and are not sexual in nature.

Another type of egg of the 3N ♀ is $X + X + A$; and this, fertilized by a normal XA sperm, gives a 3X,2A individual with a sex index of 1.88, which is 50 per cent. higher than that of the normal female. This constitution corresponds in fact to the "superfemales" that occur in these cultures and elsewhere. The superfemales are much delayed in development, are rarely able to live and are probably completely sterile.

Conversely, an $X + A + A$ egg, fertilized by the type of sperm that does not carry an X, gives an X,3A zygote with a sex-index of only 0.42. This type of individual was expected to be more male-like than an ordinary male; and such individuals were looked for among the offspring of 3N females. At first none were found; but presently it was discovered that very late in the cultures an occasional example of a distinct type of male occurred. These so-called "supermales" are likewise sterile. Recently, cytological proof has been secured that this type of sex has the constitution X,3A, which agrees with the genetical evidence previously secured.

It was observed that the intersexes showed considerable variation and seemed to form a bimodal class. And since the cytological investigation had showed that some intersexes had three and others only two of the small round fourth-chromosome, it was guessed that the more male-like mode corresponded to the full trio of fourth chromosomes, while the more female-like mode corresponded to the cytological type that lacked one fourth chromosome. An effort has been made to secure cytological evidence on this point. But this evidence is inconclusive; as is also that from an attempt to make a genetic test of the number of fourth chromosomes present through use of the fourth-chromosome mutant character eyeless. At present extra fourth chromosomes are being artificially inserted into the intersexes by continually crossing 3N mothers to males known to have an extra fourth chromosome. Contrariwise, in other lines of intersexes, fourth chromosomes are being diminished in number by continually mating 3N mothers to males known to lack one of the two fourth chromosomes. Similarly, the superfemaleness of the 3X,2A individuals might be reduced or be increased by matings with triplo-IV males or with haplo-IV males. For this experiment females are being used whose two X-chromosomes are permanently attached to each other (L. V. Morgan '22), and hence that give through non-disjunction a very high proportion of 3X-superfemales. Present indications, from the uncompleted experiments, are slightly contradictory, but tend to a conclusion which is the opposite of that earlier reported as probable on the basis of the slight evidence then available (Bridges '22). When the number of fourth chromosomes is three the intersexes are more female-like, and when the number is two they are more male-like. The fourth chromosome has a net female tendency, similar to that of the X and different from that of the other autosomes. By variation in the number of fourth chromosomes it is possible to have a fringe of minor sex-types about each of the major types of sex difference.

The list of sex-types has been enlarged by the discovery of tetraploids, or 4N individuals. These are females, quite identical with normal females in sex characteristics. The tetraploid arose in a stock of triploids; and was detected only by the strikingly different offspring given. A female supposed to be 3N was selected from the 3N stock and outcrossed to a normal male. All the offspring were triploid females (about 30) or triploid intersexes (about 20). There were no 2N offspring or supersexes. It was

seen that this result might be produced if the mother were
4N instead of 3N. For in that case all the reduced eggs
would be 2N; and these fertilized by X-sperm would give
3N females, and fertilized by Y-sperm would give 2X,3A
intersexes.

Before the discovery of this 4N individual, an expectation that it would occur had arisen from several facts.
Thus, in the three years following the discovery of
triploidy there had been found no less than twenty-five
instances of the new occurrence of triploidy. This very
high frequency was paralleled by cytological observations
that give the explanation of the origin of triploids. In
three separate preparations of ordinary 2N females it
was found that a portion of an ovary was constituted of
markedly larger cells; and in two of the individuals some
of the giant cells were in division, and the chromosomes
could be counted as 4N. Evidently there had been in
some oogonial cell a division of the chromosomes that had
not been followed by division of the nucleus and cytoplasm. The resulting tissue was tetraploid, and any reduced gamete would be 2N. Such a 2N gamete, fertilized
by a normal sperm, would account for each of the twenty-
five recurrences of triploidy.

Furthermore, in examining sections of intersexes, two
individuals were found in which similar cysts of even
larger cells were present. In one of these cysts divisions
were occurring; and the chromosomes were clearly 6N.
A 6N cyst in a 3N female would give, upon reduction, 3N
eggs, which, fertilized by X sperm, would give the expected 4N type of female.

Soon after this first case of 4N female a second similar
case was found. Also L. V. Morgan found a third case,
and was able to prove by genetic tests that four separate
X-chromosomes had been present (in press).

The fact that 4N individuals are females, not modified
as to sex, has important bearings on our ideas as to the
way in which genes interact to produce their effect. The
view adopted here is that in general the effectiveness is

in proportion to number of genes, and the significant point is the *ratio* between sets of genes that tend to produce alternative effects. On this view we find a ready explanation for the fact that such diverse forms as 2N, 3N and 4N individuals are precisely alike in their sexual characteristics; for in all these forms the effectiveness of both contending sets of influences has been doubled, trebled or quadrupled; and the ratio remains constant.

But a system of formulation different from the ratio type has been adopted by Goldschmidt in dealing with the intersexes produced in the course of his brilliant work on racial crosses of *Lymantia dispar*. To the male tendency of a particular race he assigns a positive value that is proportional to the strength of the male-determining gene or genes. To the female tendency he assigns another value also proportionate to the strength of the female-determining gene or genes. He then assumes that when in an individual the male value is greater than the female value by a certain number of units the individual is a male, and that, conversely, when the female value is greater than the male by this same number of units the individual is a female. The locus of the male tendency gene (M) is in the "Z-chromosome" of which two are present in the male and one in the female. The female tendency is strictly maternally inherited; and hence the locus of the F genes is in the W-chromosome that descends from mother to daughter. The F gene is supposed to exert its influence on the cytoplasm of the developing egg; and hence, although the male has no W-chromosome, he is supposed to have a definite female tendency that was impressed upon the cytoplasm of the egg and that persists throughout development. For a "weak" race the value assigned to F is 80, and to M, 60. In the WZ individual the cytoplasmic $\overline{\lfloor F \rfloor}$ of 80 exceeds the M value of 60 by "the epistatic minimum" of 20 units, and the individual is a female. Likewise, in the ZZ individual the $\lceil F \rceil$ is 80, but the net M value is twice 60 or 120, with an excess in the male direction of 40 units. For a strong race both F and M are higher, for example, 100 and 80, but the arithmetical relation between the values of F and M would still govern the sex of the individual. In a cross between a weak female and a strong male the ZW individual received an $\boxed{F}$ of 80 from the mother and an M of 80 from the father. The values are thus balanced midway between the excesses necessary for a female on the one hand or a male on the other, and the result is an intersex. This far the formulation is satisfactory; but when extensive series of crosses are compared, and an attempt is made to give values to the F and the M of each race that will hold throughout the entire range of experiments, this attempt is rather unsuccessful. It seems to me that a reformulation on the ratio instead of the algebraic basis would give a series of consistent indices without running counter to the very valuable physiological ideas that Goldschmidt has developed.

In the table of sex-types of Drosophila the haploid individual is entered with the index 1.25, the same as that of the 2N, 3N and 4N females. Unfortunately the haploid individual has *not* been discovered. But it is plain from the view just given that the expectation for a haploid *Drosophila melanogaster* is that it would be female in sex character. Accordingly, it is necessary to assume that the sex-determining mechanism here is essentially different from that in the bee and similar forms in which the haploid individual is a male. To me, sex-determination in the bee is the outstanding unsolved puzzle, although before the development of the idea of genic balance it seemed one of the clearest and simplest of cases. If it is true that the male is a haploid individual, then one would suppose that the diploid individual should likewise be a male, since the ratios among the sex-determining genes are not different in the two cases.

Schrader and Sturtevant have attempted a reconciliation of the cases of Drosophila and the bee by use of Goldschmidt's algebraic formulation. They assign a positive value, *e.g.,* $+2$, to each A and a negative value,

e.g., — 6, to each X. It is then assumed that the effective
relation is the algebraic sum of the values of X and A,
as given in the column to the right in Table I. On this
view the haploid might be a male. But this system has a
difficulty in that the intervals between successive indices
do not correspond very well with the observed differences
between the sex grades. Thus the smallest observed interval in fact, that between the 3N and 2N individuals, is
represented by a difference of 4 units, while the very
great interval between the male and the female is represented by only six units. At that time the 4N type was
not known; and when it is added to the series, the fit is
very poor on the algebraic system and very good on the
ratio system. I repeat that I do not regard the case of
the bee as interpretable on the same basis as Drosophila
so long as the present account of the mechanism for the
bee is unchallenged. At present the difference between
haploid and diploid sexes must be referred to the same
type of determination as that responsible for the larger
size, rougher texture of eyes and other slight changes
that distinguish the 3N from the 2N individual.

But outside of the cases like that of the bee, it seems
probable that the ratio type of interaction is the general
mode. In evidence of this may be cited the vast array
of monoecious plants in which the sex relations in the
triploid and tetraploid remain the same as in the diploid.
Of course there is high sterility in the triploid forms on
account of the instability of the 3N group in meiosis, and
the consequent production of inviable gametes or zygotes.
Among the most striking confirmations of this ratio view
of genic balance is seen in the mosses, through the brilliant work of the Marchals, Schweitzer and von Wettstein. For example, they find in a moss with separate
sexes that a 2N gametophyte that combines 2 female
groups of chromosomes is a pure female, like the haploid
female plant (Table II). Likewise the 2N gametophyte
that combines two male groups of chromosomes is a pure
male plant, like the haploid male plant. But a 2N gameto-

TABLE II

SEX TYPES IN A DIOECIOUS MOSS

Sex Type	X (100)	X' (50)	A (80)	Ratio (X + X') : A	Sex Index
Female $\begin{cases} 2N \\ N \end{cases}$	2	...	2	200 : 160	1.25
	1	...	1	100 : 80	1.25
Proterogynous Herm. 3N	2	1	3	250 : 240	1.04
Protandrous Herm. $\begin{cases} 4N \\ 2N \end{cases}$	2	2	4	300 : 320	.94
	1	1	2	150 : 160	.94
[2]Protandrous Herm. 3N	1	2	3	200 : 240	.83
Male $\begin{cases} 2N \\ N \end{cases}$	...	2	2	100 : 160	.63
	...	1	1	50 : 80	.63

phyte that combines a male and a female group is no longer a single-sexed plant but is a protandrous hermaphrodite. Furthermore, the 4N gametophyte that combines 2 male chromosome groups with 2 female groups (FFMM) is a hermaphrodite like the FM hermaphrodite. But a triploid form in which two groups are female and one is male is a hermaphrodite that is strongly protogynic instead of protandrous. On the other hand, in working with monoecious mosses where the haploid group is a hermaphrodite, then all haploid, diploid, triploid and tetraploid plants were hermaphrodite without distinction, as they should be from their possession of the same ratio of female to male determiners.

A series of sex-indices, similar to those for Drosophila, can be fitted to the dioecious mosses, as shown in Table II. Here it is assumed that there is a pair of chromosomes, X and X', whose difference accounts for the difference between the female and male types. It is assumed that in both these sexes the net effect of the other chromosomes, that may be represented by A, is male-determining. Then, since the X,A type is a female, the value for A must be less than that of X, *e.g.*, X = 100 and A = 80. Likewise, since X',A is a male, the value X' must be less than that of A, *e.g.*, X' = 50. Also, since the FM plant is a hermaphrodite that resembles the normal male more

[2] This type is not reported as realized.

than the normal female, $X + X' < 2A$. And, since the FFM plant is a hermaphrodite that resembles the normal female more closely, $X + X + X' > 3A$. We have thus five limiting equations for the three values X, X' and A. As the table shows, the assigned values of 100, 50 and 80 are possible, although there may be other slightly different values that would give a set of indices whose intervals would correspond even more closely to the observed differences than do those given.

The same conformity to the ratio rule seems to be true in the haploid, diploid, triploid and tetraploid daturas, as far as I can gather. But on the genic balance view each of the twelve kinds of chromosomes of Datura might have a distinctive internal unbalance of the sex-controlling genes, similar to the unbalance in the fourth chromosome of Drosophila. In Datura there is a full series of forms that differ from the 2N by the addition of a particular extra chromosome. If any of these twelve kinds of chromosomes contain more effective male tendency genes than female tendency genes, or vice versa, then one may well expect to discover that some of Blakeslee's "Apostles" and "Acolytes" have atypical sex-relations.

LITERATURE

Blakeslee, A. F.
>   1922. Variations in Datura, due to changes in chromosome number. AM. NAT. 56: 16–31.

Bridges, C. B.
>   1922. The origin of variations in sexual and sex-limited characters. AM. NAT. 56: 51–63.

Goldschmidt, R.
>   1920. Untersuchungen über Intersexualität. *Zeit. f. ind. Abst. u. Verer.* 23: 1–197.

Morgan, L. V.
>   1922. Non-criss-cross inheritance in Drosophila melanogaster. *Biol. Bull.* 42: 267–274.

Schweitzer, J.
>   1923. Polyploidie und Geschlechterverteilung bei Splachnum sphericum Schwartz. *Flora* 116: 1–72.

Wettstein, F. v.
>   1924. Morphologie und Physiologie des Formwechsels der Moose auf genetischer Grundlage. *Zeit. f. ind. Abst. u. Verer.* 33: 1–236.

32

Reprinted from *Natl. Acad. Sci. (U.S.A.) Proc.* **22**(11):27–33 (1936)

VARIEGATION IN DROSOPHILA AND THE INERT CHROMO-SOME REGIONS

By Jack Schultz

Carnegie Institution of Washington and William G. Kerckhoff Laboratories of the Biological Sciences, California Institute of Technology

Communicated November 12, 1935

At the time of their discovery, the variegated races of Drosophila (D. virilis, Demerec ('26); D. melanogaster, Muller ('30)) seemed to be promising material for the study of mutation. They have since developed rather into a puzzle *sui generis*. Members of the group have been discussed variously as mutable genes (Demerec ('28)), unstable translocations (Muller ('30), Patterson and Painter ('31, '32)) and, more recently, as a rather special type of duplication undergoing frequent somatic crossing-over (Stern ('35)). The discussions have not been very satisfactory; partly because the data themselves have been inadequate for the formulation of general rules, excepting only Muller's correlation of the "eversporting" types in D. melanogaster with the occurrence of chromosome rearrangement.

This paper summarizes the results of experiments with thirteen different variegations in Drosophila melanogaster. All of these belong to Muller's "eversporting displacements"—they are associated with chromosome rearrangements. Five involve the region around the white locus in chromosome 1; two the yellow region of the same chromosome; and the six remaining, the brown region of the second chromosome. I have attempted to determine what relation there might be between the nature of the rearrangements and the production of variegation. To this end, I have studied the characteristics of the rearrangements cytologically, in the salivary gland chromosomes; and the characteristics of the variegations, genetically, in relation to the different affected genes. Both series of data show a relation between variegation and the so-called "inert" regions. The first evidence of the sort came from the work of Gowen and Gay ('34), on the suppression of white-variegation by a supernumerary *Y* chromosome. This may now be extended; extent of variegation, that is, the proportion of "mutant" to wild type tissue, depends upon a quantitative relation between active and "inert" chromosome regions. In addition, however, the "inert" regions are involved in the rearrangements themselves. Thus both the chromosome structure associated with variegation and the extent of variegation are related to the "inert" chromosome regions.

The variegated races form abnormal configurations of the salivary gland chromosomes as a direct result of their relation to the "inert" regions. In a nucleus with normal chromosomes, the "inert" regions are aggregated to form a "chromocenter" (Heitz ('33, a, b,); Painter ('35)). Since in Drosophila melanogaster these regions are normally located near the spindle attachments of the chromosomes (Heitz ('33), Kaufman ('34)), a configuration is formed in which all spindle attachments are approximated. In the chromosome rearrangements where "inert" regions are transferred from the spindle attachment to other loci, the coalescence to a chromocenter persists in the salivary gland nucleus. The result is that, in nuclei of such types, rings, lateral attachments and more complex configurations may occur.

Ring configurations are regularly found when an "active" chromosome region is intercalated into an "inert" region. For example, both ends of the normal fourth chromosomes of Drosophila melanogaster are regularly part of the chromocenter (Bridges ('35)). Three of the variations involving the white locus turn out to be intercalations, into the fourth chromosome, of the section of the X from yellow to white, or in one case, from yellow to echinus. These form, then, part of a ring configuration. A similar situation exists for the yellow-silver variegation of Sturtevant ('34), scute 10-2. Another variegated white forms a ring as the result of intercalation of a small portion of the X chromosome, including white, into the inert region of the left limb of chromosome 3.

Effective "lateral" attachments occur when "inert" region is intercalated between two "active" regions. After chromocenter formation, the "active" region distal to the inert material is attached to the chromocenter by its proximal end; the other "active" region, between the inert material and the spindle attachment, is bent into a loop. Figures of this kind are found in the scute-8 inversion, in two of the variegated brown allelomorphs, and in one of the white variegations. It is readily seen that the principle is the same as in the cases of ring formation. Still more complex configurations are found in other cases, the four remaining variegated brown allelomorphs. They also result from the breakage and rearrangement of inert regions, which by their aggregation to form a chromocenter may then join the broken chromosomes. An apparently similar case has been described by Dubinin and his collaborators ('35), and interpreted as a fusion of specific intergenic bonds. Evidently on the basis of the present discussion this assumption is not yet necessary. Their "reconstruction of a normal chromosome" presumably followed disruption of the chromocenter aggregate when the preparation was smeared.

In addition to the thirteen cases I have studied, those already described by other workers lend themselves to a similar interpretation (see Patterson ('32, '33), Gowen and Gay ('34), Glass ('34a, b), Van Atta ('32), Painter ('34), Mackensen ('35), Patterson and Stone ('35), Stone and Thomas ('35)). There are some variations, that I am studying, which may be of a different type; and I have seen similar abnormal configurations unaccompanied by obvious variegation. These exceptions are, however, in a minority. The evidence seems adequate to establish the rule, that variegations are associated with abnormal configurations of the salivary gland

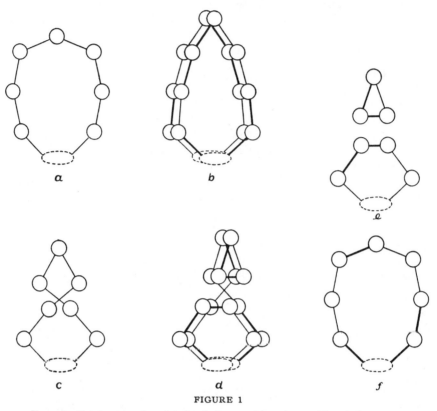

FIGURE 1

Reproduction in normal and twisted ring configurations. The circles represent chromomeres (genes), the ellipses the spindle attachment. Light lines are the old connecting fibres, heavy lines the new ones.

(*a, b*) Reproduction in a normal ring; (*c*) a twisted ring; (*d*) reproduction in a ring where distances are such as to permit abnormal linkages to form; (*e, f*) the products of reproduction: normal ring, deficient ring and fragment without spindle attachment.

By variation of the type of overlap or the plane of formation of the new fibres, rearrangements, duplications, double or interlocked rings may result.

chromosomes, caused in these cases by the aggregation of inert regions to a chromocenter.

The variegation in all of the rearrangements concerns those genes close to the locus of the break. All the available mutants located in the regions showing abnormal configurations were tested for the display of variegation. The evidence from these tests shows clearly that only those characters whose genes lie close to the locus of rearrangement as determined in the salivary gland chromosomes are affected. No unequivocal instance has yet been found of the total loss of a fragment as the cause of variegation. One such case has been reported to have been seen cytologically (Patterson and Painter ('31), Patterson ('32)). But the genetic data (Patterson ('32)) indicate variegation only for the white-Notch loci, although tests were made for other genes. If the whole fragment were lost, all the genes that are included in the region should show variegation. It would seem, then, that in this case, as in those studied and recognized as being of this nature by Gowen and Gay, and by myself, only a part of the fragment is involved in the variegation.

Changes in the extent of variegation occur when the normal balance of active and inert regions in the nucleus is upset. The suppression of variegation by an extra Y chromosome is not confined to the white-Notch cases studied by Gowen and Gay. A similar phenomenon occurs for the brown variegations (Schultz and Dobzhansky ('34), Dubinin and Heptner ('34)) and for the others as well. I have now determined the effects of changing the number of Y chromosomes, on variegation for most of the affected genes in the different stocks. In addition, I have found that not only the inert region from the Y, but also its relation to the active regions, is important. The evidence for this comes from the study of many different combinations of variegations, in types with different balance of active and inert region. The different fragments of X and Y chromosomes have been studied; triploids with different numbers of Y chromosomes; intersexes also, and superfemales; and particularly certain duplications for active regions, some of which annul, others of which simulate, the effects of the Y chromosome. Here I will discuss the effects on variegation of changing the number of Y chromosomes in the diploid.

The variegated brown allelomorph Plum-2 may be taken as an example, since it illustrates most of the relations found also in other cases. Flies containing a Plum-2 second chromosome display, when appropriately tested, four distinct types of variegation: a dominant brown (eye color), a recessive brown (eye color and testis sheath), a "minus" bristle and "light" (eye color). Of these, the first three named lie close to one side of the breakage point; the fourth ("light") is at the other of the two breaks, within the inert region of 2L. It will be recalled here, that in Plum-2 homozygotes, all the broken portions of the chromosome are joined at the chromocenter.

The characteristics of these variegations have been studied in all types from the XO male to the $XXYY$ female. Minus, which is to the right of brown, shows an extent of variegation in the XO male comparable to that shown by brown in the XY male. The susceptibility to variegation decreases as the number of Y chromosomes increases. In the case of the "light" variegation, an apparently converse relation holds.   With increase in the number of Y chromosomes, the extent of "light" areas in the eye increases; in the XO male, however, the eye is completely wild-type as regards the "light" character.   This is to be interpreted as meaning that the product of variegation in this case is wild type, and the original form "light."   Support for such an interpretation comes from the study of several other cases involving the "light" locus, in which a similar break in $2L$ has produced a similar "mutation" (position effect) to "light."   Thus the variegation in this case is a "reversion" to the wild type.   On this basis, the effect of addition of Y chromosomes in all cases is to decrease the frequency of variegation.

All the variegations—those for yellow and achaete, for white, split and Notch, and the others for loci in the brown region—exhibit similar characteristics.   They cannot be discussed here due to limitations of space.   In all cases, however, it appears that the extent of variegation increases as the Y chromosome number is lowered.   The result is that the apparent direction of "mutation," as judged by relative proportions of mutant and wild type tissue, may appear to change with change in Y chromosome number.   These relations suggest that the origin of variegation may be considered as the result of some type of unstable configuration which produces different types of "mutation" in the different Y chromosome types. The presence of the abnormal chromosome configurations in the salivary gland nuclei permits an attempt to formulate an hypothesis based on consideration of the reproduction of genes in these configurations.

The basic postulate for a theory of the variegations involves the nature of the so-called "mutations."   Either they are intra-genic, the orthodox mutation types; or they involve structural changes of the chromosomes. In one of the present cases—scute-8—Beadle and Sturtevant (in the press) have shown that germinal deficiency for the variegating portion of the chromosome may occur.   Other data, on different variegation types, may be interpreted similarly.   These belong then to the simplest type of structural change—a loss which occurs without a complementary production of duplications.   There are, in addition, a number of instances of variegation which are definitely different from known deficiencies for the same loci.   Examples are the dominant brown variegations, and certain white variegations.   These may be interpreted as due to position effects, effectively mutations in appearance, resulting from local rearrangements produced by a mechanism similar to that which produces the deficiencies.

It has long been obvious that losses or rearrangements of genes might occur as a result of the difficulties of division in abnormally shaped chromosomes (see especially the results of McClintock ('32) on ring-shaped chromosomes in Zea mays).   It is possible to consider such a process simply, following Belling's ('33) analysis of gene (chromosome) reproduction into

[*Editors' Note:* The scute-8 inversion referred to above in Beadle and Sturtevant (in press) is described in their 1936 paper in *Genetics* **21**:554–604. Part of the story is also in Patterson's 1933 paper (see reference list).]

two components. These are the synthesis of a new gene, and the mainte-
nance of the genes in linear order. According to his view, the mechanism
which maintains linear order operates after the genes have reproduced.
Thus, whenever gene reproduction occurs in crossed chromatids, the new
linear connections are formed between the closest genes—hence, according
to Belling, crossing-over. This makes no postulate as to the nature of the
new "fibre"; what is discussed is the time of its formation.

The application of such an analysis to the variegations is simple. The
reproduction of a ring-shaped chromosome will offer a model for consider-
tion. From a normal ring, two identical daughters result (Figs. 1*a*, *b*).
But if the ring is twisted so that two non-homologous portions come to lie
one over the other, the reproduction is abnormal. When the new fibres
form across the shortest way (Fig. 1*d*), a normal ring, a small ring and a
fragment lacking a spindle attachment, result under the conditions shown
in the figure. Evidently the formation of rings of different sizes, observed
by McClintock in Zea mays, might be explained in this way. Moreover,
only rings would ordinarily be produced from rings. In addition, by variations of details in the place of twist, or the plane in which the new intergenic connections are formed, a variety of other new rings are possible.
These may be gene rearrangements within the ring, or duplications which
would appear as larger rings. Further discussion, and the detailed application of the hypothesis to the various cases, is deferred until the full presentation of the data.

The effect of the Y chromosome and other inert regions on the extent of
variegation may be derived from this point of view, with an additional consideration. I have found, in comparative studies of larvae having different
numbers of Y chromosomes, differences in the "turgor" of the salivary
gland chromosomes. With the increase in number of Y chromosomes, all
the others show an increase both in the sharpness of stain and the plump-
ness of the chromosomes. This result may possibly be related to the development of accessory chromosome materials, in a manner similar to the
suggestion of McClintock that the nucleolar center in chromosome 6 of Zea
mays, has to do with the development of chromosome matrix. Here also a
nucleolar center—that of the Y chromosome—may be concerned. It is
clear that any factor changing the "turgor" of the chromosomes would have
an influence on the frequency of cross contacts of non-homologous genes.
This would appear as a change of the sort found in the extent of variegation.

The hypothesis that I have indicated for variegation is to be regarded as
a pictorial summary of the facts, cytological and genetic, about the variegations. A variety of other hypotheses are possible; differences in rate of
gene division, or of formation of intergenic connections; or a whole variety
of intragenic changes may be invoked. None that I have considered are

so simply related to the general body of cytogenetic theory as the one I have discussed, although even it has its difficulties. It is perhaps worth noting, that in its more general aspects, the theory presents the study of variegations as a study of the mechanism whereby the genes are maintained in linear order.

[1] A bibliography will be given with the detailed publication of the data.

[2] The terminology is obviously outworn: "inert" is certainly a misnomer. Heitz's use of "heterochromatin" and "euchromatin" for "inert" and active regions is probably better. For the present, however, I have used the older term because of its familiarity.

[*Editors' Note:* The following references were researched from other sources. The references for Glass 1934b, Painter 1934, Patterson and Painter 1932, and Patterson and Stone 1935 could not be located.]

REFERENCES

Belling, J. 1933. Crossing over and gene arrangement in flowering plants. *Genetics* **18**:388–413.

Bridges, C. B. 1935. Cytological data on chromosome four of *Drosophila melanogaster. Trud. Dinam. Razvit.* **10**:463–474 (Russian summary).

Demerec, M. 1926. Reddish—a frequently "mutating" character in *Drosophila virilis. Proc. Nat. Acad. Sci. U.S.A.* **13**:249–253.

_____. 1928. Mutable characters of *Drosophila virilis.* I. Reddish-alpha body character. *Genetics* **13**:359–388.

Dubinin, N. P., N. N. Sokolov, and G. G. Tiniakov. 1935. (Cyto-genetic analysis of position effect) *Biol. Zh. (Mosc.).* **4**:707–720 (Russian and English text).

_____ and M. A. Heptner. 1934. A new phenotypic effect of the Y chromosome in *Drosophila melanogaster. J. Genet.* **30**:423–446.

Glass, H. B. 1934a. A study of dominant mosaic eye-color mutants in *Drosophila melanogaster.* I. Phenotypes and loci involved. *Am. Nat.* **68**:107–114.

Gowen, J. W., and E. H. Gay. 1934. Chromosome constitution and behavior in eversporting and mottling in *Drosophila melanogaster. Genetics* **19**:189–208.

Heitz, E. 1933a. Über totale und partielle somatische Heteropyknose, sowie strukturelle Geschlechts-chromosomen bei *Drosophila funebris.* (Cytologische Untersuchungen an Dipteren II.) *Z. Zellforsch.* **19**:720–742.

_____. 1933b. Die somatische Heteropyknose bei *Drosophila melanogaster* und ihre genetische Bedeutung. (Cytologische Untersuchungen an Dipteren III. *Z. Zellforsch.* **20**:237–287.

Kaufmann, B. P. 1934. Somatic mitoses of *Drosophila melanogaster. J. Morph.* **56**:125–155.

Mackensen, O. 1935. Locating genes on salivary chromosomes. Cyto-genetic methods demonstrated in determining position of genes on the X chromosome of *Drosophila melanogaster. J. Hered.* **26**:163–174.

McClintock, B. 1932. A correlation of ring-shaped chromosomes with variegation in *Zea mays. Proc. Nat. Acad. Sci. U.S.A.* **18**:677–681.

Muller, H. J. 1930. Radiation and genetics. *Amer. Nat.* **64**:220–251.

Patterson, J. T. 1932. A new type of mottled-eyed *Drosophila* due to an unstable translocation. *Genetics* **17**:38–59.

———. 1932. The mechanism of mosaic formation in *Drosophila*. *Proc. Sixth Int. Congr. Genet.* **2**:153–155.

———. 1933. The mechanism of mosaic formation in *Drosophila*. *Genetics* **18**:32–52.

——— and T. S. Painter. 1931. A mottled-eyed *Drosophila*. *Science* **73**:530–531.

Schultz, J., and T. Dobzhansky. 1934. The relation of a dominant eye color in *Drosophila melanogaster* to the associated chromosome rearrangement. *Genetics* **19**:344–364.

Stern, C. 1935. The behavior of unstable genic loci-an hypothesis. *Proc. Nat. Acad. Sci. U.S.A.* **21**:202–208.

Stone, W. S., and I. Thomas. 1935. Crossover and disjunctional properties of X-chromosome inversions in *Drosophila melanogaster*. *Genetica* **17**:170–184.

Sturtevant, A. H. 1934. Preferential segregation of the fourth chromosomes in *Drosophila melanogaster*. *Proc. Nat. Acad. Sci. U.S.A.* **20**:515–518.

Van Atta, E. W. 1932. Genetic and cytological studies on X-radiation induced dominant eye colors of *Drosophila*. *Genetics* **17**:637–659.

33

Reprinted from *Nature* **190**(4773):372–373 (1961)

GENE ACTION IN THE X-CHROMOSOME OF THE MOUSE (MUS MUSCULUS L.)

Mary F. Lyon

Medical Research Council Radiobiological Research Unit, Harwell, Didcot.

Ohno and Hauschka[1] showed that in female mice one chromosome of mammary carcinoma cells and of normal diploid cells of the ovary, mammary gland and liver was heteropyknotic. They interpreted this chromosome as an X-chromosome and suggested that the so-called sex chromatin was composed of one heteropyknotic X-chromosome. They left open the question whether the heteropyknosis was shown by the paternal X-chromosome only, or the chromosome from either parent indifferently.

The present communication suggests that the evidence of mouse genetics indicates: (1) that the heteropyknotic X-chromosome can be either paternal or maternal in origin, in different cells of the same animal; (2) that it is genetically inactivated.

The evidence has two main parts. First, the normal phenotype of XO females in the mouse[2] shows that only one active X-chromosome is necessary for normal development, including sexual development. The second piece of evidence concerns the mosaic phenotype of female mice heterozygous for some sex-linked mutants. All sex-linked mutants so far known affecting coat colour cause a 'mottled' or 'dappled' phenotype, with patches of normal and mutant colour, in females heterozygous for them. At least six mutations to genes of this type have been reported, under the names mottled[3,4], brindled[3], tortoiseshell[5], dappled[6], and 26K[3]. They have been thought to be allelic with one another, but since no fertile males can be obtained from any except, in rare cases, brindled, direct tests of allelism have usually not been possible. In addition, a similar phenotype, described as 'variegated', is seen in females heterozygous for coat colour mutants translocated on to the X-chromosome[7,8].

It is here suggested that this mosaic phenotype is due to the inactivation of one or other X-chromosome early in embryonic development. If this is true, pigment cells descended from cells in which the chromosome carrying the mutant gene was inactivated will give rise to a normal-coloured patch and those in which the chromosome carrying the normal gene was inactivated will give rise to a mutant-coloured patch. There may be patches of intermediate colour due to cell-mingling in development. The stripes of the coat of female mice heterozygous for the gene tabby, *Ta*, which affects hair structure, would have a similar type of origin. Falconer[9] reported that the black regions of the coat of heterozygotes had a hair structure resembling that of the *Ta* hemizygotes and homozygotes, while the agouti regions had a normal structure.

Thus this hypothesis predicts that for all sex-linked genes of the mouse in which the phenotype is due to localized gene action the heterozygote will have a mosaic appearance, and that there will be a similar effect when autosomal genes are translocated to the X-chromosome. When the phenotype is not due to localized gene action various types of result are possible. Unless the gene action is restricted to the descendants of a very small number of cells at the time of inactivation, these original cells will, except in very rare instances, include both types. Therefore, the phenotype may be intermediate between the normal and hemizygote types, or the presence of any normal cells may be enough to ensure a normal phenotype, or the observed expression may vary as the proportion of normal and mutant cells varies, leading to incomplete penetrance in heterozygotes. The gene bent-tail, *Bn*[10], may fit into this category, having 95 per cent penetrance and variable expression in heterozygotes. Jimpy, *jp*, is recessive, suggesting that the presence of some normal cells is enough to ensure a normal phenotype, but Phillips[11] reported one anomalous female which showed the jimpy phenotype. Since it showed the heterozygous phenotype for *Ta* this animal cannot be interpreted as an XO female; it is possible that it represents an example of the rare instance when by chance all the cells responsible for the jimpy phenotype had the normal gene inactivated.

The genetic evidence does not indicate at what stage of embryonic development the inactivation of one X-chromosome occurs. In embryos of the cat, monkey and man sex-chromatin is first found in nuclei of the late blastocyst stage[11,12]. Inactivation of one X at a similar stage of the mouse embryo would be compatible with the observations. Since an XO female is normally fertile it is not necessary to postulate that both X-chromosomes remain functional until the formation of the gonads.

The sex-chromatin is thought to be formed from one X-chromosome also in the rat, *Rattus norvegicus*[14], and in the opossum, *Didelphis virginiana*[15]. If this should prove to be the case in all mammals, then all female mammals heterozygous for sex-linked mutant genes would be expected to show the same phenomena as those in the mouse. The coat of the tortoiseshell cat, being a mosaic of the black and yellow colours of the two homozygous types, fulfils this expectation.

[1] Ohno, S., and Hauschka, T. S., *Cancer Res.*, **20**, 541 (1960).

[2] Welshons, W. J., and Russell, L. B., *Proc. U.S. Nat. Acad. Sci.*, **45**, 560 (1959).

[3] Fraser, A. S., Sobey, S., and Spicer, C. C., *J. Genet.*, **51**, 217 (1953).

[4] Lyon, M. F., *J. Hered.*, **51**, 116 (1960).

Dickie, M. M., *J. Hered.*, **45**, 158 (1954).

[6] Phillips, R. J. S., *Genet. Res.* (in the press).

[7] Russell, L. B., and Bangham, J. W., *Genetics*, **44**, 532 (1959).

[8] Russell, L. B., and Bangham, J. W., *Genetics*, **45**, 1008 (1960).

[9] Falconer, D. S., *Z. indukt. Abstamm. u. Vererblehre*, **85**, 210 (1953).

[10] Phillips, R. J. S., *Z. indukt. Abstamm. u. Vererblehre*, **86**, 322 (1954).

[11] Austin, C. R., and Amoroso, E. C., *Exp. Cell Res.*, **13**, 419 (1957).

[12] Park, W. W., *J. Anat.*, **91**, 369 (1957).

[13] Ohno, S., Kaplan, W. D., and Kinosita, R., *Exp. Cell Res.*, **18**, 415 (1959).

[14] Ohno, S., Kaplan, W. D., and Kinosita, R., *Exp. Cell Res.*, **19**, 417 (1960).

34

Reprinted from *Science* **164**(3882):955–957 (1969)

Visualization of Nucleolar Genes

O. L. Miller, Jr. and Barbara R. Beatty

Biology Division, Oak Ridge National Laboratory

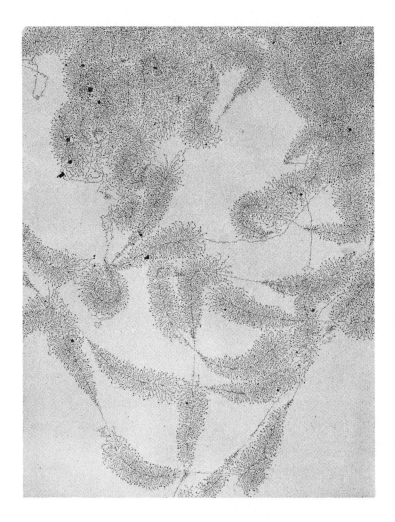

Nucleolar genes from an amphibian oocyte. These genes, which code for ribosomal RNA, repeat along the DNA axis and are visualized because approximately 100 enzymes are simultaneously transcribing each gene. The gradient of fibrils extending from each gene contains ribosomal RNA precursor molecules in progressive stages of completion (electron micrograph, × 25,000). See page 955.

During early growth of the amphibian oocyte, the chromosomal nucleolus organizer is multiplied to produce about a thousand extrachromosomal nucleoli within each nucleus (*1*). There is convincing evidence that these nucleoli function similarly to chromosomal nucleoli in the synthesis of rRNA precursor molecules (*2*). In thin sections of fixed oocytes, each extrachromosomal nucleolus typically shows a compact fibrous core surrounded by a granular cortex (Fig. 1). Previous studies have shown that only the core region contains DNA, whereas both components contain RNA and protein (*3*). The large size of the amphibian oocyte nucleus (*4*) allows rapid isolation and manipulation of the extrachromosomal nucleoli before extensive denaturation and cross-linking of proteins occurs. If saline of low molarity or deionized water is used as the isolation medium, nucleolar cores and cortices can be separated and the DNA-containing cores dispersed for electron microscopy (*5*).

Each unwound isolated nucleolar core consists of a thin axial fiber, 100 to 300 Å in diameter, that is periodically coated along its length with matrix material (Figs. 2 and 3). The axial fiber of each core forms a circle, and treatment with deoxyribonuclease breaks the core axes. The diameter of trypsin-treated axial fibers (about 30 Å) suggests that the core axis is a single double-helix DNA molecule coated with protein (*6*). The matrix segments along a core axis exhibit thin to thick gradations, and show similar polarity along the axial fiber. Each unit is separated from its neighbors by matrix-free axis segments.

Nucleolar core axes are stretched to variable degrees depending on prepara-

tive procedures. For example, drying preparations out of deionized water before staining causes little or no stretching of axial cores (Fig. 2), whereas precipitating preparations with acetone staining solution before drying stretches the core axes to varible degrees over the grid surface (Fig. 3). When regions of core axes appear unstretched or uniformly stretched, the matrix units along a specific region are similar in length; unstretched matrix units are 2 to 2.5 μ long but can be 5 μ long after severe stretching. The matrix-free segments between matrix units also show variations in length due to stretching, but, in addition, exhibit differences in length independent of stretching (Fig. 3). Most matrix-free segments are about one-third the length of adjacent matrix units, but bare regions up to ten times as long as neighboring matrix

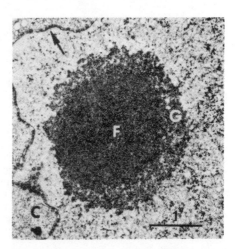

Fig. 1. Thin section of extrachromosomal nucleolus from *Triturus viridescens* oocyte. A granular cortex (*G*) surrounds a compact fibrous core (*F*). Portions of the nuclear envelope (arrow) and cytoplasm (*C*) are visible. Conventional osmium tetroxide fixation, Epon embedding, and uranyl acetate staining. Scale, 1 μ.

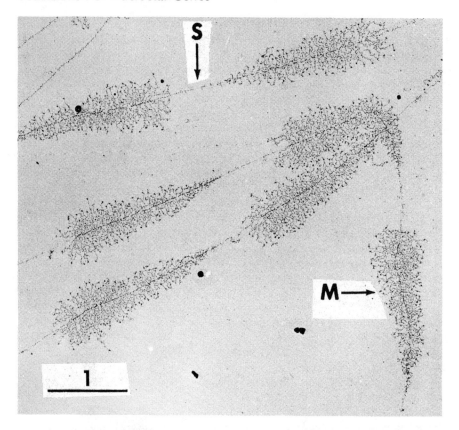

Fig. 2. Portion of a nucleolar core isolated from *Triturus viridescens* oocyte showing matrix units (*M*) separated by matrix-free segments (*S*) of the core axis. The axial fiber can be broken by treatment with deoxyribonuclease, whereas the matrix fibrils can be removed by ribonuclease, trypsin, or pepsin. The specimen was prepared by placing the contents of an oocyte nucleus in deionized water, thus causing dispersal of nucleolar components; the unwound cores were centrifuged through a neutral solution of $0.1M$ sucrose with 10 percent formalin onto a carbon-covered grid; the grid was rinsed in 0.4 percent Kodak Photo-Flo before drying; the preparation was then stained for 1 minute with 1 percent phosphotungstic acid in 50 percent ethanol at pH 2.5 (unadjusted) rinsed in 95 percent and then in 100 percent ethanol, and dried with isopentane. The matrix units and intermatrix segments apparently are unstretched by this procedure. Scale, 1 μ.

units have been observed in both *Xenopus laevis* and *Triturus viridescens*. There appears to be no pattern to the distribution of the longer matrix-free regions along the core axis.

Detailed examination of matrix units shows that each consists of about 100 thin fibrils connected by one end to the core axis and increasing in length from the thin to the thick end of the unit (Fig. 2). Treatment with ribonuclease, trypsin, or pepsin removes the matrix fibrils from the core axis. After labeling of RNA in intact oocytes for

30 to 60 minutes with tritiated ribo-
nucleosides, electron microscopic autoradiography of unwound cores shows
silver grains only over the matrix units.
This initial incorporation corresponds
in time to the appearance of labeled
40S precursor rRNA molecules in nuclear fractions from amphibian oocytes (7). Furthermore, the length of
unstretched to slightly stretched units
(2 to 3 μ) is in close agreement with
the length of DNA required to code
for the precursor molecule synthesized
in amphibians (8). Therefore, we believe that each matrix-covered DNA
segment is a gene coding for rRNA
precursor molecules.

The mechanism of RNA polymerase
action in DNA-dependent synthesis of
RNA involves a polymerization of

monomer ribonucleotides into a poly-
ribonucleotide chain that is immediately dissociated from the template
DNA strand (9). The structural arrangement of the fibrils in each matrix
unit is consistent with a model in which
numerous RNA molecules are sequen-
tially initiated before completion of the
first. Thus, visualization of the genes
coding for rRNA precursors is possible
because many molecules are simultaneously synthesized on each gene. In
that either ribonuclease or proteases
remove matrix fibrils from the core
axis, each fibril probably consists of a
growing rRNA precursor molecule
coated with protein.

If each ribonucleoprotein matrix
fibril contains one growing rRNA precursor molecule, the number of fibrils

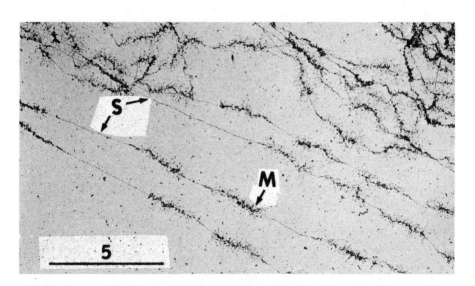

Fig. 3. Portion of nucleolar core isolated from *Triturus viridescens* oocyte, showing
matrix units (*M*) separated by matrix-free segments (*S*) of the core axis. Intermatrix
segments of various lengths are present. Specimen preparation was similar to that
described in Fig. 2, except that the centrifuged specimen was rinsed in water and
dipped without drying into 1 percent uranyl acetate in 80 percent acetone and stained
for 5 minutes. With this procedure, the matrix units and matrix-free segments appear
stretched—slightly to severely—depending on their location on the grid and on how
tightly the centrifugation step has compressed them to the grid surface. Scale, 5 μ.

per matrix unit (about 100) and the dimension of the RNA polymerase molecule in the axis of transcription (about 100 Å) (*10*) indicate that about one-third the length of each gene coding for rRNA precursors is covered with polymerase molecules (*8*). A high concentration of RNA polymerase on rRNA genes also has been reported for *Escherichia coli* (*11*).

In equilibrium gradients, the nucleolar DNA of *X. laevis* can be separated from most of the DNA of the cell as a heavier peak (G·C satellite) (*12–14*). Saturation hybridization (*12, 14*) indicates that about 40 percent of the G·C satellite codes for rRNA (*15*). Annealing experiments with fractionated satellite DNA show that the stretches of DNA coding for 18*S* and 28*S* rRNA are alternating, closely adjoining, but separated by stretches of DNA higher in G·C content and not homologous to rRNA (*13, 14*). The latter observations agree with the evidence that each rRNA precursor molecule consists of one 18*S* and one 28*S* rRNA molecule, plus a portion that is degraded during the formation of the two rRNA molecules (*13, 16*).

We propose that the redundant structural arrangement of the rRNA precursor genes and intergene segments seen in isolated nucleolar cores visually confirms the biochemical nature of nucleolus organizer DNA in amphibians. Thus, the DNA axis of the matrix-covered segments corresponds to the satellite portion that is homologous to the entire precursor rRNA molecule (that is, homologous to one 18*S* and one 28*S* rRNA molecule plus the degraded part of the precursor rRNA molecule), and the DNA in the intergene regions corresponds to the remaining portion of the satellite. Mea-surements of relative lengths of matrix-free and adjacent matrix-covered units in *X. laevis* show that the mean length of intergene segments is about two-thirds the length of a precursor rRNA gene. This indicates that approximately 40 percent of the G·C satellite is inactive nucleolar DNA and about 60 percent consists of genes coding for precursor molecules.

Although the structure of chromosomal loci synthesizing RNA has already been documented, we believe ours are the first observations of the structure of individual genes and associated transcription products whose specific function is known—namely, the extrachromosomal nucleolar genes on which rRNA precursor molecules are synthesized.

References and Notes

1. J. G. Gall, *Proc. Nat. Acad. Sci. U.S.* **60**, 553 (1968); H. C. Macgregor, *Quart. J. Microscop. Sci.* **106**, 215 (1968).
2. D. D. Brown, *Nat. Cancer Inst. Monogr.* **23**, 297 (1966); E. H. Davidson and A. E. Mirsky, *Brookhaven Symp. Biol.* **18**, 77 (1965).
3. O. L. Miller, Jr., *Nat. Cancer Inst. Monogr.* **23**, 53 (1966).
4. The nuclei in mature oocytes of some amphibia are near 1 mm in diameter. Mature oocytes of *Xenopus laevis* are about 1.5 mm, and their nuclei are near 0.6 mm in diameter. Mature oocytes of *Triturus viridescens* are near 1.75 mm and their nuclei are about 0.8 mm in diameter.
5. Details of techniques are given in legends for Figs. 2 and 3. Oocytes of *Xenopus laevis*, the African clawed toad, and *Triturus viridescens*, the spotted newt of eastern North America, were used in these studies. Limited examinations in two other genera, *Rana* and *Plethodon*, indicate that these observations probably extend to all amphibians. Earlier reports of these results are found in: O. L. Miller, Jr., and B. R. Beatty, *J. Cell Biol.* **39**, 156a (1968); ———, in *Handbook of Molecular Cytology*, A. Lima-de-Faria, Ed. (North-Holland, Amsterdam, in press); ———, *Genetics*, in press.
6. The diameter of double-helix DNA determined by electron microscopy of shadow-cast molecules [C. E. Hall, *J. Biophys. Biochem. Cytol.* **2**, 625 (1956)] and uranyl acetate–stained molecules [W. Stoeckenius, *J. Biophys. Biochem. Cytol.* **11**, 297 (1961); M. Beer and C. R. Zobel, *J. Mol. Biol.* **3**, 717 (1961)] is approximately 20 Å.

7. J. G. Gall, *Nat. Cancer Inst. Monogr.* **23**, 475 (1966).

8. For double-helix DNA in the B conformation: 2×10^6 daltons $= 1 \mu$; and 1μ of DNA length codes for 1×10^6 daltons of single-stranded RNA [A. R. Peacocke and R. B. Drysdale, *The Molecular Basis of Heredity* (Butterworths, Washington, D.C., 1965), p. 34]. The molecular weight of the $40S$ precursor rRNA in *X. laevis* has been estimated by acrylamide-gel electrophoresis to be 2.5×10^6 daltons (*14*) and by sedimentation coefficient to be 3.5×10^6 daltons (*13*). These molecules would require, respectively, 2.5μ and 3.5μ of double-helix DNA for synthesis.

9. E. K. F. Bautz, in *Molecular Genetics*, J. H. Taylor, Ed. (Academic Press, New York, 1967), pt. 2, p. 213.

10. E. Fuchs, W. Zillig, P. H. Hofschneider, A. Preuss, *J. Mol. Biol.* **10**, 546 (1964); H. S. Slayter and C. E. Hall, *ibid.* **21**, 83 (1966).

11. H. Bremer and D. Yuan, *ibid.* **38**, 163 (1968).

12. J. G. Gall, *Genetics*, in press.

13. D. D. Brown and C. S. Weber, *J. Mol. Biol.* **34**, 681 (1968).

14. M. Birnstiel, J. Spiers, I. Purdom, K. Jones, U. E. Loening, *Nature* **219**, 454 (1968).

15. The saturation hybridization value is about 20 percent. Since only one strand of the double-helix DNA is copied in transcription, the amount of double-helix DNA containing the sequences homologous to rRNA is 40 percent of the total DNA.

16. R. A. Weinberg, U. Loening, M. Willems, S. Penman, *Proc. Nat. Acad. Sci. U.S.* **58**, 1088 (1967).

17. Sponsored by the AEC under contract with Union Carbide Corporation.

35

CHROMOMERE CONSTANCY AND SPECIFIC MODIFICATIONS OF THE CHROMOSOME STRUCTURE DURING DEVELOPMENT AND ORGAN DIFFERENTIATION IN <u>CHIRONOMUS TENTANS</u>

Wolfgang Beermann

Zoological Institute of Göttingen University and the Max Planck-Institute for Marine Biology, Wilhelmshaven

These exerpts were translated expressly for this Benchmark volume by C. R. Burnham, R. L. Phillips, and Patrick Buescher, University of Minnesota, from pp. 140, 170, 171–177, 178, 179–181, 194, 195–196 of "Chromomerenkonstanz und spezifische Modifikationen der Chromosomenstruktur in der Entwicklung und Organdifferenzierung von Chironomus tentans," Chromosoma **5***:139–198 (1952) with the permission of the publisher, Springer-Verlag, Berlin, Heidelberg, New York.*

It has been known since the investigation of Heitz and Bauer (1933) that the so-called loop nuclei (Schleifenkerne) of the Diptera contain giant chromosomes, which in number and relative dimensions correspond to the mitotic chromosomes, and whose structural organization in the long axis—the cross-banding pattern—is constant and species specific. In *Drosophila* the genetic chromosome map could be placed along the cytological (Painter 1941; Bridges 1935), and through suitable experiments individual genes could be placed in definite cross-bands (Mackensen 1935 and others). It was assumed as self-evident that the cross-banding pattern of the most investigated salivary-gland chromosomes reflected directly the arrangement of the genes also in mitotic chromosomes. The giant chromosomes were described as many-stranded bundles that had arisen through replication and the remaining together of the chromonemata of normal chromosomes (Koltzoff 1934; Bridges 1935; Bauer 1935). Bauer's (1935, 1936) investigations of the fibrillar structure of the salivary-gland chromosomes of chironomids support the polyteny hypothesis. Recently, nevertheless, Kosswig and Sengün (1947a, b, c), Sengün and Kosswig (1947), and Sengün

(1948) have expressed another concept of the development and structure of the giant chromosomes of Diptera based on their findings for *Chironomus* and *Drosophila*. This is that during development of the salivary-gland chromosomes, there is fragmentation followed by a complicated rearrangement of the "chromatic material" which provides the cross-structures. Therefore the cytological localization of genes there would have only limited value, as in the concept of R. Goldschmidt (1950), who assumed that each genetic locus corresponded to a larger chromosome segment comprising several cross-bands.

This new investigation was undertaken because of the differences between the concepts postulated by Kosswig and Sengün and previous concepts.

[*Editors' Note:* Section B. "Materials and Methods" is not translated. This section is followed by a detailed description (pages 142–169) of the general morphology of the salivary-gland chromosomes; then the cross-banding pattern, the cytological maps of the 4 chromosomes (Plates 1 and 2); and the number of cross-bands in each chromosome; the development of the salivary-gland chromosomes of *C. tentans*, the chromosomes of other tissues compared with the salivary-gland chromosomes.]

F. TISSUE AND FUNCTIONALLY SPECIFIC STRUCTURAL MODIFICATIONS.

It has been known for a long time that the salivary-gland chromosomes of the Diptera do not possess along their entire length a uniform clear cross-banding pattern.

Included in the chromosome sections of heterochromatic nature can be, for instance, the proximal part of the X-chromosome of *Drosophila melanogaster* (differing greatly in appearance from the euchromatic areas). Structural disturbances in connection with the nuclei development can also appear.

Finally, at definite places there are swellings of different sizes that have been described by different authors as "bulbs," "puffs," or "secondary nucleolus forming places" and in which the cross-banding pattern is not distinguishable. Such structures are also called "Balbiani-rings" after their discoverer (Balbiani 1881). Comparative studies on the giant chromosomes of *C. tentans* have shown that these structural modifications of euchromatic chromosome regions have a meaning hitherto unknown.

Balbiani Rings of the Salivary-gland Chromosome

a. Structure and Function: In salivary-gland chromosome 4 of *C. tentans*, 3 Balbiani-rings can occur (see Bauer and Beermann 1952). In larvae growing under natural conditions, a large Balbiani-ring appears regularly at about the midpoint of the chromosome and a considerably smaller one farther to the right, almost terminal. An additional occasional Balbiani-ring in the left third of the chromosome is not found in wild animals; in its place is a large variable loosened-up zone.

The Balbiani-rings are characterized by their regular rotational symmetry. Their typical appearance is like nearly spherical rotational ellipsoids whose main axis is that of the chromosome; the spaces between the Balbiani-rings and the adjacent chromosome segments on both sides are cone-shaped [see Fig. 30b, c]. Whether in live or in fixed preparations, the surface possesses a sharply laid down boundary that seems lightly wavy or bumpy in outline. The interior of the Balbiani-rings in live observations (phase contrast) is somewhat like protoplasm in tone and so is not so dark as the main nucleoli. After a fine granulation the main branches of the split chromosomes usually can be recognized in vivo.

In good acetocarmine preparations, one can follow further the ramification of the chromosome in the region of the Balbiani-rings. It occurs in unpaired, "haploid" chromosomes in the same way as in those completely paired. The chromosome splits in a small number of branches of the same thickness which, during strong regeneration, diverge externally from the Balbiani-ring. In the typical case the main branches coming from both sides intersect; the adjacent pieces of the chromosome seem to be shoved into each other. That extends also to a comparison of the relative cross-band intervals in the anterior end of salivary-gland chromosome 4 at one time in typical expression of the number 1 Balbiani-ring lying farthest to the left and at another time by its absence (Table 4).

Table 4. **Relative intervals of Band 1C1, 2C1, and 4A1 from left end of the salivary-gland chromosome 4 of 2 individuals (a and b). (Relative intervals in percent of total length; 3-fold average error; $n = 10$ in both cases.)**

	1C1	2C1	4A1
Balbiani-Ring			
a. not developed	10.7 ± 1.0	27.1 ± 2.4	48.1 ± 2.2
b. fully developed	10.2 ± 0.8	21.8 ± 1.2	46.6 ± 1.8

In smaller nuclei with indefinite cross-band structure, the mutual approach of the adjacent segments can give the impression that the chromosome is unchanged throughout the Balbiani-ring, while thin threads appear in radial directions from the cross-bands just as Poulson and Metz (1938) have described. Normally at least the main branches possess a clear cross-band structure. The interval of the cross-bands is greater on the inside of the branch than on the outside and increases with their regeneration. The division into fibrillar bundles does not occur very regularly; the main branches divide about on the same level into a small number of fine strands, again running into each other, and the process repeats itself manyfold up to the periphery of the Balbiani-ring where the finest branching is not resolvable by light optics. The branching is accompanied by increased elongation of the interbands that result from the interval relationships of the cross-bands. Thereby a large number of very fine bands becomes recognizable. From time to time in short stretches dichotomous cleavage occurs; however, it is not the rule. The fibrils and fibrillar bundles are embedded in a Feulgen negative substance in the Balbiani-ring, whose concentration seemed to be greatest in the region of the finest peripheral branching—after subjective observation in acetocarmine stained preparations.

The Balbiani-rings are places of special metabolic activity that can be demonstrated easily experimentally: One brings full-grown larvae that were kept at least 1 hour in the cold at about 5°C back into water at room temperature; then after about 1½ hours at the earliest, a clearly recognizable (in vivo, phase contrast) accumulation of material precipitates. The periphery of the Balbiani-ring seems at the beginning of the reaction to be impregnated with many dark little drops, which later run together into larger balls. These little drops clearly arise from the above mentioned fibril covering substance. The different stages of the reaction must be followed at each particular point of time from freshly prepared salivary glands. They may, however, occasionally be followed further under the coverglass. At the latest, after 8 hours the Balbiani-rings again show the normal picture. One can treat these "cold effects" as a disturbance of the metabolic balance in the area of the nucleus: The Feulgen negative substances formed at the Balbiani-rings are transported away more slowly than the synthesis itself; thereby the extra material precipitates as coacervate.

b. Variability: Balbiani-ring 1 of the salivary-gland chromosome 4 is extraordinarily variable in its expression. A series of variants is assembled in Fig. 30. There, only "*f*" is to be viewed as a typical Balbiani-ring: In contrast to that, "*e*" sets itself apart through progressive splitting and overlapping of the main branches. Chromosomes with forms *a* to *e* or entirely without formation of Balbiani-ring 1 can appear in wild animals in the same glands next to each other. The extreme expression (*f*) was observed only in larvae raised in the laboratory; in these, on the other hand, forms *a–d* do not appear.

In contrast to the great variability of Balbiani-ring 1, the 2nd and especially the 3rd, show a remarkable constancy. Under laboratory conditions Balbiani-ring 2 occasionally does not attain its extreme development; however, among many more than 100 preparations only 2 were found in which in a single cell, it was completely absent (Fig. 31).

The occurrence of intraindividual variability in the expression of the Balbiani-ring allows the possibility that part of the differences from animal to animal is also of a phenotypic nature. Experiments, to be sure, to attain a comparison with the situation described above in wild animals through changes in the growing conditions (temperature, food)

during larval development are lacking up to now. The reciprocal experiment—growth in the wild of animals from the laboratory—is not
yet done. Against this, the appearance of heterozygotes (cf. Section 3)
gives proof that part of the variants, even if they are very rare, are genetic; Fig. 32 shows 2 of the 3 hitherto observed cases.

By comparing the individual transitional forms, one can pinpoint exactly the places from which the Balbiani-rings begin. Both large Balbiani-rings can be traced back to cross-bands designated as BR 1 and
BR 2 in the map of salivary-gland chromosome 4; in the case of the
small Balbiani-ring 3, this was not possible.

In the homologous segments of chromosome 4 in the other tissues
investigated, Balbiani-rings or their comparable structures do not appear. Only in number 4 chromosomes in Malpighian tubules- and mid-
gut chromosomes does one find in the region of the well-identifiable
BR 1 band merely a slight swelling (Fig. 33). This fact cannot be due to
less polyteny of the malpighian tubule, rectum, and mid-gut chromosomes since Balbiani-rings are found in the salivary-gland chromosome
4 in much younger larvae. Also the assumption of a stronger torsion of
the individual fibrils being somewhat correlated with the compact construction of salivary-gland chromosomes does not explain the exclusive
appearance of the Balbiani-rings in the salivary glands since at certain
other places in the chromosome the Malpighian tubule chromosomes
can form loosened-up zones similar to those of the Balbiani-rings (Fig.
40b). Therefore the Balbiani-rings of salivary-gland chromosome 4 are a
salivary-gland specific structural modification of certain chromosome
positions that can be traced back to a single cross-band.

Occasionally the positions where nucleoli form take on a Balbiani-
ring-like structure; this was up to now ascertained with certainty only
once. In that particular animal the nucleolus-forming zone of chromosome 3 was expressed as a Balbiani-ring in every salivary-gland nucleus
(Fig. 34). The nucleolar substance is clearly defined at the periphery of
the rings.

Modifications of Other Kinds

a. Presence, structure, and function: The Balbiani-rings represent
only a special case of frequently occurring changes in structure of individual points in the chromosome connected with increased functional
activity. On the other hand, there are such positions of increased metabolism in other chromosome segments which, as expected, after cold
treatment are clearly points for the deposit of thymonucleic acid-free
material. At other positions in the chromosome, one observes only occasionally also without pretreatment the presence of achromatic drops.

b. Variability: Also during development of the individual, specific differences in the expression of single chromosome loci appear: In a
comparative investigation of 20 larvae and 20 prepupae from 3 inbred

lines, it was established that some structural modifications occurred at times in the larvae, others only in the prepupae, and a third group in both stages. Thus, for example, a swelling in segment 19A of chromosome 1 was found in the larval salivary gland, which was completely reduced in the prepupae; conversely, there appeared here a new swelling in segment 18C. Similar changes are observed in the Malpighian tubule and rectum chromosomes at the setting in of metamorphosis.

According to expectations the comparison of the chromosomes of different organs shows, in conclusion, that the striking differences in expression of structural modifications are a reflection of functionally far-reaching different parts of the organism; some examples of that are described below in the discussion of Fig. 39 (the salivary-gland specificity of the Balbiani-rings of chromosome 4 was indicated on page **282**).

> Unless noted otherwise, only chromosomes fixed and stained in the same way and from the same individual were compared, and the individual patterns in the chromosome segments concerned were repeated in all cells of the investigated organs.

Fig. 39 shows the relationships in the region of the already mentioned inversion in Region 14 of chromosome 3. The segment includes 4 cross-bands (in the mid-gut) and can be seen in the 4 tissues, altogether 6 positions (marked in the diagram), which appear swollen or diffuse. An individual organ shows at times only part of these structurally changed positions; in the mid-gut those generally do not appear in the segments described. Individual cross-bands in different tissues may appear modified at the same time, others only in one organ. Between those which show in different tissues modified positions at the same time, there are moreover differences in the degree of modification (compare point 4 in rectum and malpighian tubule). The general scarcity of structural modifications is typical for the mid-gut chromosomes, to a certain extent also for the salivary-gland chromosomes.

SUMMARY

1. *Chironomus tentans* has (haploid) three long V-shaped and one short rod-shaped chromosomes. Two of the long ones carry a nucleolus.

2. The elongated condition of the giant chromosomes in developing larvae (measured as the diameter/length relationship) is tissue-specific and individually different.

7. The structure of the cross-banding may be modified secondarily. Such local structural modifications are the Balbiani-rings and the larger number of observed swellings and diffuse zones ("puffs" and "bulbs" of earlier authors) whose origin in all cases can be traced back to individual cross-bands. The structurally modified positions are probable phases of especially high metabolism, as shown experimentally, for example, for the Balbiani-rings.

8. Appearances and expression of the different structural modifica-

tions are variable, only minor variations within the cells of an organ; on the other hand, specific differences of that type are found from tissue to tissue, such that one can speak of a tissue-specific pattern of structural modifications; also there are differences between larvae and prepupae: Some structural modifications recur, others appear anew.

9. The results are a further confirmation of the polyteny hypothesis; coiled longitudinal elements are the basis for the different structural conditions of the salivary-gland chromosomes; also the structure of the Balbiani-rings is traceable to the coiling structure. The findings about the structural modifications are the first direct cytological indication that the individual elements of the genome react differentially to internal as well as external conditions.

REFERENCES

Balbiani, E. G. 1881. Sur la structure du noyau des cellules salivaires chez les larves de *Chironomus. Zool. Anz.* **4**:637–641.

Bauer, H. 1935. Der Aufbau der Chromosomen aus den Speicheldrüsen von *Chironomus Thummi* Kiefer. *I. Z. Zellforsch.* **23**:280–313.

_____. 1936. Beiträge zur vergleichenden Morphologie der Speicheldrüsen-chromosomen II. *Zool. Jb., Abt., allg. Zool. u. Physiol.* **56**:239–276.

_____ und W. Beermann. 1952. Die Polytänie der Riesenchromosomen. *Chromosoma.* **4**:630–648.

Bridges, C. B. 1935. Cytological data on chromosome four of *Drosophila melanogaster. Trud. Dinam. Razvit.* **10**:463–475. [Russian summary]

Goldschmidt, R. B. 1950. Marginalia to McClintock's work on mutable loci in Maize. *Am. Nat.* **84**:437–455.

Heitz, E., und H. Bauer. 1933. Beweise für die Chromosomennatur der Kernschleifen in den Knäuelkernen von *Bibio hortulanus* L. *Z. Zellforsch.* **17**:67–82.

Koltzoff, N. K. 1934. The structure of the chromosomes in the salivary glands of *Drosophila. Science.* **80**:312–313.

Kosswig, C., und A. Sengün. 1947a. Neuere Untersuchungen über den Bau der Riesenchromosomen der Dipteren. *Rev. Fac. Sci. Univ. Istanbul,* Ser B. **12**:107–121.

_____ und A. Sengün, 1947b. Vergleichende Untersuchungen über die Riesenchromosomen der verschiedenen Gewebearten verschiedener Dipteren. *C.r. Ann. et Arch. Soc. Turque Sci. Phys. et Nat.* **13**:94–101.

_____ and A. Sengün. 1947c. Intraindividual variability of chromosome IV of *Chironomus. J. Hered.* **38**:235–239.

Mackensen, O. 1935. Locating genes on salivary chromosomes. *J. Hered.* **26**:163–174.

Painter, T. S. 1941. An experimental study of salivary chromosomes. *Cold Spring Harb. Symp. Quant. Biol.* **9**:47–54.

Poulson, D. F., and C. W. Metz. 1938. Studies on the structure of nucleolus-forming regions and related structures in the giant salivary gland chromosomes of Diptera. *J. Morph.* **63**:363–395.

Sengün, A. 1948. Vergleichend-ontogenetische Untersuchungen über die Riesenchromosomen verschiedener Gewebearten der Chironomiden. *I Comm. Fac. Sci. Univ. Ankara* **1**:187–248.

Sengün, A., und C. Kosswig. 1947. Weiteres über den Bau der Riesenchromosomen in verschiedenen Geweben von *Chironomus*-Larven. *Chromosoma* **3**:195–207.

[*Editors' Note:* The figures mentioned in the translation appear on the following pages.]

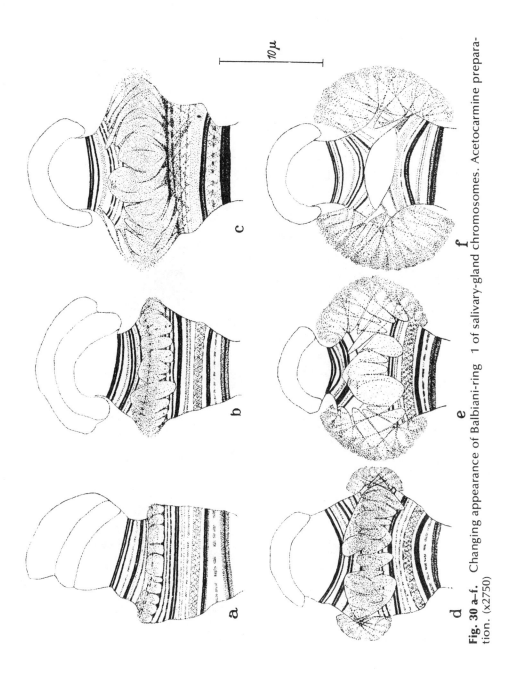

10 μ

Fig. 30 a–f. Changing appearance of Balbiani-ring 1 of salivary-gland chromosomes. Acetocarmine preparation. (x2750)

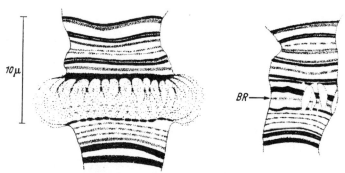

Fig. 31. The region of salivary-gland chromosome 4 in which normally the Balbiani-ring 2 appears; individual patterns in 2 cells of the same gland. Acetocarmine-euparal. (x 2660)

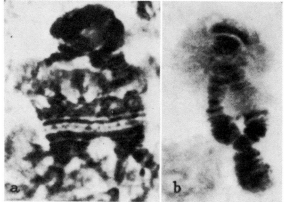

Fig. 32 a and b. Heterozygosity for the formation of the Balbiani-zones. a. In region of the Balbiani-ring 1 (in one partner the cross-band structure is slightly distorted; in the other interrupted due to being greatly spread out). b. In the zone of the Balbiani-ring 2 (here the Balbiani-ring 1 is completely formed in both homologues). (a) Acetocarmine-euparal; (b) Toluidine-blue-canada balsam. (x1500). (a) phase contrast.

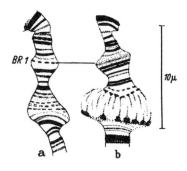

Fig. 33 a and b. Homologous segments of chromosome 1 from the mid-gut (a), and malpighian tubules (b). BR 1 cross-band, from which in the salivary gland probably the formation of Balbiani-ring 1 occurs. In (b) a loosening-up zone distal from BR 1, which often is also observed in salivary glands. Acetocarmine. (x2660)

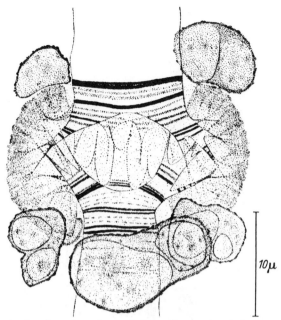

Fig. 34. Nucleolus-forming place of the salivary-gland chromosome 4 in the expression of a Balbiani-ring. The nucleolar material lies in the form of irregular balls at the surface of the Balbiani-ring. Acetocarmine-euparal. (x2000)

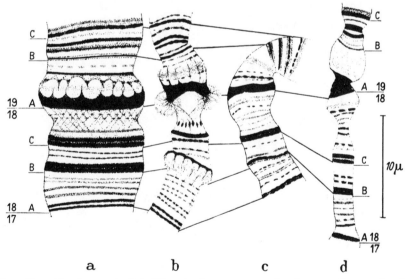

Fig. 40 a–d. Interval 18A-19B of chromosome 1, from salivary glands (a), malpighian tubule (b), rectum (c) and mid-gut (d) of the same individual. Acetocarmine-euparal. (x2660)

287

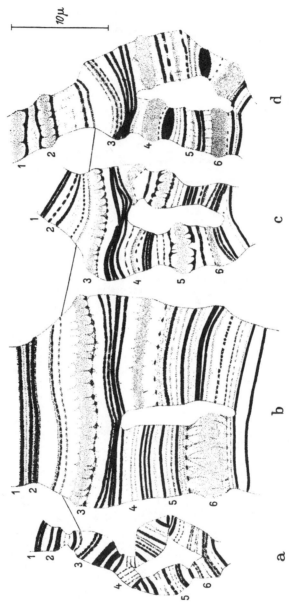

Fig. 39 a–d. Tissue specificity in the expression of individual positions in the region of the small inversion in Region 14 of chromosome 3. a. mid-gut. b. salivary gland. c. malpighian tubule. d. rectum. Homologues, in individual organs, the diffuse expressed positions marked with numbers. From the same animal. Acetocarmine-euparal. (x2660)

36

Reprinted from pp. 294, 296–303, 322, and 323 of *Z. Zellforsch. Mikrosk. Anat.*
21(2):293–328 (1934)

THE RELATION OF A PARTICULAR CHROMOSOMAL ELEMENT TO THE DEVELOPMENT OF THE NUCLEOLI IN ZEA MAYS.

By

BARBARA MCCLINTOCK [1].

California Institute of Technology, Pasadena, California.

[*Editors' Note:* Plates VIII–XIV, showing photographs 1 through 48, are not reproduced here.]

Introduction.

The study described in this paper was undertaken to show that the nucleolus is organized in the telophase by an enlarged, morphologically distinct, deep-staining chromosomal body which appears at a definite position in one chromosome of the monoploid complement of *Zea mays*. It was known that only one nucleolus appeared in each sister nucleus at the telophase of the first nuclear division in the microspore. The microspore contains the monoploid complement of ten chromosomes, one of which possesses the deep-staining body mentioned above. In diploid somatic tissue, two such chromosomes are present. The telophase nuclei in such diploid tissue show two nucleoli. In both the diploid somatic nuclei and the haploid spore, the deep-staining body mentioned above was always observed to be adjacent to the nucleolus. An interchange, produced through X-rays treatment, (ANDERSON, in press) divided this body into two parts, each interchanged chromosome possessing a section. It was found through examination of plants homozygous and heterozygous for this interchange that both sections of this body could function independently to produce a nucleolus.

[*Editors' Note:* Material has been omitted at this point.]

---

[1] National Research Council Fellow in the Biological Sciences.

Structure of the satellited chromosome.

The structure of the satellited chromosome (chromosome 6 of the monoploid complement) in *Zea mays* is best seen at the mid-prophase of meiosis (pachytene) [1]. This is shown photographically in fig. 1. For descriptive purposes it has been divided into six parts. Part *1*, fig. 1, is the satellite proper. It is usually composed of four distinct chromomeres. The basal chromomere is adjacent to the nucleolus. A thin, practically colorless thread or ribbon, part *2* (not clear in this photograph), running across the surface of the nucleolus joins the basal chromomere of the satellite with a large deep-staining body, part *3*, (and arrow photo. 10, plate IX) which is conjoined to the nucleolus. The structure of this deep-staining body is sometimes definitely reticulate. It will be shown in the following pages that this particular body, possessed only by chromosome 6 of the normal maize complement, is responsible for the orderly organization of the telophase nucleoli. In this paper it will be referred to as the nucleolar-organizing body. This conspicuous body is followed by a chromatic thread, part *4*, composed of chromomeres, which extends to the spindle fiber attachment region, part *5*. Part *6* is the long arm of the satellited chromosome. Besides the distinctive features of chromosome 6 already described it may possess, toward the end of the long arm, one or two knobs. These knobs have characteristic physical features and definite locations in the chromosome. It has been pointed out previously (CREIGHTON and McCLINTOCK, 1931) that such knobs can be followed through generations with the same precision as genes. Whether a particular satellited chromosome (chromosome 6) has none, one or two knobs depends, therefore, upon the particular culture which is used.

At somatic telophases, in diploid plants with two normal satellited chromosomes, two nucleoli, usually of similar size [2], are formed. Frequently, fusion of the two nucleoli takes place in the early telophase to form one. Thus, in diploids, somatic nuclei are characterized by one or two nucleoli.

---

[1] For variations in the types of satellited chromosomes, see page 307.

[2] For variations in size see page 316.

The attachment of the satellited chromosome to the nucleolus in somatic prophases is similar to that described for the microspore stage given below.

The microspore is very convenient for such a study. Only one chromosomal complement is present. Each chromosomal complement is isolated in a separate spore. The methods allow the prophase figures to be somewhat flattened thus affording a better view of the satellited chromosome.

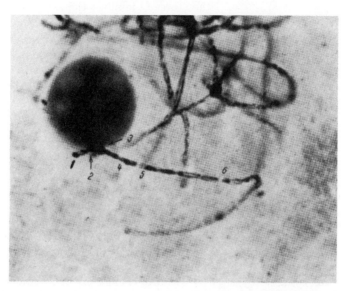

Fig. 1. Photomicrograph of two homologously associated chromosomes 6 (satellited chromosome) in the mid-prophase of meiosis showing the relationship to the nucleolus. The parts of the chromosome have been numbered; *1.* the satellite proper; *2.* the region joining the satellite to the, *3,* deep-staining nucleolar-organizing body; *4.* the chromatic thread of the short arm; *5,* the spindle fiber attachment region; *6.* the chromatic thread of the long arm. Magnification approx. 1400 ×.

Large numbers of prophase figures are available. In the prophase of a microspore with a normal monoploid complement there are ten identifiable chromosomes. There is but one nucleolus. The satellited chromosome is attached to the nucleolus, fig. 17. The attachment of this chromosome to the nucleolus is similar to that in the prophase of meiosis with the exception that the satellite, with its basal part adjacent to the nucleolar surface, is usually some distance removed from the nucleolar-organizing body. In most cases it is not possible to see a connection joining the satellite with the nucleolar-organizing body. In the drawings, therefore, a connection has been indicated by a dotted line. As the prophase approaches the metaphase conditions, the nucleolus decreases in size. As it does so the satellite begins to approach the nucleolar-organizing body. However, before this is concluded, the rapid dissolution of the nucleolus completely releases the satellited chromosome. The thread join-ing the satellite with the nucleolar-organizing body is then clearly visible. The length of this thread varies greatly in different figures and appears to depend upon the distance the satellite was removed from the nucleolar-organizing body in the previous prophase and also, upon the stage at which the satellited chromosome is released from the nucleolus in the late prophase. This variation in length of the thread joining the satellite to the nucleolar-organizing body is visible in the metaphase chromosome. The microspore is particularly useful for telophase studies since the two nuclei in a spore are known to be sister nuclei. A single nucleolus is formed in each sister nucleus. They lie in corre-sponding positions.

Fig. 2. Diagram of the chromo-somes involved in the interchange. *a* chromosomes 6 and 9. The short arm of chromsome 9 termi-nates in a knob. The breaks in the chromosome connected by slighthly bulging light lines indi-cate the position of the spindle fiber attachment region. The arrows point to the position in each chromosome where the inter-change occurred. *b* The two chromosomes resulting from the interchange. Chromosome 9⁶ has the small section of the nucleolar organizing body; chromsome 6⁹ has the larger section of the nucleolar-organizing body.

Description of interchange.

Proof that the deep-staining body of chromosome 6 which lies adjacent to the nucleolus is associated with the orderly organization of the telophase nucleoli was obtained from an interchange which di-vided this region into two parts. Each part was found capable of organizing a separate nucleolus.

The interchange involved chromosome 9 and chromosome 6 (satellited chromosome). Chromosome 9, fig. 2, is characterized by the relative lengths of its two arms. The long arm is almost twice the length of the short arm when measured in the mid-prophase of meiosis. In most cultures, the end of the short arm possesses a deep-staining knob. Chromosome 6 has been described on page 296. The interchange occurred at the position of the arrows, *a*, fig. 2, to produce the two chromosomes 9⁶ and 6⁹, *b*, fig. 2. From the diagram it can be seen that the interchange divided the deep-staining body adjacent to the nucleolus into two unequal parts. Chromosome 9 lost two-thirds of its long arm and received in its place a small section of the deep-staining body of chromosome 6, plus the satellite of chromo-some 6. Since it retained all of the short arm, the spindle fiber attach-ment region and one-third of the long arm of chromosome 9, this new chromosome has been designated chromosome 9⁶. Chromosome 6, on the other hand, retained a large portion of the deepstaining body. The segment from the long arm of chromosome 9 attached itself at its broken end to the broken end of the deep-staining body of chromosome 6. The resulting chromosome has been designated chromosome 6^9 since it possesses the long arm, the spindle fiber attachment region and most of the short arm of chromosome 6.

Somatic and meiotic nucleolar conditions in plants homozygous for the interchange.

It has been stated on page 294 that two nucleoli develop in the somatic telophase of normal diploid plants. There are two chromosomes 6 in each somatic nucleus. The two nucleoli are developed in conjunction

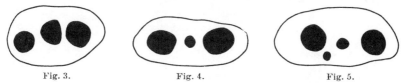

Fig. 3.   Fig. 4.   Fig. 5.

Fig. 3. Outline sketch of a nucleus of a root-tip cell in a plant trisomic for chromosome 6. Three nucleoli of approximately similar size are present.

Fig. 4. Outline sketch of a nucleus of a root-tip cell in a plant heterozygous for the interchange. There are two large nucleoli and one small nucleolus. The two large nucleoli resulted from the functioning of the nucleolar-organizing elements of chromosomes 6 and 6^6, the small nucleolus from the functioning of the nucleolar-organizing elements of chromosome 6^9.

Fig. 5. Outline sketch of a nucleus from a root-tip cell in a plant homozygous for the interchange. The two large nucleoli were formed through the functioning of the nucleolar-organizing elements of the two chromosomes 9^6; the two small nucleoli were formed through the functioning of the nucleolar-organizing elements of the two chromosomes 6^9. (The small nucleolus to the left has been displaced in the drawing since it was lying in the same plane as the large nucleolus to the left.)

with the nucleolar-organizing bodies of each chromosome 6, respectively. The union between the nucleolar-organizing body and the nucleolus is retained through the resting period and into late prophase of the next nuclear division.

The nucleolar condition in diploid plants homozygous for the interchange is strikingly different from that observed in normal diploids. The maximum number of nucleoli present in a nucleus is not two, as in normal diploids, but four. Of these four nucleoli, two are large and two are small, fig. 5.[1] The relation between these four nucleoli and the interchanged chromosomes will be brought out in the description which follows. Plants homozygous for the interchange possess two each of chromosomes 9^6 and 6^9. Homologous associations at the mid-prophase of meiosis gave three main types of figures. The first type is shown in photos. 1, a and b, and 2 a, b and c, and sketches of same, fig. 6 and 7,

[1] Due to fusions between several or all of these four nucleoli other nucleolar conditions, with reference to number and size were observed.

respectively. There is one large nucleolus. Chromosome 9^6 bivalent is attached to the nucleolus by its segment of the nucleolar-organizing body. The satellite is usually a short distance removed from this body and joined to it by an almost unstained thread or ribbon which is likewise attached to the nucleolus. This association is similar to that described

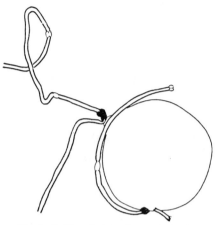

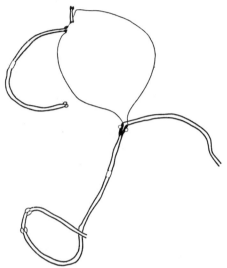

Fig. 6. Outline sketch of a pachytene configuration in a plant homozygous for the interchange. See photograph of the same, photo. 1a and b. Both chromsome pairs are attached to the nucleolus by their deep-staining nucleolar-organizing bodies. Chromosome pair 6^9 is to the left, chromosome pair 9^6 to the right. The spindle fiber attachment region is indicated by the slightly bulging light lines. This region of chromosome 9^6 is clearly visible in the photograph.

Fig. 7. Outline sketch of a pachytene configuration in a plant homozygous for the interchange. See photograph of same, photo. 2a, b and c. Chromosome 9^6, upper left, chromosome 6^9 below. Both are attached to the single nucleolus by their nucleolar-organizing bodies.

for the normal chromosome 6, page 296. Chromosome 6^9 is also attached to the nucleolus by its section of the nucleolar-organizing body. However, in this case, the interchanged piece belonging to the long arm of chromosome 9 is almost always joined directly to the deep-staining segment of the nucleolar-organizing body of chromosome 6^9, i. e., it is not removed from this body by an unstained thread. This can be seen in photos. 1 and 2. In these photos. the difference in size between the two segments of the nucleolar organizing body possessed by each interchanged chromosome is evident.

The second type of configuration is illustrated in photos. 3 and 4. These cells were photographed at a lower magnification than those shown in photos. 1 and 2. The two interchanged chromosomes with their respective nucleoli, shown in photo. 3, are sketched in fig. 8. In these sporocytes there are two nucleoli, one large and one small. The chromosome bivalent 9^6 is always associated with the large nucleolus, the chromosome bivalent 6^9 with the small nucleolus. It is necessary to emphasize this correlation since it is a part of the evidence which has lead to the conclusion that the small segment of the nucleolar-organizing body possess-cd by chromosome 9^6 produces a large nucleolus whereas the large segment of the nucleolar-organizing body possessed by chromosome 6^9 produces a small nucleolus.

A third type of configuration is sometimes found. In these figures the nucleolar-organizing bodies of each of the four chromosomes are

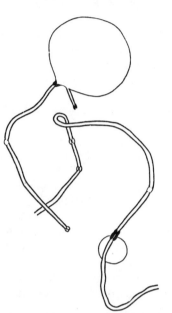

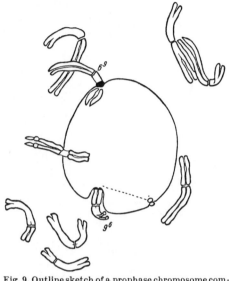

Fig. 8. Outline sketch of a pachytene configuration of a plant homozygous for the interchange. See photograph of same, photo. 3. Chromosome 9^6 is associated with the larger nucleolus; chromosome 6^9 with the smaller nucleolus.

Fig. 9. Outline sketch of a prophase chromosome complement in the microspore of a plant homozygous for the interchange. See photograph of same, photo. 28. Chromosomes 6^9 and 9^6 are associated with the single nucleolus. Note that the satellite of chromosome 9^6 is distantly removed from the small but deeply-staining nucleolar-organizing body of this chromosome and that the translocated segment of chromosome 6^9 lies close to its deep-staining nucleolar-organizing body. The "breaks" in chromosomes are the achromatic spindle fiber attachment regions.

together. The resulting configuration resembles those found in plants heterozygous for the interchange.

The microspores of these plants were studied. The prophase complements were of two kinds, those with one nucleolus and those with two nucleoli. In microspores with one nucleolus, chromosomes 9^6 and 6^9 are both attached to the single nucleolus, fig. 9 and photograph of same, photo. 28. The satellite on chromosome 9^6 is always some distance removed from the small segment of the nucleolar-organizing body of this chromosome (dotted line, fig. 9). Chromosome 6^9 is attached to the nucleolus by its section of the nucleolar-organizing body. The section of the long arm of chromosome 9 in chromosome 6^9 usually lies close to the nucleolar-organizing body of this chromosome, fig. 9. Occasionally, however, it is removed a short distance from the nucleolar-organizing body. In this case, its basal part remains adjacent to the nucleolus. In an occasional spore chromosome 6^9 appeared entirely free from the nucleolus. Chromosome 6^9 in these spores can be readily identified by its deep-staining knob and also its deep-staining nucleolar-organizing body. In these cases, the section of the long arm of chromosome 9 possessed by chromosome 6^9 is always directly adjacent to the nucleolar-organizing body.

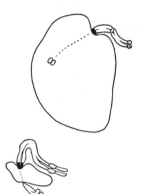

The second type of spore possessed two nucleoli, one large and one small, fig. 10. In this case, chromosome 9^6 is always associated with the large nucleolus, chromosome 6^9 with the small nucleolus.

The metaphase chromosome complement in a spore of a plant homozygous for the interchange is shown in fig. 11 and photograph of same, photo. 31. It will be noted that there is a small satellited chromosome and a longer chromosome with two "constrictions". The small satellited chromosome is chromosome 9^6, the longer chromosome with two "constrictions" is chromosome 6^9. The distal "constriction" on chromosome 6^9 is associated with the previous nucleolar attachment, whereas the more median "constriction" is the spindle fiber attachment region.

Fig. 10. Chromosomes 6^9 and 9^6 from a prophase of a microspore in a plant homozygous for the interchange. Chromosome 9^6 is associated with the larger nucleolus. The satellite is some distance removed from the nucleolar-organizing body of this chromosome. Chromosome 6^9 is associated with the smaller nucleolus. The translocated segment is removed from the nucleolar-organizing body of this chromosome during the growth of the small nucleolus.

The telophases of the first division of the nucleus in the microspore are characterized by the production of two nucleoli in each sister nucleus, one large and one small. The positions of the nucleoli correspond in the two nuclei, photos. 18, 19 and 20. Fusion of the two nucleoli frequently takes place very early in the telophase, upper nucleus photo. 20. The difference in size between the two nucleoli is quite striking in most telophase nuclei. However, there is some variation. In a few spores the two nucleoli in the telophase nucleus were more nearly the same size. This however, is rare and will be considered on page 316.

Examination of nuclei of plants homozygous for the interchange in diploid somatic tissue, in meiosis and in the microspores has been particulary instructive (1) in indicating that the interchange divided the nucleolar-organizing body into two unequal parts, (2) in showing the relative amounts of the nucleolar-organizing body that each chromosome possessed, (3) in proving that each section of the divided nucleolar-organizing body can function to produce a nucleolus and (4) in showing that the nucleoli produced by each section of the nucleolar-organizing body bear a definite size relationship to one another: the small section of the nucleolar-organizing body carried by chromosome 9^6 produces a large nucleolus whereas, the large section of the nucleolar-organizing body carried by chromosome 6^9 produces, in contrast, a small nucleolus. The following interpretation is placed upon this latter observation. The usual satellited chromosome (chromosome 6) possesses a large, somewhat elongated nucleolar-organizing body (*3*, fig. 1), which is confluent with the nucleolus mainly at one region. The rest of the nucleolar-organizing body frequently is free from the nucleolus. Examination of plants homozygous for the interchange (and also those heterozygous for the interchange) indicates that the point of interchange is near that end of the nucleolar-organizing body which is directly adjacent to the nucleolus. In consequence, chromosome 9^6 obtained the small section of the nucleolar organizing body which in the normal chromosome 6 is usually confluent with the nucleolus, and chromosome 6^9 retained that part of the nucleolar-organizing body of chromosome 6 which, in the normal chromosome 6, is usually free from the nucleolus. This evidence suggests that the functional capacities of different sections of the nucleolar-organizing body vary, i. e. that certain regions may possess greater functional capacities than other regions. On this basis, it can be assumed that in the normal chromosome 6 the distal region (farthest from the spindle fiber attachment region) has a greater functional capacity than the proximal region. That this differential capacity for functioning is a matter of speed reaction will be brought out later (page 316).

Fig. 11. Metaphase chromosomes in the microspore of a plant homozygous for the interchange. See photograph of the same, photo. 31. The small chromosome with the satellite (upper left) is chromosome 9^6; the chromosome with the secondary "constriction" (to the right of chromosome 9^6) is chromosome 6^9.

[*Editors' Note:* Material has been omitted at this point.]

Summary.

1. The nucleolus is organized in the telophase through the activity of a distinct deep-staining body having a definite position in one chromosome (the satellited chromosome) of the monoploid complement. Correlated with the number of satellited chromosomes present, the telophases of somatic tissue of haploids show one nucleolus, diploids, two nucleoli and triploids, three nucleoli. That the nucleolus develops through the activity of this body (refered to as the nucleolar-organizing body or clement) was obtained from a reciprocal translocation which broke this body into two parts. Both interchanged chromosomes possessed a section. Nucleoli developed from *each* of these two segments. Thus, plants homozygous for .the interchange developed four nucleoli in their somatic telophases; plants heterozygous for the interchange developed three nucleoli in their somatic telophases. Similarly, the telophase nucleoli resulting from the first division within the monoploid microspore of normal diploids show only one nucleolus, whereas, those of plants homozygous for the interchange are characterized by the development of two nucleoli.

2. The functional capacity to develop a nucleolus is not the same for both segments of the severed nucleolar-organizing body. This is evident when the two interchanged chromosomes are present in the same nucleus. The segment of the nucleolar-organizing body possessed by one interchanged chromosome produced a large nucleolus, whereas, the segment of the nucleolar-organizing body possessed by the other interchanged chromosome produced a small nucleolus. When this latter chromosome, with the nucleolar-organizing element of slower rate of functional capacity is present without the former (i. e. without a competing nucleolar-organizing element) it produces, in contrast, a large nucleolus.

3. The activity of the nucleolar-organizing clement is hindered by certain genomic deficiencies. When this occurs, many small nucleolar-like bodies are produced and remain associated with the other chromosomes of the complement. These small nucleoli appear to develop from a swelling and later collection into droplets of the matrix substance of the chromosome.

[*Editor's Note:* Material has been omitted at this point.]

REFERENCES

Anderson, E. G. 1934. A chromosomal interchange in maize involving the attachment to the nucleolus. *Am. Nat.* **68**:345–350.

Creighton, H. B., and B. McClintock. 1931. A correlation of cytological and genetical crossing-over in *Zea mays. Proc. Nat. Acad. Sci. U.S.A.* **17**:492–497.

37

Reprinted from *Natl. Acad. Sci. (U.S.A.) Proc.* **28**(10):433–436 (1942)

ON THE ANAPHASE MOVEMENT OF CHROMOSOMES

By M. M. Rhoades and Hilda Vilkomerson

Department of Botany, Columbia University

Communicated August 22, 1942

The kinetic properties of chromosomes are controlled by the centromere or kinetochore. In somatic mitoses it is the centromeres which become oriented on the equatorial plate while the two arms of each chromosome may extrude from the spindle figure into the cytoplasm. At metaphase a thread-like structure, descriptively known as the spindle fiber, can be seen in properly fixed and stained material extending from the centromere to the pole. The centric region plays a decisive rôle in congression, orientation on the spindle and anaphase separation. Acentric fragments behave in an irregular fashion. The centric region becomes "attached" to the spindle and leads the way to the pole at anaphase with the arms of the chromosome apparently being moved as passive bodies. A sub-terminal centromere imparts a J-shaped appearance to an anaphase chromosome while a chromosome with a median centromere assumes a V-shape. In either case the centromere (i.e., centric region) is at the apex of the configuration.

At the first meiotic metaphase the bivalent chromosomes undergo congression and orientation. When the bivalent (tetrad) moves upon the spindle the two homologous centromeres become coöriented and lie symmetrically on either side of the equatorial plane and directed toward opposite poles. The two homologous centric regions lead the way to opposite poles in anaphase disjunction. Inasmuch as two chromatids are attached to each centromere, the disjoining dyads have the appearance of double V's or J's, depending upon the position of the centromere in the chromosome. The double V's or J's characterizing the first anaphase are later transformed into X-shaped configurations by the repulsion between the four constituent arms. The undivided centromere lies at the center of the X and holds the two chromatids together until the second anaphase. The dyads at MII have their undivided centromeres lying on the equatorial plate as in a somatic mitosis. The effective split of the centromere occurs and the two chromatids comprising each dyad pass to opposite poles, with the centric region advancing foremost. In maize there is no relational coiling at the second division to delay disjunction. This fact coupled with the marked contraction of the chromosome arms at MII leads to the two chromatids moving somewhat parallelly to the poles, and the pronounced V's and J's characteristic of other mitoses are not always seen, although it is clear that the centric region is in the front. The above outline is true for maize chromosomes, and with minor exceptions holds for those plants and animals possessing chromosomes with localized centromeres. Recently Hughes-Schrader and Ris[1] have shown that the hemipterous insects have a diffuse type of spindle attachment region and consequently in these insects the movement of the chromosomes follows another pattern. There is then for chromosomes no universality in type of spindle behavior. This is emphasized by the following preliminary report of an anomalous situation in maize.

The unorthodox behavior is limited to the two meiotic divisions. The somatic mitoses are normal. Studies of meiosis have been limited to microsporogenesis. The first meiotic division is normal up to metaphase, when the bivalents congress upon the spindle figure. The pachytene chromosomes are of normal appearance with clearly defined centromeres. The bivalents become oriented on the spindle in a regular manner. The first indication of an unusual behavior occurs when structures similar to the primary centric region arise from distal portions of the chromosomes before the beginning of anaphase. These newly arisen structures will be tentatively called secondary centric regions inasmuch as they, like the primary centric region, become attached to the spindle and affect anaphase movements of the chromosomes. These secondary centric regions move poleward more rapidly than does the primary centric region, so that the distal ends of the chromosome, instead of facing the center of the spindle as is commonly true at anaphase, are pulled ahead and, overtaking the true centric region, come to lead the way to the poles. These secondary centric regions may be formed by one or more of the four arms comprising each dyad. The activity of these regions superimposed on the anaphase movement due to the primary centric region produces great complexity in the types of configurations observed. At the end of AI ten dyads usually are found at each pole, there being a surprisingly high regularity in disjunction.

Cytological conditions in the second division are much more favorable for observation and diagrammatic configurations are often found. Second prophase stages do not noticeably differ from normal but the onset of metaphase occurs before the usual contraction of the chromosomes has taken place. However, the somewhat extended dyads usually become oriented on the metaphase plate with the undivided centromere lying on the equatorial plate. Before the centromere splits and a normal anaphase separation is initiated, secondary centric regions again arise from or near the distal ends of the chromosomes. These new centric regions become attached to the spindle and move rapidly poleward, with the result that the chromosome arms become greatly attenuated if their proximal portions were anchored on the metaphase plate by the still undivided centromere (see Fig. 1). The centromere eventually divides and the monads pass to the poles.

When both ends of the same chromosome form secondary centric regions the chromosome literally backs into the pole with the apex (representing the centromere) of the V- or J-shaped chromosome pointing toward the equatorial plate while the two distal ends lead the way. Often only the distal end of one of the two arms forms a centric region, in which case the chromosome becomes an extended rod-shaped element. Occasionally both arms of the same chromosome may form secondary centric regions which are directed toward opposite poles, thus forming a chromosome bridge. Usually one of the two opposing forces prevails and the chromosome passes to one of the poles but infrequently the chromosome becomes suspended on the spindle with a consequent breaking of the chromosome at late anaphase. It is somewhat surprising that in both the first and second meiotic divisions regular disjunction usually occurs and that plants exhibiting this decidedly aberrant behavior are very fertile.

The formation of these secondary centric regions is limited to those plants having an abnormal type of chromosome $10^{2, 3}$ with extra chromatin near the distal end of the long arm. In plants homozygous for this abnormal tenth chromosome the frequency is high for the formation of these secondary centric regions, while it is much less in plants heterozygous for this chromosome. Sister plants homozygous for a normal chromosome 10 had a completely orthodox behavior.

It is obvious that the unparalleled behavior reported here is of great interest to current theories on cell mechanics, especially that of the kinetic movement of chromosomes. A detailed account will be published elsewhere.

FIGURE 1

Second metaphase showing precocious poleward movement of secondary centric regions. The functionally undivided centromeres of some of the dyads are oriented on the metaphase plate. In some cells the secondary centric regions reach the periphery of the cell before division of the centromere occurs resulting in an extreme attenuation of the chromosome arms.

Summary.—In maize the primary centric region representing the localized centromere is responsible for the kinetic movement of the chromosome in anaphase. The concentration of the kinetic forces produces J- or V-shaped configurations in anaphase. Plants carrying an abnormal type of chromosome 10 exhibited a unique behavior in that centric regions were formed by portions of the chromosome other than the centromere.

[1] Hughes-Schrader, S., and Ris, H., *Jour. Exp. Zoöl.*, **87,** 429–456 (1941).
[2] Rhoades, M. M., *Genetics*, **27,** 395–407 (1942).
[3] Longley, A. E., *Jour. Agric. Res.*, **56,** 177–195 (1938).

Part VI

CHROMOSOME BEHAVIOR

HISTORICAL PERSPECTIVES

Independent Behavior of Chromosome Pairs

Following the rediscovery of Mendel's laws there were many reports of genetic studies that involved two or more pairs of genes. The ratios were those predicted if the gene pairs segregated independently of each other. The independent orientation of chromosome pairs which would explain that result can be demonstrated cytologically if an organism has two or more chromosome pairs that are heteromorphic. This approach was used first by Carothers (1913) for the single sex or accessory chromosome and a heteromorphic pair of autosomes (mainly in *Brachystola*). A more extensive study was reported in 1917 (Paper 38). Carothers made collections of two species of grasshoppers, *Trimerotropis* and *Circotettix*, in the wild in southern and western United States and then analyzed the chromosome garniture in the individuals. She found that individuals varied in the number of heteromorphic pairs and also that all but one of the 12 chromosomes in *Trimerotropis* were subject to this change. For a particular chromosome that was heteromorphic in one individual, there were other individuals that were homomorphic for either of the two types. In a particular collection area, the ratio of homomorphic of one type: heteromorphic: homomorphic for the other type was 13:31:18. This comes very close to the 1:2:1 ratio that we might expect on Mendelian principles if the two types occur with equal frequency in nature and there is chance fertilization. In addition, segregation data for three pairs of heteromorphic autosomes and the sex chromosomes at metaphase I in the spermatocytes of two individuals fit closely that expected from independent orientation of each bivalent (see Paper 38).

Chromosome Segregation in Circles

In *Oenothera* and *Datura*, chromosomes in alternate positions in the ring usually pass to the same pole at anaphase I, the end result being the production of gametes with only parental combinations. There is no sterility except when crossing-over occurs in interstitial segments (between the breakpoints and the centromeres) which produces chromosome combinations with segments that are duplicated and deficient and usually abort. Since these also are the cross-over chromatids, recombination values within the interstitial segments are greatly reduced, as shown by Hanson (1952) in barley.

Genetic markers were used to study segregation in *Drosophila* interchange heterozygotes. In crosses between individuals heterozygous for the same interchange, the deficiency-plus-duplication (Df + Dp) combination from one parent could be recovered and recognized phenotypically when it combined with the complementary Dp + Df type from the other parent (Glass 1935). The frequency of alternate, adjacent-1, and adjacent-2 segregations could then be determined. In maize the nucleolar pattern in microspore quartets was used to determine the types and frequencies of the various types of disjunction (Burnham, Paper 39, and earlier by McClintock, unpublished). In general, the segregation from ring configurations without cross-overs in the interstitial segments was 2 alternate: 1 adjacent-1:1 adjacent-2 corresponding to 2 zigzag: 2 open-ring configurations. Since the products of adjacent-1 and -2 segregations are Dp and Df, in maize this results in 50 percent abortion of pollen and ovules, and in *Drosophila* 50 percent zygote mortality. In chain configurations—for example, those with one break in the satellite in maize—adjacent-2 segregations were missing.

For the type of open configuration with homologous centromeres passing to opposite poles (adjacent-1 segregation), there probably is a corresponding zigzag configuration (alternate-1 segregation); and for the type of open configuration with homologous centromeres passing to the same pole (adjacent-2 segregation), a corresponding zigzag configuration (alternate-2 segregation), as noted by John and Lewis (1965). Endrizzi (1974) has observed the four types of configurations and determined their frequencies in cotton.

Various suggestions have been offered to account for the fact that in certain species the chromosomes in the rings segregate in a directed manner at meiosis (alternate chromosomes usually passing to the same pole), whereas in others they do not. In some species, directed versus random segregation may be determined genetically. In *Collinsia heterophylla*, random orientation is the rule for induced translocations that had short interchange segments whereas directed orientation occurred in those with breaks in the nonchiasma forming regions (longer interchange segments) (Garber and Dhillon 1962; Soriano 1957; Zaman and Rai 1972).

Variant Chromosome Behavior

There are a number of cases in which heterozygotes of certain constitutions fail to produce the two kinds of gametes in equal numbers. Sandler and Novitski (1957) proposed the useful concept that where this "is a consequence of the mechanics of the meiotic divisions we suggest that the name meiotic drive be applied." They applied the term to cases in which *Drosophila* males heterozygous for the Segregation-distorter gene (*SD*) regularly produce a preponderance of *SD*-carrying sperm (Sandler and Novitski 1957). The degree of distortion depends on the strain. According to one report, cytological analysis demonstrated that one of the two telophase I poles produced two functional sperm, and the other pole two nonfunctional sperm. Preferential disjunction would explain the abnormal segregation (Peacock and Erickson 1965).

In maize a certain stock known as abnormal chromosome 10 (K 10) has an additional segment, half of whose pachytene length is pycnotic, added to the long arm replacing a short distal segment (Longley 1938). Plants heterozygous for this chromosome, when used as the female parent, have a little over 70 percent transmission of the longer chromosome. Megaspore competition and a lethal on the normal chromosome were ruled out as possible explanations (Rhoades 1942). Longley (1945) reported that in plants heterozygous also for a knobbed constitution in chromosomes 6 and 9, there was preferential recovery of the knobbed chromosome. This preferential segregation occurs only when heteromorphic dyads are present following crossing-over between the knob and the centromere (Rhoades and Dempsey 1966).

The mechanism leading to preferential segregation has as its basis the observation by Rhoades and Vilkomerson (Paper 37) that in plants possessing abnormal chromosome 10, chromosomal regions remote from the true centromere showed centric (now termed *neocentric*) activity. Those regions apparently were functionally double at metaphase because they moved precociously toward the poles before the remainder of the chromosome. Sibling plants without abnormal chromosome 10 showed no such neocentric activity. Later chromosome knobs were implicated in neocentric activity. Rhoades (1952) reported that only regions of chromosomes or chromatids with knobs possessed neocentric activity. The result was a greater recovery of knobbed chromosomes among the functional megaspores. Preferential segregation occurred for those genes linked to a chromosome knob. The knob had to be heterozygous and presumably some distance from the centromere to allow crossing-over. Rhoades studied a paracentric inversion heterozygote that produced an acentric fragment carrying a knob. The acentric fragment exhibited no neocentric activity even though abnormal chromosome 10 was present in the cell. Neocentric activity of chromosome knobs appears to require that the knob be physically associated on the same chromosome segment as the true centromere. Rhoades suggested that abnormal chromosome 10 causes the true centromere to overproduce a substance that travels along the chromosome from the centromere to a knob. The knob then exhibits neocentric activity. Since the knobbed acentric fragment produced by crossing-over in the inversion loop possessed no neocentric activity, the movement of the substance from the true centromere must occur after the time of crossing-over.

In interchanges involving a supernumerary B-chromosome in maize, there is a high rate of nondisjunction for the B-chromosome or for an interchange chromosome with the B-centromere at the division of the generative nucleus in microspores (Paper 40). This unusual behavior is useful for various cytogenetic studies, particularly in placing genes to chromosome arm (also see Paper 40 and Roman and Ullstrup 1951). Nondisjunction of the B-chromosome is believed to be due to delayed replication of the B-centromere region or adjacent heterochromatin, thus causing both sister chromatids to pass to the same pole at the second postmeiotic division. Later Rhoades noticed that certain stocks of maize were undergoing chromatin loss, but only in those with B-chromosomes (Rhoades, Dempsey, and Ghidoni 1967; Rhoades and Dempsey 1972). When two or more B-chromosomes are present in the microspores, the normal A-chromosomes that have knobs frequently behave abnormally at the second division of the microspore. The A-chromosome with a knob may be lost entirely, terminal or internal segments of the arm with the knob may be lost, or the entire arm may be lost. Transpositions between different chromosomes also occur (Rhoades and Dempsey 1975). This behavior was interpreted as being due to the B-chromosome causing a delay in DNA replication of the knob region at the second postmeiotic division, thus leading to a bridge at anaphase and subsequent breakage. This finding will be useful in future studies of chromosome behavior and in studying interactions between chromosomes in relation to chromosome replication.

Another cytological cause of nonrandom segregation is found in *Drosophila*, heterozygous for a paracentric inversion. Single cross-overs within the inversion result in a dicentric chromatid that forms a tie that orients the two cross-over chromatids toward the two inner, non-functional cells in the linear quartet of division products (Beadle and Sturtevant 1935, and shown cytologically in *Sciara* by Carson 1946). The resulting reduced recombination is not accompanied by reduced egg hatch. In maize there are several paracentric inversions that produce when heterozygous 15 to 20 percent pollen abortion but fertile ears, suggesting that the chromatid tie operates in maize. In one paracentric inversion heterozygote in barley this tie may not operate since there is both pollen and ovule abortion.

Chromosome Elimination

In oats one strain had normal meiosis, but certain of the seedlings had only 40 chromosomes instead of the normal 42 and developed into dwarf plants. At some time between fertilization and maturation of the embryo, one specific pair of chromosomes was lost (Griffiths and Thomas 1953).

In barley in crosses between *Hordeum vulgare* and *H. bulbosum*, fertilization occurs, but there is a progressive loss of *H. bulbosum* chromosomes during embryo development, leaving a haploid set of *H. vulgare* chromosomes (Kasha and Kao 1970; Lange 1971).

In *Sciara* sex expression is determined after fertilization. All zygotes have identical chromosome complements, but certain sex chromosomes are lost at the fifth or sixth cleavage division in the zygote (Metz 1938) which allows for the differentiation of males and females.

Triploid Behavior

In triploid *Datura* chromosome distribution at meiosis was studied in micro- and megasporogenesis. Distributions of the 12 extra chromosomes were not at random; 9-3, 10-2, 11-1, and 12-0 were in excess. Also the frequency of lagging and loss of univalents was much higher in megasporogenesis than in microsporogenesis (Satina and Blakeslee 1937a, 1937b).

Applications of Chromosome Behavior to Genome Analysis

Studying chromosome pairing in crosses between related species differing in chromosome number is one method of determining homologies between polyploid and diploid species.

Meiotic chromosome behavior in wheat in 21 II × 14 II and 14 II × 7 II F_1-hybrids and in subsequent generations was reported by Kihara (Paper 42). He offered several different possibilities for the origin of these species with different chromosome numbers—for example, one of these involved two tetraploid species, one having 7 IIa + 7 IIc, the other 7 IIb + 7 IIc. From the hybrid between them a species with 7 IIa + 7 IIb + 7 IIc might then be derived. Thus he focused attention on the concept of 7 as the basic number and on polyploids being comprised of different basic sets. In his genetic studies of the inheritance of various characters in these hybrids, he used A, B, and D to represent all the genes carried by the chromosomes in each particular genome. The first diagrammatic representation of the different sets of 7 chromosomes showing the chromosome relationships in the 7 II, 14 II, and 21 II wheat species was published by Gaines and Aase (1926) in a report of their studies of a 21-chromosome haploid.

Studies of genome relationships in *Triticum* and *Aegilops* were continued by Kihara and co-workers for many years (see review by Lilienfeld and Kihara 1951).

Speltoid variants had been found in various wheat varieties, and their cytology and breeding behavior studied by numerous workers (for a summary see Huskins 1946). Many were found to be aneuploids. Kihara (Paper 42) also obtained constant 40-chromosome wheat plants (lacking one chromosome pair) among the progeny of pentaploid (21 II × 14 II) hybrids. These were dwarf or semidwarf. A systematic study of aneuploids in *Triticum vulgare*, 21 II wheat, was begun by Sears (Paper 43). In subsequent studies, he developed the complete set of monosomics and identified the chromosomes that belong in each of the three sets of 7 chromosomes and which ones were partially homologous (homoeologous). Monosomics are useful not only in studying and locating genes but also in breeding for crop improvement. As mentioned earlier (see Part III), one of the important discoveries was that the pairing in *T. vulgare* between chromosomes that are partially homologous is eliminated by a genetic factor located on one chromosome. Hence essentially diploid pairing might evolve in an allo- or autopolyploid as a result of gene mutation.

In *Nicotiana* Goodspeed and Clausen (Paper 44) showed that the tetraploid *N. tabacum* species might have arisen by chromosome doubling of the F_1 between two different diploid species. Hybrids of *N. sylvestris* × *N. tabacum* and *N. tomentosa* × *N. tabacum* had 12II plus 12 univalents in the F_1 hybrids. The *N. sylvestris* × *N. tomentosa* hybrid had 24 univalents. Doubling the chromosome number of this hybrid yielded a plant resembling *N. tabacum* but differing in many respects because of changes that had occurred in *N. tabacum* since its origin and in the diploid species subsequent to the original hybridization.

A hybrid between *N. rustica* (*n* = 24) and *N. paniculata* (*n* = 12) was used in two series of backcrosses, one to the *N. paniculata* parent, the other to the *N. rustica* parent (Lammerts 1934, 1935). From the former Lammerts selected true breeding lines that had 12II but with distinct differences from *N. paniculata*. When crossed with *N. paniculata*, the differences behaved as simple recessives in most cases, but a type with spurred corolla depended on the interaction of four recessive genes. The explanation Lammerts offered was that the *N. rustica* chromosomes that paired with *N. paniculata* chromosomes are partially rearranged. Crossing-over would result in deficiencies and duplications, the smaller ones of which might be viable and behave as recessives. An alternative explanation is that mutants had occurred in the tetraploid species that would not be seen until the partially homologous chromosome in the other genome had been lost. From the backcrosses to *N. rustica*, true-breeding derivative lines with 24II were established that also differed from the recurrent *N. rustica* parent. Seven of these behaved as simple dominants, two as recessives, and two were more complex. Only the first explanation would seem to apply here.

Similar studies of species relationships were made in cotton. The cultivated American species are allopolyploids with *n* = 13 + 13. The chromosomes belonging to one set of 13 are considerably larger than those belonging to the other set of 13 (Skovsted 1934; Webber 1939). Studies of homologous genetic characters within two of the *n* = 26-chromosome species (*Gossypium hirsutum* and *G. barbadense*), showed simple Mendelian segregation. In crosses between the two species, there was a wide range of expression of the particular character in the F_2. Harland (1936) in a summary of his and others' work interpreted the results as meaning that under long isolation a different coordinated system of modifiers was built up in each species. When crossed, segregation for the two sets of modifiers resulted in a wide range of expressions of the character as well as many off-type plants.

Introgressive Hybridization

Anderson and Hubricht (1938) showed, in a study of species closely related to *Tradescantia virginiana*, that interspecific hybridization is comparatively frequent. Herbarium specimens, field work, and breeding tests showed that certain species had incorporated characters from another species. Presumably this had been the result of an original hybridization, followed by backcrossing to one of the parent species. Anderson termed this *introgressive hybridization*. Anderson's methods of pictorial and graphic analysis have been useful in providing clues as to the species that may have been the parents of naturally occurring allopolyploids.

Breakage-fusion-bridge Cycle (Chromatid and Chromosome Types)

Early evidence from X-irradiation and ring chromosome studies had suggested that broken chromosome ends may fuse. McClintock (1938) decided to study experimentally the behavior of broken chromosomes in a system where the break could be produced at a known cell division and where the broken chromosome could be followed in the immediate successive divisions. A paracentric inversion that had been discovered in maize satisfied these requirements; crossing-over within the inversion loop of a heterozygous paracentric inversion leads to bridges and subsequently to broken chromosomes that can be followed throughout the meiotic and postmeiotic divisions. She demonstrated that the chromatid ends of a broken chromosome can fuse at the position of breakage and generate a new bridge which itself will break. In 1941 (Paper 41) McClintock showed that this sequence of events can occur over and over (cyclic) in the postmeiotic mitoses and in the endosperm of a kernel receiving a broken chromosome. The cycle does not continue in the sporophytic part of the plant; that is, the broken chromosome "heals" in the zygote. This cycle is the result of fusion of broken sister chromatids; they apparently heal in the zygote and are incapable of fusion. Later McClintock (1942) demonstrated that two broken chromosomes, brought together in the same zygote by the fusion of male and female gametes carrying a broken chromosome, could fuse and initiate the chromosome-type breakage-fusion-bridge cycle. Also interesting to note is that the important discovery of controlling elements in maize was made while studying the chromosome-type breakage-fusion-bridge cycles. As noted earlier, mutable loci were detected in progeny from self-pollination of plants that had undergone this cycle.

Editors' Comments
on Papers 38 Through 44

Paper 38 reports the first extensive test of a correlation between the cytological behavior of several chromosome pairs and genetic segregation ratios. Carothers' 1913 paper analyzed a case in which there was one heteromorphic chromosome pair in addition to the accessory chromosome. By that time, segregation ratios obtained for two or more independent gene pairs were those predicted if the chromosome pairs oriented on the metaphase plate independently of each other. In several species of grasshoppers, certain individuals had, in addition to the accessory or sex chromosome, one or more chromosome pairs that were heteromorphic. Carothers showed that these cytologically recognizable autosomal bivalents and the accessory or sex chromosome all segregated independently of one another. This furnished a firm cytological basis for the genetic ratios obtained for independent pairs of genes.

Paper 39 was the first report of extensive studies in maize on the frequencies of the different kinds of chromosome segregation from the associations of four chromosomes at meiosis in interchange heterozygotes. The interchanges used involved chromosome 6 and differed greatly from each other with respect to: length of interstitial segments (between the centromeres and the interchange breakpoints); frequencies of ring-of-four versus chain-of-four configurations. The analysis depends on the fact that chromosome 6 has the nucleolus organizer. Nondisjunction for it produces one cell with no organizer (resulting in microspores with only scattered, diffuse nucleoli) and one cell with two organizers (resulting in microspores with two separate nucleoli which sometimes fuse). By scoring the microspore quartet types and the amount of pollen abortion in the various interchange heterozygotes, the frequencies of the different kinds of segregation, alternate: adjacent-1: adjacent-2, were determined.

Paper 40 is an example in maize of chromosome nondisjunction that occurs with a very high frequency at only one cell division in the life cycle, namely, the second postmeiotic division in the development of the male gametophyte. An understanding of what determines this behavior might add greatly to our knowledge about the mechanics of chromosome movements at cell division. As described by Roman the nondisjunction was discovered by using plants heterozygous for an interchange between a normal (A) and a supernumerary (B) chromosome. Only the interchange chromosome with the B-centromere and the A-chromosome segment underwent nondisjunction at the second postmeiotic division to form the two sperm nuclei. One sperm was therefore deficient for this chromosome, including the A-chromosome segment, and the other sperm was duplicate for it. Roman presents data showing how genes can be placed to chromosome from F_1 data. This also furnishes a method for comparing phenotypic effects of a duplication and a deficiency for the same A-chromosome segment.

McClintock's 1938 paper (Paper 41) was the first demonstration of the fusion of broken ends of sister chromatids; but the broken chromosomes that came from dicentric chromatids produced by crossing-over in a paracentric inversion heterozygote were deficient and not transmitted regularly to the next generation. The two methods used in the 1941 paper produced broken chromosomes, many of which had a duplication but no deficiency which enabled them to survive at least through the ovules in female gametophytes. Thus she could study the behavior of broken chromosome ends not only at meiosis and in the divisions to form gametophytes but also in the endosperm and embryo. One of the methods used and the results are described in the reprinted excerpts. This report is a significant contribution to knowledge about chromosome replication and behavior. Molecular explanations of chromosome replication must accommodate these findings. The methodology itself is an important contribution.

The excerpts from the original paper by Kihara (Paper 42) include the first report of the concept of basic chromosome sets in an allopolyploid. Included are Kihara's cytological observations on the F_1-hybrids between wheat species with 7 chromosome pairs and a species with 14 pairs, termed *triploid hybrids*: also included are the discussion of how the species with different numbers might have originated and his concept of the genetic constitution of the basic sets. Not included in the excerpts are his cytological observations on the pentaploid hybrids from crosses of species having 21 II with ones having 14 II and his observations on offspring up to the fifth generation derived from them. Also omitted are the results of his inheritance studies.

Sears (Paper 43) discusses possible ways of utilizing aneuploids in polyploid *Triticum vulgare* wheat for a broad cytological and genetical analysis. Up to the time of his studies, very little progress had been made. In this paper Sears also reports the chromosome numbers for plants obtained by crossing a haploid from the Chinese Spring variety of *T. vulgare* with normal *T. vulgare* pollen. Among the progeny from selfing those plants, there were monosomics (20 II + I), trisomics (20 II + III), and some nullisomics (20II, one pair missing). This paper initiated extensive cytogenetic studies in wheat. Similar studies have followed in polyploid oats and cotton.

Paper 44 furnished cytological evidence that the allotetraploid *Nicotiana tabacum* ($2n = 48$) may have originated from a hybrid between two diploid species *N. sylvestris* and *N. tomentosa*, each with $2n = 24$. Cytological observations on meiotic behavior are included for F_1's of each diploid crossed with *N. tabacum* and also for F_1's between the two diploids. This represents the first attempt to resynthesize an allopolyploid using species determined to be ancestral by the genome analysis method (i.e., the study of pairing in hybrids between the putative ancestors and also between them and the polyploid species).

REFERENCES

Anderson, E., and L. Hubricht. 1938. Hybridization in *Tradescantia* III. The evidence for introgressive hybridization. *Am. J. Bot.* **25**:396–402.

Beadle, G. W., and A. H. Sturtevant. 1935. X chromosome inversions and meiosis in *Drosophila melanogaster*. *Proc. Nat. Acad. Sci. U.S.A.* **21**:384–390.

Carothers, E. E. 1913. The Mendelian ratio in relation to certain Orthopteran chromosomes. *J. Morph.* **24**:487–511.

Carson, H. L. 1946. The selective elimination of inversion dicentric chromatids during meiosis in the eggs of *Sciara impatiens*. *Genetics* **31**:95–113.

Endrizzi, J. E. 1974. Alternate-1 and alternate-2 disjunctions in heterozygous reciprocal translocations. *Genetics* **77**:55–60.

Gaines, E. F., and H. C. Aase. 1926. A haploid wheat plant. *Am. J. Bot.* **13**:373–385.

Garber, E. D., and T. S. Dhillon. 1962. The genus Collinsia XVII. A cytogenetic study of radiation-induced reciprocal translocations in *C. heterophylla*. *Genetics* **47**:461–467.

Glass, B. 1935. A study of factors influencing chromosomal segregation in translocations of *Drosophila melanogaster*. *Missouri Agric. Exp. Stn. Res. Bull.* **231**:1–28.

Griffiths, D. J., and P. T. Thomas. 1953. Genotypic control of chromosome loss in *Avena*. Proc. 9th Int. Congr. Genet. *Caryologia* **6**, Sup. II: 1172–1175.

Hanson, W. D. 1952. An interpretation of the observed amount of recombination in interchange heterozygotes of barley. *Genetics* **37**:90–100.

Harland, S. C. 1936. The genetical conception of the species. *Biol. Rev.* **11**:83–112.

Huskins, C. L. 1946. Fatuoid, speltoid and related mutations of oats and wheat. *Bot. Rev.* **12**:457–514.

John, B., and K. R. Lewis. 1965. The meiotic system. *Protoplasmatologia* **6**(F1):1–335.

Kasha, K. J., and K. N. Kao. 1970. High frequency haploid production in barley (*Hordeum vulgare* L.). *Nature* **225**:874–876.

Lammerts, W. E. 1934. Derivative types obtained by back-crossing *Nicotiana rustica-paniculata* to *N. paniculata*. *J. Genet.* **29**:355–366.

_____. 1935. Derivative types obtained by backcrossing *N. rustica-paniculata* to *N. rustica*. *Z. indukt. Abstamm.-Vererbungsl.* **68**:417–435.

Lange, W. 1971. Crosses between *Hordeum vulgare* L. and *H. bulbosum* L. II. Elimination of chromosomes in hybrid tissues. *Euphytica* **20**:181–194.

Lilienfeld, F. A., and H. Kihara. 1951. Genome-analysis in *Triticum* and *Aegilops* X. Concluding review. *Cytologia* **16**:101–123.

Longley, A. E. 1938. Chromosomes of maize from North American Indians. *J. Agric. Res. (Washington, D.C.)* **56**:177–195.

_____. 1945. Abnormal segregation during megasporogenesis in maize. *Genetics* **30**:100–113.

McClintock, B. 1938. The fusion of broken ends of sister half-chromatids following chromatid breakage at meiotic anaphases. *Missouri Agric. Exp. Stn. Bull.* **290**:1–48.

_____. 1942. The fusion of broken ends of chromosomes following nuclear fusion. *Proc. Nat. Acad. Sci. U.S.A.* **28**:458–463.

Metz, C. W. 1938. Chromosome behavior, inheritance and sex determination in *Sciara*. *Am. Nat.* **72**:485–520.

Peacock, W. J., and J. Erickson. 1965. Segregation-distortion and regularly nonfunctional products of spermatogenesis in *Drosophila melanogaster*. *Genetics* **51**:313–328.

Rhoades, M. M. 1942. Preferential segregation in maize. *Genetics* **27**:395–407.

_____. 1952. Preferential segregation in maize. Chapt. 4 in *Heterosis*. (J. W. Gowen, Ed.) Iowa State College Press, Ames, Iowa.

_____ and E. Dempsey. 1966. The effect of abnormal chromosome 10 on preferential segregation and crossing over in maize. *Genetics* **53**:989–1020.

_____. 1972. On the mechanism of chromatin loss induced by the B chromosome of maize. *Genetics* **71**:73–96.

_____. 1975. Stabilization of freshly broken chromosome ends in the endosperm mitoses. *Maize Genet. Coop. News Letter* **49**:53–58.

_____ and A. Ghidoni. 1967. Chromosome elimination in maize induced by supernumerary B chromosomes. *Proc. Nat. Acad. Sci. U.S.A.* **57**:1626–1632.

Roman, H., and A. J. Ullstrup. 1951. The use of A-B translocations to locate genes in maize. *Agron. J.* **43**:450–454.

Sandler, L., and E. Novitski. 1957. Meiotic drive as an evolutionary force. *Am. Nat.* **91**:105–110.

Satina, S., and A. F. Blakeslee. 1937a. Chromosome behavior in triploids of *Datura stramonium*. I. The male gametophyte. *Am. J. Bot.* **24**:518–527.

_____. 1937b. Chromosome behavior in triploids of *Datura*. II. The female gametophyte. *Am. J. Bot.* **24**:621–627.

Skovsted, A. 1934. Cytological studies in cotton. II. Two interspecific hybrids between Asiatic and new world cottons. *J. Genet.* **28**:407–424.

Soriano, J. D. 1957. The genus *Collinsia*. IV. The cytogenetics of colchicine-induced reciprocal translocations in *C. heterophylla*. *Bot. Gaz.* **118**:139–145.

Webber, J. M. 1939. Relationships in the genus *Gossypium* as indicated by cytological data. *J. Agric. Res.* **58**:237–261.

Zaman, M. A., and K. S. Rai. 1972. Cytogenetics of thirteen radiation-induced reciprocal translocations in *Collinsia heterophylla*. *Cytologia* **37**:629–638.

38

Reprinted from pp. 445, 447–448, 449, 450–451, 452–456 of *J. Morphol.*
28(2):445–520 (1917)

THE SEGREGATION AND RECOMBINATION OF HOMOLOGOUS CHROMOSOMES AS FOUND IN TWO GENERA OF ACRIDIDAE (ORTHOPTERA)

E. ELEANOR CAROTHERS

From the Zoological Laboratory of the University of Pennsylvania[1]

[*Editors' Note:* In the original, material precedes this excerpt.]

As is well known, in the Orthoptera and many other animals as well, the first maturation division determines which of the derivative cells will, in fertilization, produce a male and which a female, since the accessory chromosome—the sex determinant—passes undivided into one of the daughter cells at this time. While probably it is purely a matter of chance which of the daughter cells the accessory enters—that is, it parallels any Mendelian character in the matter of segregation—nevertheless, it marks unalterably, after it has passed to one pole, the male from the female producing spermatozoon. If one of the homologues of any pair is recognizably different from its mate and these homologues should segregate from each other at the first maturation division their manner of segregation in relation to sex would be apparent. The only instances of this sort to be

---

[1] Contribution from the Zoological Laboratory of the University of Pennsylvania.

reported, so far, are those in which the homologues differ markedly in size. The first was the work of the writer which appeared in 1913, giving data showing the alternative distribution of the unequal dyads of one tetrad in relation to sex in certain Oedipodinae. Three hundred first spermatocytes were counted; in 51.3 per cent the larger dyad was going into the same second spermatocyte as the accessory, and in 48.7 per cent the smaller dyad was accompanying the accessory. Shortly afterward a paper by Voïnov ('14) appeared, giving like results and an essentially similar conclusion from a study of Gryllotalpa vulgaris.

Wenrich, in 1914, reported similar results from a count of four hundred and seventy-two first spermatocytes of Phrynotettix magnus where a tetrad with unequal dyads is also present. And lastly, Robertson ('15), in an addendum to a paper in which he reports an unequal pair of chromosomes in Tettigidea parvipennis and in Acridium granulatum, states that these unequal pairs agree with the one in my material in regard to the distribution of their dyads in relation to the accessory. These I believe are the only instances so far reported where the chromosomes derived from different parents could be followed in the germ cells and their distribution determined.

[*Editors' Note:* Material has been omitted at this point.]

2. Nomenclature

The unusual conditions of the chromosomes in this group have made advisable the introduction of four new terms.

1. Homomorphic—used to designate those tetrads made up of morphologically similar homologues.

2. Heteromorphic—used to designate those tetrads made up of morphologically different homologues.

3. Telomitic—a term used to indicate terminal fiber attachment.

4. Atelomitic—a term used to indicate nonterminal fiber attachment.

The two latter are extensions of the ideas involved in the terms 'Hippiscus type' and 'Stenobothrus type' as used by McClung ('14). This has seemed desirable since the work of Robertson ('16) and some recent work of McClung indicates that the chromosomes of Stenobothrus may be fundamentally different from those of the typical Acrididian complex. I shall also apply the terms telomitic and atelomitic to spermatogonial and somatic chromosomes as well as to those of the first spermatocyte. The term Hippiscus type will be used interchangeably with telomitic for those first spermatocyte chromosomes which are comparable in structure to the Hippiscus rings. However, in this material the tetrads of the Hippiscus type are usually transformed into rods in the metaphase as described by McClung ('14). In the same way Stenobothrus type and atelomitic may be used interchangeably.

3. Special fitness of material

The genus Trimerotropis is confined to the American continent and contains over thirty species, the extreme members of which merge with those of four other genera. The present paper deals with the fallax group of the genus and its relation to the genus Circotettix, with especial reference to the heteromorphic tetrads found in each.

[*Editors' Note:* Material has been omitted at this point.]

Dr. McClung has shown ('14) that the point of spindle fiber attachment is normally constant, but at some period in the history of this species there must have been a reorganization to the extent of a shifting of the attachment in certain chromosomes. The most striking result has been to produce J-shaped tetrads in the first spermatocyte. Such shapes are due to one homologue being rod-shaped; that is, it has terminal fiber attachment, while its mate has nonterminal fiber attachment. Of the twelve first spermatocyte chromosomes, ten tetrads and the accessory are affected by this reorganization. If the female is similarly involved and if there is free fertilization, bringing about chance recombination, one might expect to find in a given individual all ten of the tetrads represented by two rod-shaped homologues, giving rod-shaped tetrads, or one might find all of the homologues with nonterminal fiber attachment, resulting in tetrads of the Stenobothrus type (McClung '14). Between these two extremes there might be present every possible combination in J's, rod's and Stenobothrus-like rings.

These expectations have been realized in part; and, furthermore, in two of these ten tetrads there is a third type with both homologues having nonterminal fiber attachment, but with a secondary shifting so that one of the free arms is longer than the other. One finds, besides such heteromorphic tetrads (plate 14, fig. 22c), homomorphic tetrads of the two expected types, one with both arms long (plate 14, fig. 23a) and the other with both arms short (plate 14, fig. 10a).

Another peculiarity, that of a single constriction occurring near the proximal ends (that is, the ends directed towards the poles), may mark certain tetrads. An example may be seen on plate 14, figure 21a. One dyad of such a tetrad (Chromosome number 3, plate 4) may be of three types; V-shaped (plate 5, fig. 31), plain rod (plate 5, fig. 32) or constricted rod (plate 5, figs. 31 and 32); these occur in various combinations, furnishing a visible mechanism for possible triple allelomorphs.

[*Editors' Note:* Material has been omitted at this point. Also, Plates 2 through 14 are not reproduced here.]

1. First spermatocytes

a. Segregation of homologues of J-shaped tetrads in individual no. 1, Trimerotropis (?) suffusa (?). This is one of two individuals belonging to what I designate as form 'B.' They were collected on Orcas Island in Puget Sound by a party from Kansas University in the summer of 1909. Both individuals are alike.

Of the twelve first spermatocyte chromosomes (plate 1), four (nos. 9 to 12) are atelomitic—Stenobothrus type (McClung, '14)—four (nos. 2, 3, 5, and 6) are telomitic—Hippiscus type, the accessory (no. 4) is V-shaped, atelomitic, while the remaining three tetrads (nos. 1, 7 and 8) have one dyad of the Stenobothrus type and the other of the Hippiscus type.

Several important questions at once present themselves. 1) Since these heteromorphic but homologous dyads segregate in the first maturation division, what is their distribution in relation to the accessory (that is, to sex)? 2) As it is obvious that at least sixteen sorts of spermatozoa are formed in these two individuals, would the well known constancy of the complex—as shown by numerous workers on Orthoptera—hold for this species or would a large number of individuals give all possible combinations? Or would certain combinations result and others fail? 3) Would the somatic complex of the female be constant? (Using this as an index of the oogonial complex). 4) If the complex is constant for the species, what is the mechanism by which it is regulated? Is there selective fertilization of a most complex sort, or is there free fertilization with regulation occurring at the time of the maturation of the egg? (Since copulation occurs some twenty-four hours before the polar bodies are formed the latter seemed quite possible.)

The first of these questions was a matter to be determined by a study of a given individual and was at once worked out from the material in hand. The others required a considerable number of individuals and it was largely for the purpose of obtaining these that a collecting trip through the southern and western states was undertaken during the summer of 1915.

For a determination of the method of segregation of the heteromorphic homologues in relation to the accessory, one hundred camera lucida drawings of entire complexes were made at random. As stated above, the two individuals in the collection were alike; fifty-seven of the drawings were from one individual and forty-three from the other. Plate 1 is based on only one individual, however, but is representative of both.

These drawings are from sections. The chromosomes from one cell are always in two and sometimes three sections. The sections were all present, in order and in straight rows, so that the problem of identifying in successive sections these large clear cells with sharply formed spindles is much simpler than it may appear to those accustomed to less favorable material. The chromosomes were first outlined under the camera lucida in their actual relation to each other in the sections, then a careful study was made and details filled in when necessary. Later, the chromosomes represented in plate 1 were arranged roughly according to size and placed on the plate so that the transverse rows represent the chromosomes found in one cell while the vertical rows represent the corresponding chromosomes in different cells. The eight complexes shown on this plate are typical of the conditions found in the hundred cells. They are so placed that the accessory is always passing to the upper pole.

Taking up first the small one at the right (no. 1'), which is one of the heteromorphic pairs, we find in five of the cells (b, c, d, f, h) the atelomitic dyad going to the cell which lacks the accessory, while in three (e, g, i) it is going into the same cell as the accessory. The other two heteromorphic chromosomes, numbers 7 and 8, respectively, are so nearly identical in size and behavior that no attempt was made to distinguish between them. Instead, the segregation of their homologues in relation to each other and in relation to the accessory was noted and gave all of the information desired. If these two chromosomes are compared with each other in the first four cells (b, c, d, e), it will be observed that the dyads with nonterminal fiber attachment are going to opposite poles; in the remaining four cells (f, g, h, i) they are going to the same pole. But in cells f and g they approach the pole which lacks the accessory, while in the last two they will enter the same cell as the accessory. Now, if we compare chromosome number 1 with numbers 7 and 8, it will be seen that its dyads also segregate independently of either of the others. For instance, in the last four cells (f, g, h, i) its atelomitic dyad passes, either to the same second spermatocyte as the similar dyads of the larger chromosomes (f and i), or to the cell which receives the telomitic dyads (g and h). It is evident then that here are four chromosomes (nos. 1, 4, 7, 8) for which this is the segregation division and that they are distributed more or less without regard to each other or to the second spermatocytes.

If there is free segregation, the number of equally possible combinations in the gametes of a single individual is represented by the formula 2^n in which n represents the number of chromosomes in the reduced series; that is, the number of pairs of homologues. In this instance $n = 4$ since we are considering only the accessory and the three heteromorphic tetrads. Then 2^4, or 16, is the possible number of combinations of these chromosomes in the gametes of this individual. While the number of morphologically different gametes formed as a result of the segregation of any *given* three is 8 (2^3), of any given two is 4 (2^2) of a given one 2 (2^1). The occurrence of *any* combination is shown by the coefficients of the expanded binomial raised to the nth power in which case n again represents the number of homologous pairs. In this instance the series of coefficients is 1–4–6–4–1.

From the two formulae given above we should expect to find

Any *given* 4 V's in $\frac{1}{16}$ of the gametes	Any 4 V's in $\frac{1}{16}$ of the gametes
Any *given* 3 V's in $\frac{1}{8}$ of the gametes	Any 3 V's in $\frac{4}{16}$ of the gametes
Any *given* 2 V's in $\frac{1}{4}$ of the gametes	Any 2 V's in $\frac{6}{16}$ of the gametes
Any *given* 1 V in $\frac{1}{2}$ of the gametes	Any 1 V in $\frac{4}{16}$ of the gametes
and 0 V in $\frac{1}{16}$ of the gametes	and 0 V in $\frac{1}{16}$ of the gametes

The difference—it will be noted—between these two series is in regard to *any* one, two or three as opposed to a *given* one, two or three; that is, in the latter case, we must distinguish between the V's with which we are dealing. A given V (as the accessory) would be found in one-half of the gametes, but, on the other hand, one-quarter of the gametes would contain only one V.

Perhaps both may be better shown graphically. Let A, B, C, D, represent the V-shaped homologues and the accessory and a, b, c, d, the rod-shaped homologues of the tetrads and the absence of any homologue in the case of the accessory. There are then the foregoing sixteen combinations any one of which is equally probable; *all* four V's together, e.g., A B C D, one-sixteenth of the time; *any* three, as in the second division, four-sixteenths of the time; any *given* three, e.g., A, B, C, two-sixteenths of the time; *any* two, as in the third division, six-sixteenths of the time; any *given* two, four-sixteenths of the time; while any *given* one occurs eight-sixteenths of the time and *only* one four-sixteenths of the time.

A B C D   a B C D   a b C D   a b c D   a b c d
  A b C D   A b c D   a b c d
  A B c D   A B c d   a B c d
  A B C d   a B c D   A b c d
                a B C d
                A b C d

To be specific if A = accessory, B the V-shaped dyad of number 1 and C and D the indistinguishable ones of numbers 7 and 8, then, out of sixteen second spermatocytes, we should have the accessory in eight-sixteenths, the accessory and V of tetrad number 1 in four-sixteenths, any two V-shaped dyads six-sixteenths, any three in four-sixteenths. As the one hundred spermatocytes counted represent two hundred derivative cells, we have:

	Expected		Actual count
A given V (Accessory or no. 1)	$\frac{1}{2}$ x 200 =	100	100
Only one V................................	$\frac{1}{4}$ x 200 =	50	48
Two given V's (Accessory and no. 1 or those of 7 and 8)	$\frac{1}{4}$ x 200 =	50	46 and 47
Any two V's................................	$\frac{3}{8}$ x 200 =	75	84
Three given V's (nos. 1, 7, 8 or accessory, 7, 8)	$\frac{1}{8}$ x 200 =	25	22 and 21
Any three V's................................	$\frac{1}{4}$ x 200 =	50	48
All four V-shaped................................	$\frac{1}{16}$ x 200 =	$12\frac{1}{2}$	8

Considering that these figures are based on only one hundred first spermatocytes, the results are probably as near the expectations as could be anticipated for any objects on the basis of chance distribution for a like number of trials.

[*Editors' Note:* Material has been omitted at this point.]

REFERENCES

Carothers, E. E. 1913. The Mendelian ratio in relation to certain Orthopteran chromosomes. *J. Morph.* **24**:487–511.

McClung, C. E. 1914. A comparative study of the chromosomes in Orthopteran spermatogenesis. *J. Morph.* **25**:651–749.

Robertson, W. R. B. 1915. Chromosome studies. III. Inequalities and deficiencies in homologous chromosomes; their bearing upon synapsis and the loss of unit characters. *J. Morph.* **26**:109–141.

_____. 1916. Chromosome studies. I. Taxonomic relationships shown in the chromosomes of Tettigidae and Acrididae: V-shaped chromosomes and their significance in Acrididae, Locustidae, and Gryllidae: chromosomes and variation. *J. Morph.* **27**:179–331.

Voinov, D. 1914b. Recherches sur la spermatogenèse du *Gryllotalpa vulgaris* Latr. *Arch. Zool. Exp.* **54**:439–499.

Wenrich, D. H. 1914. Synapsis and the individuality of the chromosomes. *Science* **41**:441 (Abstract).

Explanation of Plates

The drawings were made with the aid of the camera lucida at a magnification of 2400 diameters and have been reduced one-third in reproduction.

Plates 1 to 10 show side views of first spermatocyte metaphases, arranged as nearly as possible according to size. Each horizontal row represents one cell, the vertical rows, corresponding chromosomes in different cells. They are so arranged that the accessory, number 4, is always passing to the upper pole.

PLATE 1

Lateral views of first spermatocyte complexes from a single individual (1), showing mode of segregation of three heteromorphic tetrads (chromosomes 1, 7, and 8).

b to e. Alternate distribution of atelomitic dyads of chromosomes 7 and 8.

f and g. Concurrent distribution of same homologues segregating opposite the accessory, number 4. In (f) the atelomitic dyad of chromosome 1 is accompanying the similar dyads of chromosomes 7 and 8, while in (g), it is passing to the opposite pole.

h and i. Same as (f) and (g) except that atelomitic dyads of 7 and 8 are passing to the same pole as the accessory.

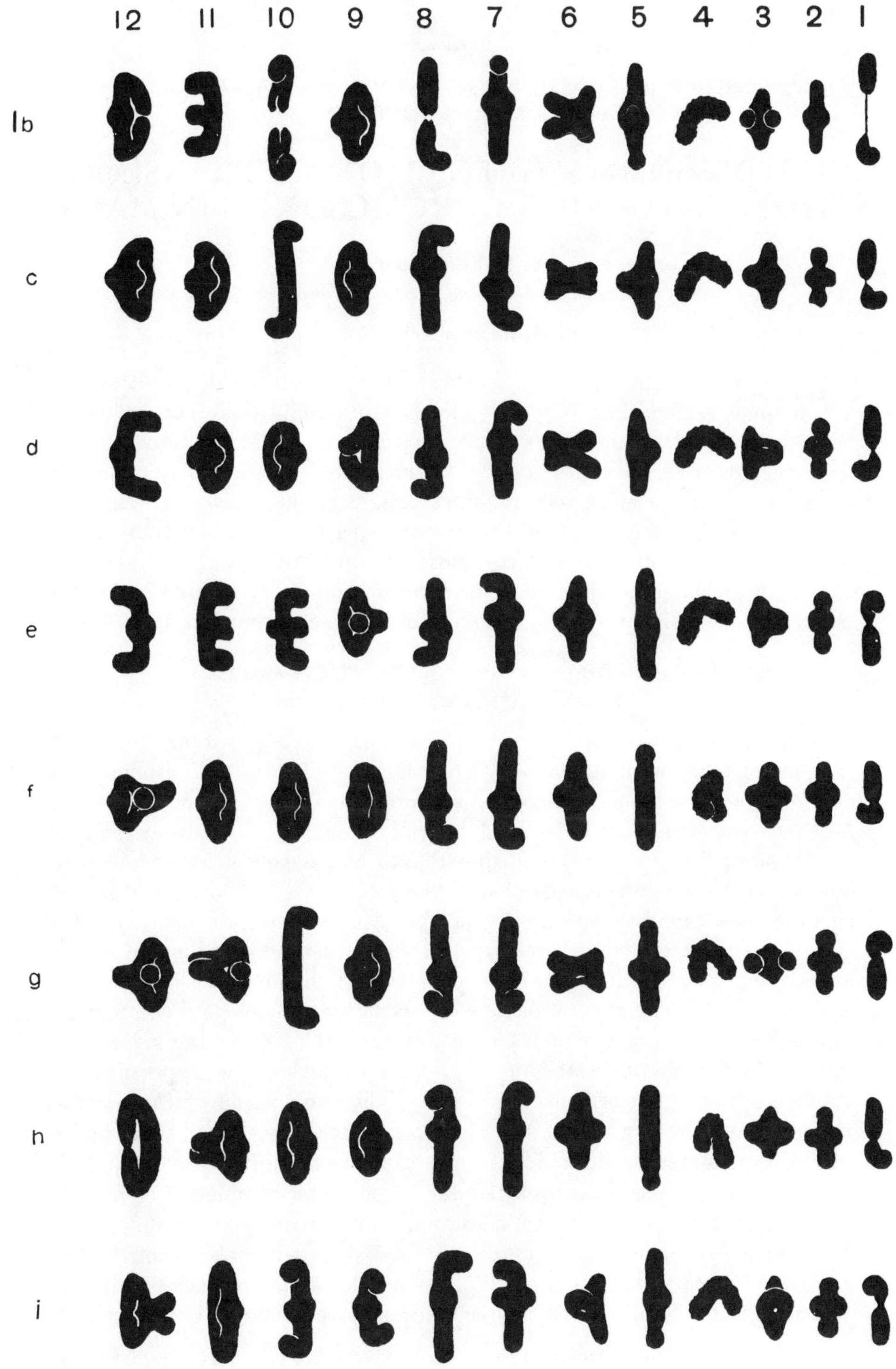

PLATE 1

39

Reprinted from pp. 446–451, 452–456, 459–461, 463, 464, 478–480 of
Genetics 35(4):446–481 (1950)

CHROMOSOME SEGREGATION IN TRANSLOCA-
TIONS INVOLVING CHROMOSOME 6 IN MAIZE[1]

C. R. BURNHAM

University of Minnesota, St. Paul, Minnesota

Received December 30, 1949

O NE of the unsolved problems concerning translocations deals with chromosome segregation. Why is it mostly alternate in Oenothera, while in maize both alternate and adjacent segregations occur? Maize translocation heterozygotes involving chromosome 6 which carries a nucleolar organizer region are useful for such a study, since certain of the types of segregation result in recognizably different microspore quartets (McCLINTOCK 1934, 1945). The results of a study of chromosome segregation using 27 translocations with breakages at different loci in chromosome 6 are presented in this paper, the theoretical results first, followed by the experimental data.

THEORETICAL RESULTS IN THE SPORE QUARTETS AND POLLEN OF
TRANSLOCATION HETEROZYGOTES

Considering only the theoretically possible 2-2 segregations from the complex of four chromosomes (fig. 1A) alternate or adjacent chromosomes may pass to the same pole, there being two types of adjacents, designated by McCLINTOCK (1945) as: adjacent 1 in which homologous centromeres pass to opposite poles at division I (disjunction of homologous centromeres, nondisjunction of the translocated pieces) and adjacent 2 in which they pass to the same pole (nondisjunction of homologous centromeres, disjunction of the translocated pieces). These three types of segregation may be thought of also as resulting from different planes of separation at the first division, alternate and adjacent 1 being along a plane which separates homologous centromeres and adjacent 2 along a plane at right angles to this. When the segregation involves nondisjunction of the nucleolar organizer regions, two types of spores result, one having two organizers and potentially two nucleoli which may fuse, the other type having no organizer and consequently scattered small bodies of nucleolar material, referred to in this paper as a "diffuse" spore type. In maize the spores remain in quartets for a time after completion of meiosis, and with the proper acetocarmine staining any spore quartets differing in their nucleolar makeup may be recognized. It is possible also to distinguish the two division planes, since a cell division follows the first meiotic division.

The expected results of these three kinds of segregation fall into three main

[1] Contribution from the Division of Agronomy and Plant Genetics, UNIVERSITY OF MINNESOTA, St. Paul, Minn. Paper No. 2479 of the Scientific Journal Series, MINNESOTA AGRICULTURAL EXPERIMENT STATION. Assistance was furnished from research funds of the UNIVERSITY OF MINNESOTA GRADUATE SCHOOL. A sabbatical leave, together with a GOSNEY FELLOWSHIP at the CALIFORNIA INSTITUTE OF TECHNOLOGY (1947–48), made its completion possible. I wish to acknowledge the assistance of GERTRUD JOACHIM, C. H. LI, and H. R. HIGHKIN in gathering the pollen sterility and quartet data. I am indebted to C. M. ARNDT for drawing figures 1A and B.

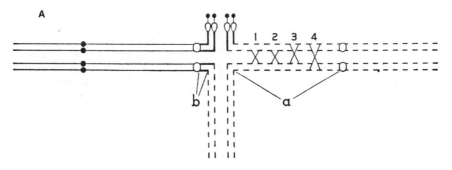

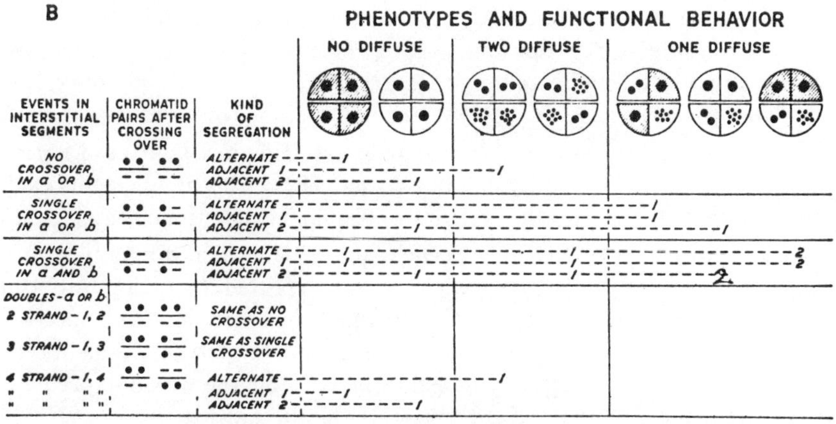

FIGURE 1A. Pachytene diagram of a 5–6 translocation heterozygote with the break in chromosome 6 (solid line) in the short arm, that in 5 (dotted line) in the long arm, *b* and *a* being the interstitial segments in the respective chromosomes. At the short arm end of chromosome 6 are the satellite and next to it the nucleolar organizer region (stippled). The clear circles represent the centromeres, the dense ones the pycnotic knobs.

FIGURE 1B. The spore quartet types and their frequencies resulting from various events (crossovers or non crossovers) in the interstitial segments in figure 1A. The second column shows the nucleolar organizer constitutions of the resulting chromatid pairs (with sister centromeres) before orientation for Division I; and — indicate chromatids with and without the nucleolar organizer respectively. The shaded spores in the resulting quartets are normal, the empty ones abortive due to chromosomal deficiency. The first division plane is indicated by the wider separation. The two abnormal spores in the first quartet type in the "one diffuse" class, and the two diffuse-nucleolate spores in the second quartet type in the "two diffuse" class should be by chance contiguous but on opposite sides of the first division plane as frequently as in the diagonal position shown.

groups depending on the position of the breakage point in chromosome 6: (1) those with the break in the short arm between the centromere and the nucleolar organizer, (2) those with the break in the long arm or in the satellite, and (3) those with the break in the nucleolar organizer. Segregation of two chromosomes to each pole will be considered first, then the results from 3-1 segregation.

*Group 1, those with one break in the short arm of chromosome 6
between the centromere and the nucleolar organizer.*

The expected results for this group have been presented (BURNHAM 1949), but are summarized here in figure 1B. Three main types of spore quartets may be recognized cytologically based on the number of spores with "diffuse" nucleolar material, i.e. "no diffuse," "two diffuse," and "one diffuse." The first division is assumed to be reductional for homologous centromeres (in adjacent 2 segregations the first division is nondisjunctional for them but the second division is again equational). By referring to figure 1A and to column 2 in figure 1B, the kinds of quartets expected from each type of segregation, with and without crossing over[2] in the interstitial segments, may be seen to be those shown in the last three columns of figure 1B. For example with no crossing over in the interstitial segments, at the end of the first division following alternate segregation one pair of chromatids at each pole has two nucleolar organizers (indicated by dots) and one pair has no organizers (indicated by dashes): ∴⁻⁻. At the second division, one chromatid from each of the two pairs passes to the same pole, resulting in a quartet of normal functional spores (indicated by shading in fig. 1B) with one nucleolus each: ⁞. Adjacent 2 segregation results in a quartet indistinguishable in appearance from this (one nucleolus in each spore) but which differs from it in having spores all of which are deficient and expected to abort. Adjacent 1 segregation results in the quartet type which has two "diffuse" spores, a "two diffuse" quartet. With no crossing over in the interstitial segments, the difference between the observed pollen abortion and the amount predicted from the frequency of "two diffuse" quartets should indicate the frequency of adjacent 2 segregation and hence the frequencies of the three types of segregation could be determined.

A single crossover in either interstitial segment, followed either by alternate or adjacent 1 segregation results in the third recognizable spore quartet type having two spores with one nucleolus each (functional), one with potentially two nucleoli, and one with diffuse nucleolar material, the latter two being on opposite sides of the first division plane and abortive (fig. 1B). This will be referred to as a "one diffuse" or a crossover type quartet. The segregation in it has been referred to in certain Oenothera and Pisum studies on translocations as "half disjunction." A high frequency of such crossovers in a species having mainly alternate segregation in rings should result in sterility approaching 50 percent. If adjacent 2 segregation followed such a crossover, a "no diffuse" as well as this "one diffuse" type would be expected, in both of which all four spores would be abortive. However, if chromosomes that crossover pass to opposite poles, as evidence reported here and by McCLINTOCK (1945) indicates, this type of segregation would not occur. In that case, in a translocation with only one interstitial segment that is short enough to preclude

[2] Throughout this paper the terms "crossing over" or "crossovers" refer to the cytological process, contrasted with "genetic crossing over" or "recovered crossovers" for the observed genetic result. Unless stated otherwise, the crossovers referred to are those occurring in interstitial segments.

double crossovers it would be possible to measure the frequency of single crossovers in this segment, and of the different kinds of segregation from the non crossover meiocytes.

Simultaneous single crossovers in the two interstitial segments result in the three types of quartets from each kind of segregation (fig. 1B). Translocations in which this is frequent would not be as satisfactory for an analysis of chromosome segregation. Note that the two abnormal spores in the "crossover" type and the two "diffuse" spores in the "two diffuse" quartet are on the same side and on opposite sides, respectively, of the first division plane; whereas the reverse is true for single crossing over in one segment and for no crossing over.

The expected results from double crossing over in either region a or b are also shown in figure 1B. The 2-strand doubles give the same quartet types as with no crossing over, and the 3-strand doubles give the same types as does a single crossover. Following a 4-strand double, adjacent 1 segregation gives rise to a normal quartet with functional spores while alternate segregation results in a quartet having two spores with diffuse nucleoli (all four spores abortive), just the reverse of those following no crossing over. Only if alternate and adjacent 1 segregations following 4-strand double crossing over are equally frequent would the frequency of quartets containing two spores with diffuse nucleoli still be an accurate measure of adjacent 1 segregation. Although adjacent 2 segregation may not be expected following the 2- and 4-strand doubles, if it did occur it would result in normal-appearing quartets with four abortive spores and hence would be included in any observed difference between predicted and observed sterility.

A very important point, previously overlooked, has been presented by LAMM (1948) based on TOMETORP'S unpublished analysis (called to my attention by DR. H. H. KRAMER as this was being completed), (see also HANSON and KRAMER 1949) i.e. that the two crossover chromatids resulting from a single crossover in the interstitial segment are carried by the abortive spores when the segregation is alternate; and are carried by the normal spores only when it is adjacent 1. Double crossover chromatids following 2-strand and 3-strand doubles, and single crossover chromatids following 4-strand doubles may be recovered when the segregation is alternate. In species having mostly alternate segregation the observed genetic crossing over should be greatly reduced. In corn where alternate and adjacent 1 segregations are probably equally frequent following the crossovers, reduction in observed recombination would be expected from that source.

In presenting the data, it is assumed that the first division is reductional at the centromere. The evidence from other sources makes this probable, although certain infrequent quartet types to be described later could be explained by rare equational separation.

Where the breaks are fairly close to the centromeres, non-homologous pairing may separate homologous centromeres. This would not necessarily result in equational separation of these centromeres at division 1, but might affect the action of the centromeres on orientation from those figures.

It should also be noted that in the quartets containing functionally normal spores a ratio of 1 with the standard normal:1 with the translocation chromosome complex is expected.

Group 2. Translocations with one break in the long arm or in the satellite of chromosome 6.

In plants heterozygous for a translocation in which the break in chromosome 6 was in the long arm, the pachytene configuration is represented in figure 2; c and d indicating the interstitial segments. In this type of translocation, a study of the diagram will show that adjacent 2 segregation results in

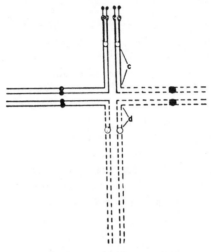

FIGURE 2. Pachytene diagram of a 5–6 translocation heterozygote with the breaks in 6 (solid line) and 5 (dotted line) in the long arms, c and d being the interstitial segments in the respective chromosomes. Adjacent 2 segregation (when homologous centromeres pass to the same pole) results in the "two diffuse" type, alternate or adjacent 1 segregation results in the "no diffuse" type, with or without crossing over in the interstitial segments.

"two diffuse" quartets, while alternate and adjacent 1 segregations result in "no diffuse" quartets, with or without crossing over in either or both interstitial segments. The frequency of "two diffuse" quartets is the frequency of adjacent 2 segregation. Single crossovers and 3-strand doubles again result in "half disjunction" quartets in which two spores are normal and two abortive, but they have one nucleolus each and cannot be recognized cytologically. If the interstitial regions are so short that little or no crossing over occurs in them, the excess of observed over predicted pollen abortion would measure the frequency of adjacent 1 segregation (quartets which appear normal but in which all four spores abort). The remaining normal, i.e. "no diffuse," quartets would result from alternate segregation. In those having either or both interstitial segments long, only if there is independent information on the frequency of single and multiple crossing over in them can the frequency of adjacent 1 and alternate segregations be estimated. Translocations of this type, group 2, are useful since in all of them the frequency of adjacent 2 segregation can be determined. No assumptions regarding the kind of segregation following crossing over are necessary, the only difficulty coming from 3-1 segregation to be discussed later.

The problem of segregation frequencies might be solved also by cytological determinations of the relative frequencies of open and zigzag rings at metaphase if it could be assumed that only the zigzag rings lead to alternate segregation. However, KOLLER (1944) has reported in mice that the zigzag rings may be oriented to give adjacent segregation. If differences in morphology are great enough, the two types of adjacents may be recognized at early anaphase I. Such a cytological study would not be complete unless it were possible to recognize those figures in which crossing over had occurred in an interstitial segment.

For *translocations with one break in the satellite of chromosome 6*, the quartet types are the same as for the preceding ones in which the break in chromosome 6 was in the long arm. One difference in this group is that spores deficient only for a portion of the satellite are not abortive (BURNHAM 1932b), hence allowance must be made for this in comparing observed with predicted pollen abortion.

Group 3. Translocations with one break in the nucleolus organizer region of chromosome 6.

In the translocations with the break in chromosome 6 in the nucleolus organizer region, if both pieces of that region (the translocated and the remaining piece) retain the nucleolar organizing function there will be a potential total of six nucleoli in the normal quartet, only two of which should be normal in size. The nucleoli organized by the pieces of the organizer region may be of similar size or unequal (McCLINTOCK 1934). Various combinations of these in the spores of a quartet are possible with different segregations and with crossing over. Since fusion of these nucleoli often occurs, an accurate determination of the relative frequencies of different segregations would be difficult or impossible from the quartet study. A quartet in which one spore has diffuse nucleoli is theoretically possible if adjacent 2 segregation occurs following crossing over in an interstitial segment. A cytological determination of the frequencies of zig-zag and open rings may be the best means of studying segregation in translocations belonging to this group. McCLINTOCK (1934) has reported the results for one such translocation (*T6–9a*) by identifying cytologically the chromosome constitutions of the microspores.

Spore quartet types following 3-1 disjunction of chromosomes from the ring. A 3-1 segregation of the chromosomes in the ring may result in a "two diffuse" or a "no diffuse" quartet, depending on whether two or only one of the three chromosomes going to one pole carries a nucleolar organizer. In either case, two spores are normal, two abortive. If the different kinds of 3-1 segregation were at random, these two types would be equally frequent and the errors in the predicted sterility would balance each other. As a check on possible sources of error, it may be important to determine the frequency with which "two diffuse" quartets may be expected from 3-1 segregation.

[Editors' Note: Material has been omitted at this point.]

EXPERIMENTAL RESULTS FOR TRANSLOCATIONS HAVING ONE BREAK IN
THE SHORT ARM OF CHROMOSOME 6 BETWEEN THE CENTRO-
MERE AND THE NUCLEOLAR ORGANIZER (GROUP 1)

Included in this group are *T5-6c* and *T4-6* Li with one long interstitial segment; and *T2-6a*, *T6-10b* and *T5-6c* homozygous for an inversion in chromosome 5 (*In5a*), all of which have short interstitial segments.

The most extensive studies were on *T5-6c* with and without *In5a* used to shift one centromere without changing the lengths of the two axes or the four arms of the "cross" (preliminary reports, BURNHAM 1945, 1948, 1949). I am indebted to DR. BARBARA MCCLINTOCK for suggesting the problem and furnishing seed of $T5\text{-}6c$, the inversion in chromosome 5, and a stock combining them, together with spore quartet data on $T5\text{-}6c$ and $T6\text{-}10b$. The conclusion from those data that chromosomes that cross over pass to opposite poles has been reported (MCCLINTOCK 1945). The data in this paper are entirely from later material grown in the field at UNIVERSITY FARM, St. Paul, Minnesota.

Pachytene morphology

To interpret the experimental data on chromosome segregation in the crosses it is essential to know where the breakages occurred to produce the chromosomal changes. In plants heterozygous for $T5\text{-}6c$, considerable variation in the position of the "cross" was noted at pachytene. Of 21 figures observed on one

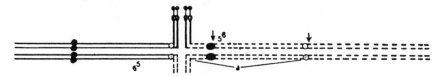

FIGURE 3. Pachytene diagram of $T5\text{-}6c/+$ with the break in 6 in the short arm adjacent to the centromere and that in 5 near the end of the long arm, 6^5 and 5^6 being the two translocated chromosomes, a the interstitial segment. The arrows show the points of breakage in inversion $5a$.

slide, ten had the "cross" in the short arm of chromosome 6 as illustrated in the diagram (fig. 3), while eleven had it in the long arm of chromosome 6, many of them at a considerable distance. Only in the former were the homologous chromosome 6 centromeres synapsed. In the latter they were on different arms of the cross. These observations indicate the breaks were in the short arm of chromosome 6 and toward the end of the long arm of 5. A study of the two translocated chromosomes in the stock homozygous for $T5\text{-}6c$ served to determine the precise breakage points, since the short arm of the normal chromosome 6 has large, darkly-staining chromomeres while the segment of the long arm of chromosome 5 distal to the knob has very small, lightly-staining ones. Since the entire short arm of the 6^5 translocated chromosome (read as "6 with a piece of 5") had only small chromomeres, the break in chromosome 6 must have been in the short arm adjacent to the centromere. The other 5^6 translocated chromosome was attached to the nucleolus, and the point where the darkly staining segment of chromosome 6 met the lightly staining one must have been the point of breakage in chromosome 5. It was considerably distal to the knob in the long arm of chromosome 5.

The breakage points in the normal chromosomes and the resulting translocated ones are shown in the diagrams, figures 4 c, d, and a, b, respectively, the average lengths of the segments in mm as measured in the *camera lucida* drawings at 5500×magnification being included. Based on these measurements, the lengths of the interchanged pieces are probably about 100 percent of the normal short arm of chromosome 6 and 11 percent of the normal long arm of chromosome 5. These observations agree with MCCLINTOCK's statement in a private communication that the breakage points were readily observable.

Pachytene stages from the stock homozygous for an inversion in chromosome 5 (*In 5a*) revealed a morphology which permitted a fairly accurate determination of the positions of these two breaks. In the inverted chromosome, a heavily stained region in the short arm adjacent to the centromere was observed in many of the figures, while in all the figures examined there was a very small knob in the long arm. This morphology would be expected if one break in the normal chromosome 5 had been in the knob present in the long arm and the other break in the short arm adjacent to the centromere. This would leave one piece of the knob adjacent to the centromere in its new position and transpose the other piece of the knob to the old position of the centromere. It is also possible that the dark region next to the centromere may have been the dark chromomeres found at that point in certain stocks

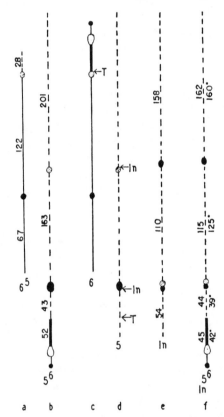

FIGURE 4. Diagrams of chromosomes 5 and 6 in homozygous normal, translocation and inversion stocks. The figures along the chromosome segments are the average lengths in mm from camera lucida drawings at about 5500×magnification. Measurements marked ● on the inverted 5⁶ chromosome were calculated from the separate *T5–6c* and *In5a* measurements.

a, b = the 6⁵ and 5⁶ chromosomes in *T5–6c*.

c, d = normal chromosomes 6 and 5, arrows indicating the points of breakage

e = chromosome 5 of *In5a*

f = inverted chromosome 5⁶ of *T5–6cIn5a*.

and not a piece of the subterminal knob. In that case one break adjacent to the knob but on the distal side, and the other a short distance from the centromere in the short arm, would result in a very similar morphology. Based on the assumption that one break was in the knob, about 67 per cent of the normal long arm plus the centromere and none of the short arm were included in the inversion (fig. 4e), the points at which breakage occurred to produce the inversion being indicated on the normal chromosome (fig. 4d; only slightly different if the break had been adjácent to the knob).

The homozygous stock which has T5-6c and the inversion combined showed all the features of chromosome morphology at pachytene expected from what was found in the separate stocks. The 6^5 chromosome was the same as in T5-6c; while the 5^6 was attached to the nucleolus, had its centromere much closer to this end due to the inversion, and showed the small knob in the long arm (fig. 4f). The darkly-stained region in the short arm adjacent to the centromere was observed only in part of the figures, similar to what was described above for In5a. The observed average lengths of these regions (based on three cells), together with those calculated from the observed values in the separate T5-6c and inversion stocks (marked with ·), are also shown in figure 4f. The agreement is very close in spite of the fact that the two are based on different slides from different plants.

A study of pachytene pairing was attempted for plants heterozygous for T5-6c and homozygous for the inversion to determine if the shift in centromere position affected the frequency or extent of non-homologous pairing. As to the "cross" arm lengths, the configuration is expected to be similar to that shown in the diagram (fig. 3) for T5-6c/+ alone. Several figures were observed in which the "cross" was in the long arm of chromosome 6 indicating non-homologous pairing. More observations must be made for an adequate comparison of frequencies between this and the stock not carrying the inversion.

In the T2-6a heterozygote at pachytene, the "cross" was frequently far out in the long arm of chromosome 6 (BURNHAM 1932a), the chromosome 6 centromeres then being on different arms indicating non-homologous association. In an occasional figure the "cross" was at a position in which these centromeres were paired, indicating the break in chromosome 6 was at or near the centromere. Since pollen abortion in the T2-6a heterozygote was 50 percent, it is probable that at least part of the short arm of 6 was translocated. Examination of the homozygous stock showed the break to be in the short arm of chromosome 6, the figures not being good enough to determine the precise point. Chromosome 2^6 now has the organizer and satellite. The break in chromosome 2 was between the knob and the centromere at about L-0.4, assuming the break in 6 to be close to the centromere.

The breakage points in the translocations used may be summarized as follows:

T2-6a	2L-0.4 (or a little less)	6S-0.0 (or up to 0.5?)
T4-6Li (C. H. LI, unpublished)	4S-0.7	6S-0.2
T6-10b (McCLINTOCK, unpublished)	6S-0.5	10L-0.58
T5-6c	5L-0.89	6S-0.0
T5-6c In 5a	(see below)	6S-0.0

In this last stock, the break in 5 is at L-0.89 in terms of the normal 5, but the centromere is now at about L-0.67; so that the distance between the new centromere position and the break point is 0.22 (0.89−0.67).

One other translocation, Connecticut 1-6, has been reported (ROBERTS 1942) to have the break in chromosome 6 in the short arm at 0.2. In the stock used for this study, only 26 percent of pollen abortion and 0.6 percent of quartets having two spores with diffuse nucleoli were observed. Examination of the homozygous stock showed the break in chromosome 6 was in the outer portion of the organizer, and both translocated chromosomes were attached to the nucleolus. The break in chromosome 1 was at about 0.05 of the long arm. Chromosome 1[6] therefore has a satellite and a small piece of the organizer region of chromosome 6 while 6[1] has retained the major portion of the nucleolar organizer region. This also accounts for the low pollen abortion since the class of spores deficient for the translocated piece of 6 probably does not abort. This translocation therefore belongs in group 3. ROBERTS has furnished his original cytological observations, including spore quartet counts. The latter showed 36 percent of "two diffuse" quartets and 4.6 percent of crossover type quartets; clearly indicating the stock I used is a different translocation.

Data on spore quartets from plants with 10 pairs of chromosomes as normal checks

For *In 5a*, based on 3,446 quartets, 0.5 percent of "two diffuse," and 0.6 percent of "one diffuse" types were observed accounting for about 0.8 percent of pollen abortion (5.4 percent observed). For *T5-6c*, based on 1,401 spore quartets, 1.1 percent of "two diffuse" and 1.8 percent with "one diffuse" were observed accounting for about two percent of pollen abortion (3.9 percent observed). One slide had six percent of abnormal quartets. In *T5-6c* homozygous for *In 5a* only 0.003 percent of "two diffuse" and "one diffuse" were observed. No data were obtained on pollen sterility on these latter plants. These cytological results may be explained on the basis of McCLINTOCK'S conclusion (1941) that a chiasma in the short arm between the centromere and the nucleolus organizer region may result in breakage of the chromatids followed by fusion and sometimes bridge plus fragment formation. The lower observed frequency of abnormal quartets in *T5-6c In 5a* as compared with *T5-6c* is in line with this explanation since the region in which this might occur is much longer in *T5-6c*. Loss of fragments carrying the nucleolar organizers could account for the observed low frequency of spores with two nucleoli in the "one" or "two diffuse" quartets (in 35 "two diffuse" quartets, none had a second spore with two nucleoli; while in 45 "one diffuse" type, only four had a spore with two nucleoli).

[*Editors' Note:* Material has been omitted at this point.]

Chromosome segregation, spore quartet data

Group 1—Translocations with one break in the short arm of chromosome 6

a. *with short interstitial segments* including *T5-6c In5a*, *T2-6a* and *T6-10b*. A summary of the data for these has been presented (BURNHAM 1949), but are presented here in greater detail. Plants heterozygous for *T5-6c* and homozygous for *In5a* were produced by crossing a stock homozygous for both *T5-6c* and *In5a* with one homozygous for *In5a*. Figure 3 represents the pachytene configuration, except that both chromosome 5 centromeres are shifted to the point near the knob indicated by the left arrow. Figure 4, e and f represent the two number 5 chromosomes. The detailed data are shown here in table 1, together with the calculated frequencies of the segregation types in the non-crossover quartets. The three groups of plants for which sub-totals are given in the table were from different crosses of related material. There was considerable variation between crosses and between plants from the same cross. With one exception, the plants in culture 6,913 showed much lower percentages of quartets having one or two "diffuse" spores than did the others. The reason for this difference is not known. One plant, 6,913-7, had only 41 percent of pollen abortion and consequently a very low calculated frequency of adjacent 2 segregation. It was not a 2n+1 plant. As shown earlier, a low percentage of the abnormal quartets, less than two percent, was observed in the homozygous controls, but since the predicted pollen abortion probably includes most of these (and in general probably more than indicated by the quartet phenotype) no correction was used. Even if the observed abortion in the translocation heterozygotes were corrected for the full amount, the conclusions would be essentially the same since this would reduce the frequency of adjacent 2 segregation by only two to five percent. The observed frequency of "crossover type" quartets, 10.6 percent for the grand total in table 1, is so low that probably no double crossing over occurred. Hence this is the percentage of single crossing over in the interstitial segment of chromosome 5, and, as shown in figure 1B, may be the result of alternate or adjacent 1 segregations following these single crossovers. Since there were probably no double crossovers, the "two diffuse" type is the result of adjacent 1 segregation in chromatid tetrads which had no crossing over in the interstitial segment. There were 27.8 percent of this type. Since the "crossover type" quartet should produce two and the "two diffuse" type should produce four aborted spores, the predicted pollen abortion for the total of all the data was 33.0 percent. The observed pollen abortion was 50.5 (table 1), an excess of 17.5 percent which should measure the frequency of adjacent 2 segregation (normal-ap-

TABLE 1

Summary of spore quartet and pollen abortion data, including frequencies of segregation types, for a group 1 translocation heterozygote (one break in the short arm of chromosome 6): T5-6c homozygous for the inversion in chromosome 5 (In5a); a short interstitial segment.

CULTURE PLANT NO.	NO. OF SLIDES	SPORE QUARTETS					TOTAL	POLLEN ABORTION, PERCENT			NON-CROSSOVER QUARTETS			
		NO DIFFUSE	TWO DIFFUSE	ONE DIFFUSE*	TOTAL	PERCENT CROSS-OVER TYPE		OB-SERVED	PRE-DICTED	PARTLY† FILLED	TOTAL	SEGREGATION TYPES IN %		
												ALTER-NATE	ADJA-CENT 1	ADJA-CENT 2
6913- 3	2	442	145	51	638	8.0	2,486	51.1	26.7	9.5	587	48.7	24.7	26.6
-29	2	694	235	48	977	4.9	2,958	53.1	26.5	15.7	929	46.7	25.3	28.0
- 7	1	210	111	64	385	16.6	3,008	41.0	37.1	11.0	321	60.7	34.6	4.7
-12	2	602	239	79	920	8.6	3,370	49.8	30.3	11.7	841	50.3	28.4	21.3
-14	2	469	137	51	657	7.8	2,223	47.1	24.7	12.5	606	53.1	22.6	24.3
-18	3	436	150	44	630	7.0	1,907	50.4	27.3	12.8	586	49.5	25.6	24.9
subtotal		2,853	1,017	337	4,207	8.0	15,952	48.7	28.2	12.2	3,870	51.4	26.3	22.3
6924- 1	2	565	314	116	995	11.7	2,074	55.8	37.4	17.3	879	43.5	35.7	20.8
-10	2	288	186	133	607	21.9	1,874	53.9	41.6	14.0	474	44.9	39.2	15.8
subtotal		853	500	249	1,602	15.5	3,948	54.9	39.0	15.7	1,353	44.2	37.0	18.8
6925- 9	2	435	282	83	800	10.4	2,672	51.6	40.4	—	717	48.1	39.3	12.6
6926- 8	2	293	197	90	580	15.5	3,187	51.1	41.7	—	490	48.6	40.2	11.2
subtotal		728	479	173	1,380	12.5	5,859	51.3	41.0	—	1,207	48.5	39.7	11.8
Total		4,434	1,996	759	7,189	10.6	25,759	50.2	33.0	12.9	6,430	49.7	31.0	19.2

* Crossover type.
† Observed.

pearing quartets but all aborted spores). Of the normal-appearing quartets 1,258 (17.5 percent of 7,189, the total number of quartets) were from adjacent 2 segregation; and the remainder were from alternate segregation. Of the non-crossover quartets, the percentages resulting from alternate, adjacent 1, and adjacent 2 segregations were therefore 49.4, 31.0 and 19.6 respectively. If alternate and adjacent 1 segregations are assumed to be equally frequent following crossing over in the interstitial segment, these values when the crossover quartets are included become 49.5, 33.0 and 17.5 percent, respectively; a small change in the ratio. The percentage of crossover quartets varied from 4.9 to 21.9 percent while the percentage of adjacent 2 segregations varied from 4.7 to 28.0. The correlation between the two is -0.77 which, for 10 comparisons, is highly significant ($P = 0.01$). In other words, the higher the frequency of crossing over, the lower the frequency of adjacent 2 segregation. In the non-crossover quartets, alternate segregations accounted for about 50 percent while adjacent 1 and 2 together accounted for the other 50 percent. This may be accounted for if for each "open" ring a "zigzag" one is equally likely. The average ovule abortion, 49.3 percent, did not differ significantly from the observed pollen abortion, 50.5.

[*Editors' Note:* Material has been omitted at this point.]

Group 1b. Translocations with at least one long interstitial segment; T5-6c, T4-6 Li. Similar data on quartets and pollen together with the calculated segregation frequencies for plants heterozygous for *T5-6c* without the inversion and for *T4-6* Li are in table 3. For *T5-6c*, figure 3 represents the pachytene configuration, fig. 4a, b the translocated chromosomes. Based on 4,229 quartets from six plants (three crosses) an average of 62.8 percent were of the "crossover type," the value ranging from 46.4 to 70.6 percent for individual plants. There were 742 quartets or 17.5 percent of the "two diffuse" type. Since the observed pollen abortion (48.1 percent) was no greater than the predicted (49.3), actually a little lower, no measurable amount of adjacent 2 segregation is indicated.

The degree of ovule abortion was 47.6 percent, practically the same as the observed pollen abortion, 48.1, indicating little if any megaspore competition. BRIEGER (1945) has presented what he considers as evidence for such competition in corn.

For *T4-6* Li, the average percent of crossover type quartets was 62.6 percent, and again the predicted abortion was close to the observed 50 percent indicating very little adjacent 2 segregation. The high percentage (63) of "crossover type" quartets for both translocations indicates that multiple crossovers were frequent in the interstitial segment. The frequency of adjacent 1 segregations cannot be calculated directly, since the "two diffuse" quartet is expected following 2-strand and 4-strand doubles as well as from non crossovers (see fig. 1B).

TABLE 3

Summary of spore quartet and pollen abortion data, including frequencies of segregation types, for two group 1 translocation heterozygotes (one break in the short arm of chromosome 6): T5-6c and T4-6Li; one very long interstitial segment.

CULTURE, PLANT NO.	NO. SLIDES	SPORE QUARTET						POLLEN				SEGREGATION TYPE
		NO. DIFFUSE	TWO DIFFUSE	ONE DIFFUSE	TOTAL	PERCENT "TWO" DIFFUSE	PERCENT CROSS-OVER TYPE	TOTAL	ABORTION			PERCENT ADJACENT 2
									OB-SERVED	PRE-DICTED	PARTLY FILLED	
T5-6c/+												
6899- 1	2	175	155	286	616	25.2	46.4	1,835	50.0	48.4	24.3	1.6
- 4	5	104	129	375	608	21.2	61.7	5,637	42.0	52.1	20.4	0.0
6900- 6	3	178	157	674	1,009	15.6	66.8	2,318	51.4	49.0	24.2	2.4
6901- 7	2	124	108	406	638	16.9	63.6	2,163	48.5	48.7	22.4	0.0
- 9	3	97	102	477	676	15.1	70.6	2,065	48.7	50.4	21.8	0.0
-10	2	131	91	460	682	13.3	67.4	2,733	47.7	47.1	20.3	0.6
Total		809	742	2,678	4,229	17.9*	62.8*	16,751	48.1*	49.3*	22.2*	0.0
T4-6Li												
15085- 3	2	163	109	319	591	18.4	54.0	2,883	48.0	45.4	15.3	2.6
15086- 1	2	83	70	378	531	13.2	71.2	1,991	51.6	48.8	—	2.8
Total		246	179	697	1,122	15.8*	62.6*	4,874	49.8*	47.1*	—	2.7

* Average percents.

[*Editors' Note:* Material has been omitted at this point.]

SUMMARY

1. This paper presents a study of chromosome segregation in 27 translocation heterozygotes involving chromosome 6 in corn. Certain segregations can be recognized by the nucleolar make-up of the spore quartets or by differences between predicted and observed pollen abortion. In translocations with the break in the short arm of chromosome 6, crossing over in the interstitial segments results in a distinctive quartet type.

2. Three translocation heterozygotes, (*2-6a*, *6-10b*, and *T5-6c* homozygous

for inversion *5a*) with short interstitial segments (1.0, 5.3 and 10.6 percent respectively of crossover type quartets), had the following percentages of alternate: adjacent 1 (homologous centromeres disjoin): adjacent 2 (non disjunction of homologous centromeres): for *T2-6a*, 48:26:26; for *T6-10b*, 58:19:23; for *T5-6cIn5a*, 50:31:19.

3. For individual plants in *T5-6cIn5a*, there was a high negative correlation (r = −0.77) between the percentage of crossover type quartets and the frequency of adjacent 2 segregation.

4. Two translocation heterozygotes, *T4-6* Li and *T5-6c*, had a long interstitial segment (63 percent of crossover quartets in each), and very little adjacent 2 segregation. The difference between *T5-6c* and *T5-6cIn5a* was only in the length of the interstitial segment, axis lengths and arm lengths of the pachytene "cross" remaining the same.

5. For translocations with the break in the long arm of chromosome 6:

a. A group of ten with short interstitial segments had 14 to 36 percent of adjacent 2 segregation, the average being 25.

b. A group of nine with at least one long interstitial segment had an average of 5.4 percent of adjacent 2 segregation. One, *T6-8a* with 17.4 percent, overlapped the other group.

6. Four chain-forming translocations, two with short and two with long interstitial segments, had very low frequencies of adjacent 2 segregation.

7. In *T5-6c/+* with a high percent of crossover quartets having two normal and two abortive spores, ovule and pollen abortion were about equal indicating no megaspore competition.

8. Relative lengths of the two axes is not a factor determining segregation. The segregations are similar whether the centromeres are on the shorter or the longer axis.

General Conclusions

1. In corn, at least for translocations involving chromosome 6, the following relationships between crossing over in interstitial segments and chromosome segregation in translocation heterozygotes are found:

a. *In rings*, low crossover frequency (short segments) is accompanied by a high frequency of adjacent 2; while high crossover frequency (long segment) is accompanied by a low frequency of adjacent 2 (little or none). The evidence indicates that when chromosomes cross over in the interstitial segment they pass to opposite poles, and that adjacent 2 segregation occurs only when there has been no crossing over.

b. In chains, on the other hand, little or no adjacent 2 segregation occurs in any of the translocations, no matter how short the interstitial segment.

2. It is suggested that directed segregation in translocation heterozygotes may be genetically controlled, the amount of spore abortion in such a species depending on the frequency of crossing over in the interstitial segment. Genetic recombination in such a species as pointed out by LAMM (1948; quoting Tometorp) may be greatly reduced, since single crossovers and certain of the multiples are not recovered following alternate segregation. This does

not account for all the reduced crossing over in corn translocations, since there is no evidence of any great amount of directed segregation. In the search for such genes in corn, translocations with short interstitial segments should be used.

REFERENCES

Brieger, F. G. 1945. Competicao entre megasporios em milho. *An. Esc. sup. Agric., Sao Paulo. Univ.* **2**(18):239–267.

Burnham, C. R. 1932a. The association of non-homologous parts in a chromosomal interchange of maize. *Proc. 6th Int. Cong. Genetics* **2**:19–20 (Abstract).

_____. 1932b. An interchange in maize giving low sterility and chain configurations. *Proc. Nat. Acad. Sci. U.S.A.* **18**:434–440.

_____. 1945. Chromosome disjunction in maize interchanges. *Genetics* **30**:2 (Abstract).

_____. 1948. Chromosome segregation in translocations involving chromosome 6 in maize. *Genetics* **33**:605 (Abstract).

_____. 1949. Chromosome segregation in maize translocations in relation to crossing over in interstitial segments. *Proc. Nat. Acad. Sci. U.S.A.* **35**:349–356.

Hanson, W. D., and H. H. Kramer. 1949. The genetic analysis of two chromosome interchanges in barley from F_2 data. *Genetics* **34**:687–700.

Koller, P. C. 1944. Segmental interchange in mice. *Genetics* **29**:247–263.

Lamm, R. 1948. Linkage values in an interchange complex in *Pisum. Hereditas* **34**:280–288.

McClintock, B. 1934. The relation of a particular chromosomal element to the development of the nucleoli in *Zea mays. Z. Zellforsch. Mikrosk. Anat.* **21**:294–328.

_____. 1941. Spontaneous alterations in chromosome size and form in *Zea mays. Cold Spring Harb. Symp. Quant. Biol.* **9**:72–80.

_____. 1945. Neurospora. I. Preliminary observations of the chromosomes of *Neurospora crassa. Am. J. Bot.* **32**:671–678.

Roberts, L. M. 1942. The effects of translocation on growth in *Zea mays. Genetics* **27**:584–603.

ERRATA

Page 449, line 10 from the bottom should read "...following the cross-overs, no reduction..."

Page 459, line 2 from the bottom should read "...pollen abortion was 50.2..."

40

Reprinted from *Genetics* **32**(4):391–409 (1947)

MITOTIC NONDISJUNCTION IN THE CASE OF INTERCHANGES INVOLVING THE B-TYPE CHROMOSOME IN MAIZE[1]

HERSCHEL ROMAN

University of Washington, Seattle, Washington

Received April 19, 1947

INTERCHANGES between the B-type chromosome and members of the basic set (A-type chromosomes) in maize provide a means of clarifying the anomalous behavior of the B-type itself. They also make possible a method whereby plants deficient for a specific A-type segment or carrying the segment as a duplication may be regularly produced for use in various cytogenetic studies.

The characteristics of the B-type have been described in detail by RANDOLPH (1941). The chromosome occurs as a supernumerary in many strains of maize. Within a strain, the number of B-types may vary considerably from plant to plant. RANDOLPH (1928) found a range of zero to eight among 43 plants of the Black Mexican variety. The chromosome does not produce a specific genetic effect even when present in relatively large numbers; a plant devoid of B-types is not noticeably different in appearance from one carrying as many as ten (RANDOLPH 1941).

The source of the variability in B chromosome number within a strain has been investigated in crosses in which only one of the parents furnished the B-types. The results of two such crosses and their reciprocals are given in table 1. On an orthodox basis, the progeny from these crosses should not have more than one B chromosome. Each of the crosses, however, yielded plants with two or more B-types and the proportion of these exceptional plants was particularly high when the B-types were transmitted by the male parent (i.e., in the oB×1B and oB×2B crosses).

These results are due only in small part, if at all, to aberrant meiotic behavior. In the microsporocytes of a 2B plant, for example, the B chromosomes usually form a bivalent in the first meiotic prophase and disjunction follows. The regularity of this procedure may be verified by chromosomal counts in the microspore nuclei formed as a result of meiosis. In 130 figures counted, RANDOLPH (1941) found 116 with one B-type and 14 with none. Since male gametes with two or more B-types are obtained from a 2B plant (as shown by the results of the oB×2B cross), these must have been produced from 1B microspores. The proportion of oB progeny from the oB×2B cross is also greater than would be expected from the microspore counts. This suggests that oB gametes are also derived from 1B microspores.

The indication that both oB and 2B gametes are obtained from 1B microspores suggests the possibility that the B chromosome is undergoing "mitotic nondisjunction" in one, perhaps rarely in both, of the divisions which produce

---

[1] This investigation was begun and largely completed at the UNIVERSITY OF MISSOURI, Columbia, Missouri.

the gametes. By this mechanism, a B chromosome would be distributed in such a way that one of the daughter nuclei would receive two B-types and the other would receive none. The result of the oB×1B cross lends support to this view. Although a sizeable number of plants produced by this cross received two or more B chromosomes, the average number of B-types per plant was approximately 0.5. This may be interpreted to mean that the production of 2B gametes from 1B microspores is balanced by the production of a numerically

TABLE 1

Distribution of B chromosomes among progeny of crosses between oB and 1B or 2B parents.

NUMBER OF B-TYPES IN PARENTS	B-TYPE NUMBER IN F_1					NUMBER OF F_1 PLANTS	AVERAGE NO. B-TYPES PER F_1 PLANT	INVESTIGATOR
	0	1	2	3	4			
o×1	76	22	15			113		Longley 1927
	32	13	5	2		52		Randolph 1941
	108	35	20	2		165	.49	
o×2	22		37		2	61	1.34	Randolph 1941
1×o	109	58	3			170		Longley 1927
	46	19				65		Randolph 1941
	155	77	3			235	.35	
2×o	8	41	3			52	.90	Randolph 1941

equivalent class of oB gametes, as would be expected from the hypothesis of mitotic nondisjunction.

A direct cytological determination of what is happening during the microspore divisions is not feasible for technical reasons. A determination by genetic methods is also precluded since the B chromosome cannot be identified by a phenotypic effect. However, through interchanges with members of the basic set (these will be referred to as A-B interchanges), it should be possible to attach A-type segments carrying known genes to B-type segments and thus to trace the behavior of the latter. This method rests on the assumption that the anomalous behavior of the B chromosome does not depend on its remaining intact. The validity of this assumption is indicated by observations of B-type fragments (RANDOLPH 1941). It should also be possible, by a comparison of the various B-segments produced by breakage at different points in the chromosome, to locate the element within it that is responsible for its aberrant performance.

MATERIALS AND METHODS

Mature pollen from plants with from four to ten B chromosomes was treated with X-rays and applied to silks of plants devoid of B-types. The radiation was produced by a Coolidge tube operated at 138 K.V.P. and was unfiltered except for the anther wall and the damp paper towelling in which the anthers were wrapped. The doses applied ranged from 600 to 1,600 r-units.

Only a small fraction of the progeny was expected to carry the A-B interchanges. The plants most likely to carry them were screened from the F_1

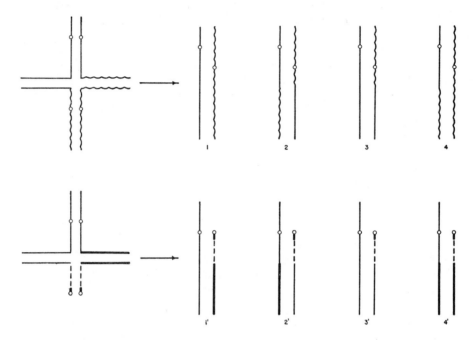

FIG. 1.—Main spore types expected from a plant heterozygous for an A-A interchange (above) compared with those expected from a plant heterozygous for an A-B interchange (below). (Only the chromosomes involved in the interchange are shown.) Note that types 3 and 4 are both deficient for A-chromatin whereas only type 4' is thus deficient. The respective pachytene configurations are given at the left. The B-type is represented as having a terminal centromere; see p. 405 for a fuller discussion of this.

population by an examination of the mature pollen of each plant. Normal maize plants in healthy condition produce pollen with nearly all grains uniform in size and well-filled with starch. There are usually a few defective grains present, seldom as high as five percent. Plants heterozygous for a deleterious deficiency produce normal and defective grains in approximately equal numbers. Plants heterozygous for an interchange between A-type chromosomes (A-A interchanges) usually yield the same pollen picture since two of the four main types of spores formed are deficient for A-chromatin (fig. 1). On the same basis, plants heterozygous for an A-B interchange would be expected to produce 25 percent of defective pollen since one of the two deficient spore types is deficient only for B-chromatin and this does not alter pollen appearance. This spore type also carries a duplication for A-chromatin but duplications ordinarily have no visible effect on pollen development.

The A-B interchanges could therefore be distinguished by pollen examination from deficiencies and most A-A interchanges. There are, however, some A-A interchanges in which one of the deficient spore types produced by the heterozygous plant develops into a visibly normal pollen grain with the result that only 25 percent of the grains are defective in appearance. Examples of this type, involving the satellite of chromosome 6, have been reported by BURNHAM (1932) and CLARKE and ANDERSON (1935). Plants heterozygous for an inversion also produce defective pollen, in proportions depending on the amount of crossing over within the inverted region. Microsporocyte samples were taken from the F_1 plants and were stored until after pollen examination. The identification of those plants heterozygous for an A-B interchange, among the 25 percent group, was accomplished by cytological examination.

The presence of an A-B interchange could be detected either in pachytene or in diakinesis. In pachytene, the interchange, in heterozygous condition, assumes the familiar cross-configuration if an intact B-type is present. If an intact B-type is not present, a T-configuration is frequently formed, as a result of the homologous pairing of the A-segments and the nonhomologous pairing of the B-segments. In either case, the interchange complex may be detected readily owing to the singular pachytene appearance of the B-type (McCLINTOCK 1933). Plants heterozygous for an A-B interchange show, in diakinesis, nine A-bivalents and the interchange complex, in contrast with plants heterozygous for an A-A interchange, in which eight bivalents are formed. The A-B interchange complex is in a chain configuration if an intact B-type is not present. If it is present, it usually associates with the complex to form a ring configuration. This offers especially convincing evidence of an A-B interchange since B chromosomes ordinarily do not pair with members of the basic set. When two or more extra B-types are present, some or all of them often associate with the complex in multivalent pairing; otherwise, they form separate univalents, bivalents, or multivalent groups.

EXPERIMENTAL RESULTS

Eight A-B interchanges were obtained from approximately 500 plants screened by the method just described. Chromosomes 1 and 7 each were involved in two interchanges and chromosomes 4, 6, and 9 each were involved once. In one of the interchanges, the A chromosome has not yet been identified.

This report deals only with the interchange involving chromosome 4, designated as TB-4a. The investigation of the other interchanges is still in a preliminary stage. It should be noted, however, that in five of the latter, the investigation has proceeded far enough to indicate that the behavior described for TB-4a applies to these as well.

In the case of TB-4a, the point of interchange in chromosome 4 is in the short arm, approximately one-eighth of the distance from the centromere to the end of the arm. In the B-type, the point of breakage is at or very near the junction of the euchromatic and the main heterochromatic segments (fig. 2). Approximately 25 percent of the pollen grains produced by a plant heterozygous for TB-4a are devoid of starch. The pollen grains of the homozygote are normal in appearance.

Evidence of mitotic nondisjunction

The hypothesis of mitotic nondisjunction, set forth to account for the anomalous transmission of the B-types, has different consequences depending on whether nondisjunction occurs in the first or the second division of the microspore. The formation of the male gametes in maize proceeds as follows. The microspore nucleus derived from the meiotic divisions divides to form a generative and a vegetative, or tube, nucleus. The generative nucleus divides again to form the two gametic nuclei. Thus the mature pollen grain is trinucleate; it carries two crescent-shaped sperm and the tube cell (WEATHERWAX 1917). The latter presumably controls the growth of the pollen tube down the

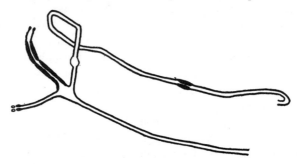

FIG. 2.—Camera lucida drawing of TB-4a in heterozygous condition in the pachytene stage of the microsporocyte.

silk and is not directly involved in the fertilization process. One of the gametic nuclei fuses with the polar nuclei in the embryo sac to form the primordial endosperm nucleus. The other gamete fertilizes the egg.

If, starting with a 1B nucleus, nondisjunction occurs only in the first division of the microspore, either the generative nucleus or the tube nucleus would receive two B-types and the other would receive none. If the generative nucleus receives the two B-types, the gametic nuclei derived from it would each have two B-types. If the generative nucleus does not receive the B-types, the gametes would also lack them. In either case, the two gametes within a single pollen grain would be identical with respect to B chromosome number. If, on the other hand, mitotic nondisjunction occurs only in the second division of the microspore, both the generative and the tube nuclei would have one B-type. The gametic nuclei of a pollen grain in this case would not be alike in B chromosome number; one would receive two B-types and the other would receive none.

In appropriate crosses, fertilization involving male gametes that are alike in B chromosome number would produce a seed with an endosperm and embryo also alike in this respect. If the gametes of a pollen grain are not alike, the endosperm and embryo would also differ. From a 0B×2B cross, for example, we would expect from nondisjunction in the first division of the microspore that the embryo and endosperm would both contain either two B-types or none. If nondisjunction occurs in the second division, we would expect some of the seeds to have a 2B endosperm and a oB embryo and others to have a oB endosperm and a 2B embryo, depending on the manner in which the oB and 2B gametes of the pollen grain are engaged in fertilization. The time of nondisjunction could therefore be fixed by an examination of these tissues; unfortunately, the endosperm does not lend itself readily to cytological study.

Crosses involving TB-4a were made to test the hypothesis of mitotic non-

TABLE 2

Results obtained when Su Su plants homozygous for TB-4a were crossed as seed parents with normal su su plants.

CROSS	TOTAL SEEDS	SUGARY SEEDS
34.3—2×6.1—1	228	0
34.4—4×6.1—1	165	0
35.1—1×3.4—4	84	0
35.1—2×6.2—6	285	0
35.2—7×6.1—1	244	0
35.2—8×6.1—1	151	0
Totals	1,157	0

disjunction and to decide between the foregoing alternatives. Plants homozygous for the interchange were intercrossed with plants that were normal and lacked B chromosomes. (These crosses are analogous to the oB×2B and 2B×oB crosses insofar as B-chromatin is concerned except that the B-types involved in the interchange are not intact.) The normal parent was homozygous for *su* (sugary endosperm), a recessive gene located in the short arm of chromosome 4 (ANDERSON and RANDOLPH 1945). The interchange parent carried the dominant allele, *Su*, in homozygous condition.

The results of these crosses were widely different depending on whether the interchange parent served as the male or the female. When TB-4a was transmitted by the seed parent, all of the F_1 seeds were nonsugary (table 2), as would be expected from normal behavior. When the interchange parent served as the pollen source, approximately half of the F_1 seeds were sugary (table 3).

The sugary seeds obtained from the latter cross were interpreted as indicating a deficiency in the endosperm of the interchange chromosome bearing the *Su* gene. These seeds were grown for a cytological examination of the microsporocytes to determine the chromosomal content of the embryo. Twenty-four plants were examined and each was found to have an extra B^4 chromosome, the interchange chromosome with a B-type centromere. They contained two B^4 chromosomes in addition to a normal chromosome 4 and a 4^B chromosome (the interchange chromosome with the chromosome 4 centromere).

It remained to be determined whether the chromosome that is absent in the endosperm of the sugary seeds is the same one that is present as a duplication in the hyperploid embryo. This was accomplished by using the hyperploid plants as seed parents in crosses with normal plants that were homozygous for *su*. It was observed in diakinesis of the hyperploids that in a majority of the cells, the normal chromosome 4 paired with the 4^B chromosome and the two

TABLE 3

Results obtained when Su Su plants homozygous for TB-4a were crossed as pollen parents with normal su su plants. (34.3-7 had no intact B-types; 34.4-7 had two intact B-types.)

CROSS	TOTAL SEEDS	SUGARY SEEDS	
		%	χ^{2*}
3.3-1×34.3-7	222	48.2	.16
6.2-3×34.3-7	166	44.0	2.4
6.2-6×34.3-7	270	55.2	2.9
78.1A-2×34.3-7	278	48.6	.23
78.1-2×34.4-7	201	53.2	.84
7.2-2×34.4-7	209	57.9	5.2
7.3-3×34.4-7	200	48.0	.32
8.4-3×34.4-7	245	57.6	5.6
Totals	1791	51.9	2.7

* Calculated for deviation from 50%.

TABLE 4

Results of crosses involving hyperploid plants for location of Su with respect to breakage point in chromosome 4. See text for description of cross.

TOTAL SEEDS	SUGARY SEEDS %
150	5.3
234	5.1
287	5.9
220	11.4
300	5.3
266	5.6
305	3.3
158	4.4
261	11.1
Totals 2,181	6.4

B^4 chromosomes formed a separate bivalent. In some cells, all four chromosomes were associated in a single complex. It would be expected from these observations that two types of spores would predominate, one containing 4^B and B^4 and the other containing chromosome 4 and B^4 (fig. 3). If *Su* is on 4^B, we would expect, since the normal chromosome 4 carries *su*, that half of the seeds obtained from the cross would be sugary. If *Su* is located on B^4, each of the two main types of spores would have this allele and only a small percentage of sugary seeds would be expected, these resulting from meiotic disjunction of the relatively rare complex association. The results of the cross (table 4) show that *Su* is on the B⁴ chromosome.

As already indicated, the noncorrespondence in the chromosomal composition of the endosperm and embryo in the sugary seeds is evidence of a parallel noncorrespondence in the composition of the gametes contained in the pollen grains furnished by the parent homozygous for TB-4a. Thus the gametic nucleus that fused with the polar nuclei to form the primary endosperm nucleus was deficient for the B⁴ chromosome and that which fertilized the egg carried this chromosome in duplicate. This means that nondisjunction of the B⁴ chromosome occurred in the division of the generative nucleus. Figure 4

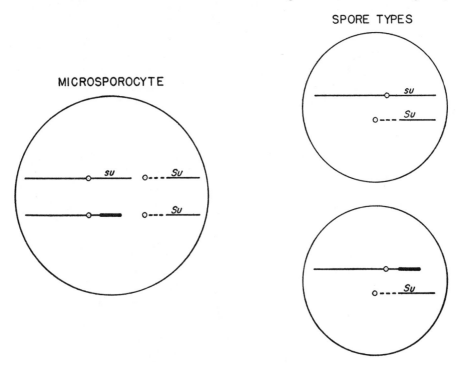

FIG. 3.—Association of chromosomes in diakinesis of microsporocyte and expected spore types.

illustrates the probable course of events in the formation of a pollen grain on the basis of nondisjunction in this division.

Another type of seed is also expected as a result of mitotic nondisjunction. If the gamete deficient for the B⁴ chromosome fertilizes the egg and the gamete carrying two doses of this chromosome fuses with the polar nuclei, we should obtain seeds that are nonsugary and have a deficient embryo. Twenty-seven plants grown from nonsugary seeds included twenty-four that were deficient for the B⁴ chromosome and otherwise carried a normal chromosome 4 and 4^B. The remaining three were simply heterozygous for the interchange, as would be expected from normal disjunction in the division of the generative nucleus.

In summary, when a plant homozygous for TB-4a and used as the pollen

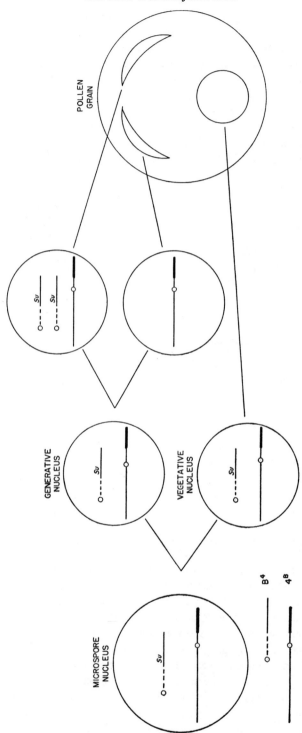

FIG. 4.—Probable course of events in the development of the pollen grain in which nondisjunction has occurred in the division of the generative nucleus.

parent is crossed with a normal plant, three types of seeds are produced. One of these has a hyperploid embryo and a deficient endosperm. Another has a deficient embryo and a hyperploid endosperm. The third type, which does not occur as frequently as the other two, has a euploid embryo, heterozygous for the interchange, and presumably also a euploid endosperm. The first two types are the result of mitotic nondisjunction of the B^4 chromosome in the second microspore division; the last, of normal disjunction.

If mitotic nondisjunction occurs in the development of the female gametophyte, its products are not transmitted to the viable eggs. When TB-4a was transmitted by the eggs in the cross of TB-4a *(Su Su)* × Normal *(su su)*, none of the F_1 seeds were sugary (table 2). Twenty-one of the seeds were grown and all were found to be simply heterozygous for the interchange. It may be that the occurrence of nondisjunction, in the development of the female gametophyte, produces an inviable gamete which prevents seed formation or results in a seed lacking an embryo. However, there was no marked evidence of the elimination of inviable gametes on the ears obtained from this cross and all of the seeds had normal-appearing embryos. If mitotic nondisjunction occurs at all in the formation of the egg, it is infrequent compared with the rate with which it occurs in the second division of the microspore.

Rate of mitotic nondisjunction

The three types of seed obtained from the cross described in the preceding section are referable to three types of fertilization (fig. 5). Types I and II are reciprocals and involve male gametes that are aneuploid as a consequence of mitotic nondisjunction. In type III, the male gametes are euploid and are the product of normal disjunction.

The rate of nondisjunction in the division of the generative nucleus may be estimated from the data in table 3 if two assumptions are made. One is that the fertilization types I and II occur in equal numbers. The second assumption is that the three types of seed are equally viable. The ears obtained from the cross carried a full set of seed and the sugary and nonsugary seeds were normal in appearance. There was no gradation in seed size to suggest a possible elimination of either kind.

The consequence of each occurrence of mitotic nondisjunction would be either a sugary or a nonsugary seed and, on the first assumption, either type would be equally probable. The frequency of sugary seeds would thus be half of the rate of occurrence of mitotic nondisjunction. If the latter were 100 percent, that is, if it occurred in all of the second microspore mitoses, 50 percent of the seeds obtained from the cross should be sugary. The sugary seeds comprised 51.4 percent of the total population obtained from eight crosses involving two different interchange parents (table 3). By the chi-square test, this value does not deviate significantly from 50 percent but the deviation from 45 percent appears to be significant. It would seem, therefore, that the rate of mitotic nondisjunction is very high, between 90 and 100 percent. The evidence that it is not 100 percent has already been presented; three of the twenty-four plants grown from nonsugary seeds were euploid and heterozygous for the interchange and were therefore the products of normal disjunction.

The calculation of the rate of nondisjunction must be made with some reservation. In two of the crosses listed in table 3, the chi-square values, 5.2 and 5.6, suggest that the deviations from 50 percent may not be due to sampling error alone. The same pollen parent was used in both crosses but it was also used in two other crosses, made under comparable conditions, which did

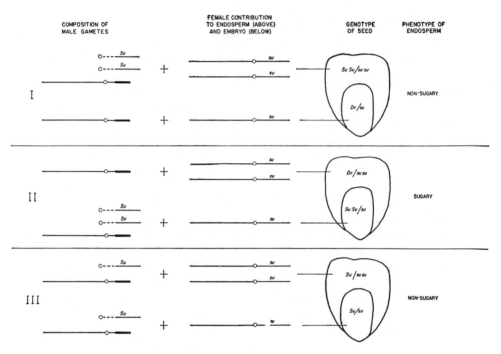

FIG. 5.—Types of double-fertilization as a result of mitotic nondisjunction (I and II) and normal disjunction (III). Types I and II are reciprocals.

not give a marked deviation. This suggests an influence of the female parent on the sugary seed frequency.

If a significant deviation from a 50 percent segregation of sugary seeds persists in additional crosses, the assumption that fertilization types I and II occur with equal frequency will require revision. Otherwise, another source of sugary seeds, not indicated by the results so far obtained, will be required. The excess of sugary seeds in some crosses suggests that fertilization type II occurs more frequently in these crosses than type I. The ratio of these two types cannot be determined directly from seed counts, since nonsugary seeds are derived from both types I and III. These could be separated if the seeds were grown and the resulting plants examined for the presence or absence of chromosome B[4]. This experiment is in progress.

Stability of the B^4 chromosome during endosperm development

In view of the high rate of mitotic nondisjunction of the B^4 chromosome in the division of the generative nucleus, it would be of interest to test for its occurrence in other divisions. Evidence has already been presented that mitotic nondisjunction is rare or absent in the divisions of the female gametophyte that lead to the formation of the egg. The study can be extended to the endosperm through the use of the sugary character.

The endosperms of the nonsugary seeds produced by the cross of Normal (*su su*) $\times$ TB-4a (*Su Su*) and its reciprocal are expected to be of three types, with respect to the sugary gene. When the interchange is transmitted by the seed parent, the expected composition of the endosperm is *Su Su su*. When it is transmitted by the pollen parent, two endosperm types, *su su Su Su* and *su su Su*, are obtained (fig. 5). The B^4 chromosome carries *Su* and the normal chromosome 4 carries *su*.

The occurrence of mitotic nondisjunction of the B^4 chromosome in the first division of a primary endosperm nucleus of composition *su su Su* would result in a sugary sector deficient for this chromosome in an otherwise nonsugary seed. Assuming uniform endosperm development, such a sector would comprise half the area of the endosperm. Smaller sectors would be obtained from mitotic nondisjunction in later divisions. In either case, the relationship between the frequency of sectors and the rate of nondisjunction would be a simple one; each occurrence of nondisjunction should yield a sugary sector.

The relationship is much more complicated in the seeds which have two B^4 chromosomes in the endosperm. The "loss" of both of these would be a requirement for the occurrence of a sugary sector. It is conceivable that the loss could occur in either of two ways as a consequence of nondisjunction. First, one of the B^4 chromosomes could undergo nondisjunction and the other disjoin normally to produce daughter cells with three and one B^4 chromosomes, respectively. The recurrence of nondisjunction in a subsequent division of the cell containing one B^4 chromosome would yield a cell line deficient for this chromosome, hence a sugary sector. Or, secondly, the B^4 chromosomes might undergo nondisjunction simultaneously, with either of two results. If the B^4 chromosomes move to one pole, the daughter cells would have four B^4 chromosomes in one and none in the other. If they move in opposite directions, the original condition would be restored in each of the daughter cells. If these alternatives are equally probable, only half of the cases of simultaneous nondisjunction would lead to sugary sectors.

A simple example will illustrate the relationship between the rate of nondisjunction and the frequency of sugary sectors when two B^4 chromosomes are present. A sugary sector comprising half the endosperm area would be expected from half the cases of simultaneous nondisjunction occurring in the division of the primary endosperm nucleus. If x is the rate of nondisjunction for each B^4 chromosome per cell division and if each acts independently, the expected frequency of sectors of this size would be given by $x^2/2$. For a rate of nondisjunction of ten percent (one occurrence per ten divisions), the frequency of

357

"1/2"-sectors would be 0.5 percent, or five per thousand seeds. Smaller sectors would be expected to occur more frequently owing to a greater opportunity for both simultaneous and recurrent nondisjunction in later divisions. Sectors comprising less than one-sixteenth of the endosperm are detectable (STADLER and SPRAGUE 1936).

The seeds obtained from the crosses between normal ($su\ su$) plants and plants homozygous for TB-4a ($Su\ Su$) were examined carefully for sugary sectors. None were found among 1157 seeds produced when the interchange was transmitted by the seed parent. The reciprocal cross yielded seven sectors among 1678 seeds examined. The production of endosperms containing a single B^4 by the latter cross (type III, fig. 5) probably accounts for the difference in the frequency of sugary sectors in the two crosses.

It should be made clear at this point that the occurrence of a sugary sector does not necessarily mean that mitotic nondisjunction of the B^4 chromosome is involved. These sectors occur spontaneously in appropriate crosses involving normal parents, with a frequency of approximately one-tenth of one percent (STADLER and SPRAGUE 1936). They could arise as a result of mutation or the production of deficiencies including the Su locus due to fragmentation of the B^4 chromosome. The B-type chromosome is subject to fragmentation (RANDOLPH 1941; DARLINGTON and UPCOTT 1941) and since the B^4 chromosome contains the proximal third of the B-type chromosome, it is not unlikely that this mechanism furnishes at least a fraction of the sugary sectors found. With regard to nondisjunction, the results indicate that the B^4 chromosome is remarkably regular in its behavior in the endosperm mitoses, in contrast with its aberrant behavior in the division of the generative nucleus of the microspore.

DISCUSSION

Mitotic nondisjunction, as demonstrated for the B^4 chromosome, explains the transmission of the intact B-chromosome in most particulars. The exceptional oB and 2B plants obtained from the oB×1B and oB×2B crosses (table 1) have their counterparts in the deficient and hyperploid plants produced by the crosses involving TB-4a as the male parent. The relatively rare 3B and 4B progeny from these crosses, however, cannot be accounted for on the basis of a single occurrence of nondisjunction. They suggest that this process may also occur occasionally during some division other than the second microspore mitosis, possibly in meiosis.

The uniformly high rate of nondisjunction, approximating 100 percent, estimated for the B^4 chromosome does not apply in all cases to the intact B-type. A high rate is evident in the results of the oB×2B cross, in which no 1B progeny, indicative of normal disjunction, were obtained. In the oB×1B cross, however, a considerable number of 1B plants were produced and a good match with the observed data can be obtained if a rate of nondisjunction of only 50 percent is assumed (table 5). The calculations in this table are made on the basis that oB and 1B microspores are produced in equal numbers by the 1B parent. This assumption affects the proportion of the oB progeny in the F_1

population. It does not affect, however, the ratio of the 1B and 2B progeny to each other, and this ratio is in itself a sufficient indicator of the rate of non-disjunction.

It is unlikely that the rate of nondisjunction is influenced by whether the parent has one B-type or two, since in either case the microspore in which nondisjunction occurs has one B-type. Different rates are found for different plants of the same B chromosome number, as is shown by the results of five oB×1B crosses, each involving a different 1B parent (LONGLEY 1927). One of these crosses yielded eight plants with two B-types and none with one (100 percent nondisjunction) and another gave none with two B-types and nine with one (no nondisjunction). The effect of environmental variation on the rate of nondisjunction is not known. There is a possibility that we are dealing

TABLE 5

Comparison of data from oB×1B crosses with distribution expected from various rates of mitotic nondisjunction.

	B CHROMOSOME NUMBER IN F_1			
	0	1	2	3
Observed*	108	35	20	2
Expected from:				
50% nondisjunction	103	41	21	
75%	114	21	31	
100%	124	0	41	

* Combined data of LONGLEY and RANDOLPH.

with two or more types of B chromosomes (a view held for different reasons by DARLINGTON and UPCOTT 1941) which differ in their capacity for normal disjunction. It is also possible that variations in the basic genotype are responsible for these differences in the rate of nondisjunction. The work with TB-4a (and several other A-B interchanges) has not shown these differences to a marked extent but the results are not extensive enough to warrant the conclusion that they do not occur.

Localization of the chromosomal element responsible for mitotic nondisjunction

Fragment chromosomes derived from the B-type chromosome by terminal deletion give transmission results that are similar to those obtained for the intact B-type itself (RANDOLPH 1941). The smallest of these is the "F" chromosome which consists "merely of a centromere with an almost insignificant amount of attached chromatin." The evidence that this diminutive chromosome undergoes nondisjunction suggests that the B-type centromere is responsible for the aberrant process.

Similar evidence has been obtained from the investigation of the A-B interchanges. In the case of TB-4a, the interchange chromosome which undergoes mitotic nondisjunction carries the centromere and the proximal third or so of the B-type chromosome. The other interchange chromosome, carrying the distal two-thirds of the B-type, disjoins normally. The study of five other A-B interchanges has proceeded far enough to show that the same relationship holds in these as well; the aberrant chromosome in each case is that one which possesses the B centromere.

In these interchanges, the points of breakage in the B-type chromosome range from one which occurred near the end of the chromosome to another well within the proximal euchromatic segment. The size of the A-segment in the B^A chromosome also varies widely. In one interchange, the A-segment consists of only the tip of the long arm of chromosome 7. In another, almost all of the short arm of chromosome 1 is involved, to form an interchange chromosome which is longer than the shorter members of the basic set. The length of the chromosome undergoing mitotic nondisjunction evidently has little or nothing to do with determining the occurrence of this process.

Other examples of mitotic nondisjunction are found in rye (HASEGAWA 1934; MÜNTZING 1946), in sorghum (DARLINGTON and THOMAS 1941), and in the fungus gnat *Sciara* (METZ 1938). In these, as in maize, nondisjunction occurs for the most part only in certain mitoses. The supernumerary in rye undergoes nondisjunction in the first microspore mitosis and also at some point in the development of the female gametophyte. In sorghum, nondisjunction occurs in a supernumerary division of the vegetative nucleus of the pollen grain. In *Sciara*, it occurs in the division of the secondary spermatocyte. In the four cases, a specific chromosome (or a derivative of this chromosome) undergoes mitotic nondisjunction; the other chromosomes of the set disjoin normally. Thus there are two conditions which are essential to the occurrence of mitotic nondisjunction: (1) a chromosome of a particular kind, and (2) a mitosis that imposes a restriction to normal disjunction that is specific for this chromosome.

A comparison of the aberrant chromosomes in the examples cited above reveals a striking relation with respect to the position of the centromere. The centromere in rye is subterminal and in maize it is terminal or nearly so. In *Sciara*, the position is different in different species. It is at or near the end of the chromosome in *S. coprophila* whereas it is median in *S. pauciseta* (SCHMUCK 1934). The aberrant chromosomes in sorghum, as reported in two plants, also have median centromeres and are of two distinct sizes (DARLINGTON and THOMAS 1941). The larger of these has been identified as an iso-chromosome. It has been shown in rye that large and small isochromosomes are derived from the standard supernumerary which has a subterminal centromere (MÜNTZING 1944). The aberrant chromosomes in sorghum (and perhaps also those with median centromeres in some species of *Sciara*) may have been similarly derived. Thus, in each case, the chromosome that is undergoing mitotic nondisjunction either is or may have originated from a chromosome with a subterminal or terminal centromere, in contrast with the normal chromosomes of the respective sets (again expecting *Sciara*) which have distinctly internal, usually median or submedian, centromeres.

There is some argument concerning the position of the centromere of the B-type in maize. McCLINTOCK (1933) described it as terminal on the basis of examination in the pachytene stage of the microsporocyte. RANDOLPH (1941) reported a diminutive short arm, seen in pachytene, which was not evident in all figures. He believes that the centromere assumes a spurious terminal appearance in some figures because the short arm folds back against the pycnotic knob adjacent to the centromere in the long arm. DARLINGTON and UPCOTT (1941) reported a sizeable short arm, long enough to permit the formation of observable chiasmata. They suggest that the B-type may vary structurally to give various positions for the centromere. The B-types involved in the A-B interchanges all appear to have terminal centromeres when examined in pachytene and no marked structural differences were found. It is however possible that a very small arm, such as RANDOLPH observed, could have been missed in the examinations.

The similarity between the aberrant chromosomes of the different organisms suggests a relationship between mitotic nondisjunction and the position of the centromere. A chromosome with a subterminal or terminal centromere could be derived from a chromosome with a more centrally located centromere in one of two ways, either by rearrangement of the chromosome or by loss of a segment in one arm. Either case would require a break at or near the centromere. An aberrant chromosome might be produced if, in the process of rearrangement or loss, there is an impairment of the effectiveness of the centromere or adjacent chromatin vital to normal disjunction. If this were the case, it should be possible to produce a chromosome showing mitotic nondisjunction as a result of such structural changes in a normal chromosome.

The telocentric chromosome studied by RHOADES (1940) in maize may represent a case of this kind. This chromosome was found in a plant obtained from a cross involving a parent trisomic for chromosome 5. It consists of the entire short arm of chromosome 5 and has a centromere that is unquestionably terminal. When plants hyperploid for the telocentric chromosome are crossed with normal diploids, the progeny include hyperploid plants of two types. One of these contains a single telocentric chromosome in addition to the diploid complement; the other has two doses of the telocentric chromosome in the form of an isochromosome. These two types are produced in about equal numbers when the male parent transmits the telocentric chromosome. When it is transmitted by the female parent, the progeny containing an isochromosome comprise only 0.57 percent of the hyperploid types. The formation of isochromosomes from telocentric chromosomes is thus much more frequent in the male germ line than in the female germ line.

This dependence on sex is remarkably similar to that found for the occurrence of mitotic nondisjunction of the B-type and the B^4 chromosomes. The tests made by RHOADES show that the formation of the isochromosome is a postmeiotic process. In the light of the behavior of the B^4 chromosome, the results obtained for the telocentric chromosome may be interpreted as another case of nondisjunction in the division of the generative nucleus. The results differ in one respect from those obtained for the B^4 chromosome. In the case of the latter, two chromosomes are distributed as individuals to one of the gametes. The telocentric chromosome, however, forms an isochromosome in the process. It is doubtful that nondisjunction occurs first and is followed by a fusion of the terminal centromeres since RHOADES found that two telocentric chromosomes could exist in one nucleus without forming an isochromosome. It would seem, rather, that the centromere of the telocentric chromosome either fails to divide or divides tranversely (DARLINGTON 1940) to produce the isochromosome. In this respect, the telocentric differs from both the B-type and the B^4 chromosomes. In all three cases, however, it is probable that the centromere is responsible for the aberrant behavior.

Cytological observations by MÜNTZING (1946) in rye suggest that the centromere may not be the only chromosomal element responsible for mitotic nondisjunction. The aberrant chromosome in rye lags in the first microspore mitosis and is usually included in the generative nucleus. The lagging is due to the failure of a region near the centromere to divide in concert with the rest of the chromosome. The activity of the centromere appears to be normal

Some other uses of A-B interchanges

The A-B interchanges provide a means of obtaining aneuploid plant types that are difficult to produce by other methods. The duplications and deficiencies that are obtained in this way are useful in the study of various cytogenetic problems. The examples that follow show some uses of the aneuploid types in problems under investigation.

It has been shown in the case of TB-4a that functional deficient and hyperploid male gametes are produced by mitotic nondisjunction in the division of the generative nucleus. As a result, progeny are obtained which carry a specific segment of chromosome 4 in one, two, and three doses. It should be possible, by appropriate crosses, to obtain the segment in even higher numbers. For example, a plant hyperploid for the B^4 chromosome forms hyperploid eggs. If these are fertilized by sperm that are also hyperploid for the B^4 chromosome, plants should be obtained that carry four doses of the segment of chromosome 4. The limit to continued accumulation would depend only on the degree of unbalance caused by the segment. Preliminary results have shown that a dosage series may also be established in the case of other A-B interchanges as well. The mechanism of mitotic nondisjunction thus provides a method for the study of the effect of specific chromosomal segments in various numbers. In the same way, it makes possible a determination of the effect of known recessive or dominant genes located in these segments.

The production of functional deficient gametes may also be utilized to locate recessive genes within the chromosome. A plant which contains the recessive gene is crossed as the seed parent with a plant carrying an A-B interchange and homozygous for the dominant allele. The deficient progeny will show the recessive character if the locus of the gene is distal to the point of breakage in the A chromosome. This method is illustrated by the use of TB-4a in locating *su*. The TB-4a parent was homozygous for *Su*. Crosses with *su su* plants gave kernels that were of the sugary phenotype. These kernels were deficient for the B^4 chromosome thus indicating that *Su* is carried on this chromosome.

Intercrosses involving different A-B interchanges may provide information as to whether or not maize is of polyploid origin. In appropriate crosses, plants can be produced that have two different A-segments in duplication. For example, a cross between a plant that has TB-4a with one that has TB-7a will yield plants in which the B^4 and the B^7 chromosomes are present as duplications. If the segment of chromosome 4 has a region within it that is a duplication of a region in the chromosome 7 segment, the homology might be revealed by the regular synapsis of these regions.

Another interesting aspect of these interchanges is the possibility of an interaction between the heterochromatin of the B-type and the adjacent A-chromatin in the interchange chromosomes. Drosophila investigations have shown that the action of a gene may be altered when, following translocation, it is brought adjacent to or relatively near a heterochromatic region. Such an effect has not been observed in maize although many interchanges between A chromosomes have been studied. No marked effect has been apparent in the case of TB-4a, but the break in the B-type in this interchange is probably not within the heterochromatic segment. In other A-B interchanges, the B-type was broken well within this segment with the result that B heterochromatin is in direct contact with A euchromatin. Interchanges of this kind provide material for a study of the heterochromatin-euchromatin relationship in maize.

SUMMARY

Eight interchanges between A-type and B-type chromosomes were obtained from B-bearing pollen treated with X-rays. The behavior of one of these, involving chromosome 4 of the basic set, is reported in this paper.

The results of crosses involving plants homozygous for this interchange (designated TB-4a) show that one of the interchange chromosomes (B^4) undergoes nondisjunction in the second division of the microspore. As a consequence, the two gametes of a pollen grain carry different chromosomal complements. One has the aberrant chromosome in duplicate; the other is deficient for this chromosome. Both gametes are functional.

Nondisjunction of this chromosome occurs in most but not all of the second microspore mitoses. The seeds obtained from a cross between a pollen parent homozygous for TB-4a and a normal seed parent are therefore of three kinds. One has an endosperm that is deficient for the B^4 chromosome and an embryo that carries the chromosome in duplicate. Another has a hyperploid endosperm and a deficient embryo. The third has a euploid endosperm and is simply heterozygous for TB-4a in the embryo.

The B^4 chromosome possesses the centromere and proximal third of the B-type chromosome. The interchange chromosome bearing the centromere of chromosome 4 disjoins normally. Preliminary studies in five other A-B interchanges have shown that it is the interchange chromosome which bears the B-centromere that is aberrant in each case.

The B^4 chromosome is relatively stable in endosperm development and undergoes nondisjunction rarely, if ever, in the development of the egg.

The anomalous male transmission of the intact B-type chromosome may also be accounted for on the basis of nondisjunction in the second microspore division.

ACKNOWLEDGMENTS

I wish to express my appreciation to PROFESSOR L. J. STADLER for his guidance during the course of this investigation; to DOCTOR BARBARA MCCLINTOCK for her valued advice and criticism; and to PROFESSOR L. F. RANDOLPH, who generously supplied me with some of the B chromosome stocks used in the experiments.

LITERATURE CITED

ANDERSON, E. G., and L. F. RANDOLPH, 1945 Location of the centromeres on the linkage maps of maize. Genetics **30:** 518–526.

BURNHAM, C. R., 1932 An interchange in maize giving low sterility and chain configurations. Proc. Nat. Acad. Sci. **18:** 434–440.

CLARKE, A. E., and E. G. ANDERSON, 1935 A chromosomal interchange in maize without ring formation. Amer. J. Bot. **22:** 711–716.

DARLINGTON, C. D., 1940 The origin of isochromosomes. J. Genet. **39:** 351–361.

DARLINGTON, C. D., and P. T. THOMAS, 1941 Morbid mitosis and the activity of inert chromosomes in Sorghum. Proc. Roy. Soc. London **130:** 127–150.

DARLINGTON, C. D., and M. B. UPCOTT, 1941 The activity of inert chromosomes in Zea Mays. J. Genet. **41:** 275–296.

HASEGAWA, NOBUMI, 1934 A cytological study on 8-chromosome rye. Cytologia **6:** 68–77.

LONGLEY, A. E., 1927 Supernumerary chromosomes in Zea Mays. J. Agric. Res. **35:** 769–784.

MCCLINTOCK, BARBARA, 1933 The association of non-homologous parts of chromosomes in the mid-prophase of meiosis in Zea Mays. Z. Zellf. Mik. Anat. **19:** 191–237.

METZ, C. W., 1938 Chromosome behavior, inheritance and sex determination in Sciara. Amer. Nat. **72:** 485–520.

MÜNTZING, ARNE, 1944 Cytological studies of extra fragment chromosomes in rye I. Iso-fragments produced by misdivision. Hereditas **30:** 231–248.

1946 Cytological studies of extra fragment chromosomes in rye III. The mechanism of nondisjunction at the pollen mitosis. Hereditas **32:** 97–119.

RANDOLPH, L. F., 1928 Chromosome numbers in Zea Mays L. Cornell Agric. Exp. Sta. Memoir **117:** 44 pp.

1941 Genetic characteristics of the B chromosomes in maize. Genetics **26:** 608–631.

RHOADES, M. M., 1940 Studies of a telocentric chromosome in maize with reference to the stability of its centromere. Genetics **25:** 483–520.

SCHMUCK, M. LOUISE, 1934 The male somatic chromosome group in Sciara pauciseta. Biol. Bull. **66:** 224–227.

STADLER, L. J., and G. F. SPRAGUE, 1936 Genetic effects of ultraviolet radiation in maize. I. Unfiltered radiation. Proc. Nat. Acad. Sci. **22:** 572–578.

WEATHERWAX, PAUL, 1917 The development of the spikelets of Zea Mays. Bull. Torrey Bot. Cl. **44:** 483–496.

41

Reprinted from pp. 234–238, 260–261, 264–267, 271–272, 276–277, 280–281 of
Genetics **26**(2):234–282 (1941)

THE STABILITY OF BROKEN ENDS OF
CHROMOSOMES IN ZEA MAYS

BARBARA McCLINTOCK
University of Missouri, Columbia, Missouri

Received November 27, 1940

I. INTRODUCTION

IF CHROMOSOMES are broken by various means, the broken ends appear to be adhesive and tend to fuse with one another 2-by-2. This has been abundantly illustrated in the studies of chromosomal aberrations induced by X-ray treatment. It also occurs after mechanical rupture of ring-shaped chromosomes during somatic mitoses in maize and is assumed to occur during the normal process of crossing-over. In a previous publication (McCLINTOCK 1938b) it was shown that following breakage of a single chromatid in a meiotic anaphase of maize, fusion occurs at the position of breakage between the two sister halves of this broken chromatid. Because of this fusion, the two sister halves cannot separate freely from one another in the following mitotic anaphase. As the two centromeres of the terminally united chromsomes pass to opposite poles in this mitotic anaphase, a chromatin bridge is produced. The tension on the bridge configuration, following the poleward migration of the centromeres, results in rupture. Once again, a chromatid with a broken end enters each sister telophase nucleus. The questions then arise: Will fusions occur at the position of breakage between the two sister halves of each of these broken chromosomes giving rise to an anaphase bridge configuration in the following mitosis? If so, will this breakage-fusion-bridge cycle continue in each successive nuclear division, or will the broken end, produced by the rupture of an anaphase bridge configuration, eventually "heal," thus discontinuing the breakage-fusion-bridge cycle? Answers to these questions were presented briefly in a preliminary publication (McCLINTOCK 1939). The results presented in this latter publication and those presented in this paper have led to the following conclusions. (1) If a chromosome, broken at the previous meiotic anaphase, is delivered to the primary endosperm nucleus through either the male or the female gametophyte, the breakage-fusion-bridge cycle will continue in the successive nuclear divisions during the development of the endosperm tissues. (2) A similarly broken chromosome delivered to the zygote nucleus by either the sperm or the egg does not give rise to bridge configurations in successive nuclear divisions in the sporophytic tissues. The broken end heals. There is a complete cessation of the breakage-fusion-bridge cycle. (3) The breakage-fusion-bridge cycle is confined to the gametophytic and endosperm tissues of the generation immediately following the initial break in the chromosome. (4) Healing of the

365

broken end in the embryonic sporophyte is permanent. When a chromosome with a healed broken end is introduced into gametophytic or endosperm tissues in succeeding generations, no fusions of broken ends result either between sister halves of the broken chromosome or between two such broken chromosomes when both are introduced into a single nucleus. It is the purpose of this paper to present the evidence for these conclusions.

II. THE TYPES OF GENETIC VARIEGATION PRODUCED BY THE BREAKAGE-FUSION-BRIDGE CYCLE

If a broken chromosome continued the breakage-fusion-bridge cycle in successive nuclear divisions, its presence should be made evident by genetic variegation in endosperm and plant tissues. This would follow when the

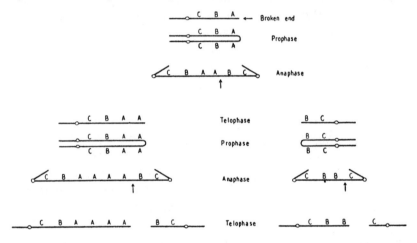

FIGURE 1.—A representative illustration of the method by which variegation may be produced in tissues carrying a chromosome with a broken end. The clear circle represents the centromere. The dominant genes *A B* and *C* are carried by the arm with the broken end, *A* being near the broken end and *C* near the centromere. The homologue of this chromosome (not diagrammed) is considered to be normal and to carry the genes *a b* and *c*. Division of this broken chromosome results in fusion at the position of breakage between the two split halves (prophase, second diagram from top). This is followed by a bridge configuration in the following anaphase (anaphase, third diagram from top). The arrow points to the position of breakage, the two broken chromosomes entering the sister telophase nuclei (telophase, right and left). This process is repeated in successive divisions. One such division is diagrammed below each of these two telophase chromatids. The diagrams illustrate how dominant genes may be deleted or reduplicated followed the breakage-fusion-bridge cycle.

broken chromosome carried dominant genes and its normal homologue carried the recessive alleles. Figure 1 illustrates the method by which variegation is produced in consequence of the breakage-fusion-bridge cycle. The line at the top of the figure represents a chromosome with a broken end. The dominant genes *A B* and *C* are carried by the arm with a broken end, *A* being close to the broken end and *C* nearest to the centromere. The diagram immediately below represents this chromosome at the following prophase, fusion having occurred between the two sister chromatids at the position of previous breakage. Separation of the two sister centromeres at anaphase results in a bridge configuration (third diagram, fig. 1). If a break occurred at the position of the arrow, a chromosome carrying the genes $A\ A\ B\ C$ and possessing a broken end would enter one telophase nucleus (left in the diagram) and one chromosome carrying the genes B and C would enter the sister telophase nucleus (right in diagram). All the cells arising from this latter cell would lack the dominant gene A. Thus, the recessive gene a would appear in all cells arising from this cell, and the process which results in variegation would have commenced. Each of these broken chromosomes could, in turn, repeat the process just outlined. In the telophase chromosome to the right, the broken ends of the two sister chromatids would again be fused at the succeeding prophase (see diagram), and a bridge configuration would result at the following anaphase (see diagram below). If a break occurred at the arrow, a broken chromosome carrying $B\ B\ C$ would enter one telophase nucleus and a broken chromosome carrying only C would enter the sister nucleus. All the cells arising from this latter cell would have lost the dominant genes A and B, and the tissues would show the character of the recessive alleles a and b. In the first telophase to the left, a similar process has been diagrammed whereby the dominant gene A is lost to one daughter nucleus and repeated duplications of A genes are introduced into the sister nucleus.

The diagram (fig. 1) is merely an example. The break in the first anaphase might have occurred between the two A genes. Several nuclear cycles might take place before a break occurred to one side of these genes resulting in the loss of the A gene to one of the daughter nuclei. On the other hand, the first or successive breaks might have occurred close to one of the centromeres resulting in the loss of all three dominant genes to one nucleus and their duplication in the sister nucleus. Variegation patterns of dominant and recessive tissue resulting from this type of behavior should be very distinctive. Some tissues could be totally recessive—that is, $a\ b\ c$. In this case, all three genes would be lost from a nucleus in a single anaphase break. Other tissues could be recessive for a but variegated for b and c. In this case A would be lost from the cell following one anaphase break, while B and C would be subsequently lost. In these tissues, patches should be found which are (1) $a\ b\ c$ where B and C have subsequently and simultaneously been lost; (2) $a\ b\ C$ from which B and not C has been lost. In this latter patch, which is wholly $a\ b$ in genetic character, still smaller patches should be found which are $a\ b\ c$. No tissues should be found which show the genetic constitution $a\ B\ c$. In other words, when variegation within variegation is present, the developmental pattern should show the loss of the terminal genes before loss of genes close to the centromere. Since variegation of endosperm tissues was the means by which individuals possessing broken chromosomes were detected, a description of the types of variegation observed will be given after the method has been described by which broken chromosomes may be obtained.

In order to secure cytological evidence of the presence of a broken chromosome in a plant and to study the genetic consequences of its behavior, two methods were employed. Both methods involve the breakage of chromosome 9 at a meiotic anaphase and its deliverance to the endosperm and the zygote after successful passage through the developmental periods of the male or the female gametophytes (the pollen grain and embryo sac, respectively). This necessitates the production of a broken chromosome

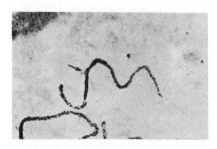

FIGURE 2.—Photomicrograph of chromosome 9 at pachytene in a microsporocyte. The centromere appears as a grey bulge (arrow). The short arm of one of the homologues terminates in a small knob. Note the deep-staining region adjacent to the centromere in the short arm and the smaller, more widely spaced chromomeres in the distal part of the arm. The proximal deep-staining segment appears relatively longer in this photograph than in the diagrams given in this paper. This is due to an overlapping of the two homologous chromosomes immediately distal to the junction of the deep-staining and lighter-staining regions.

which possesses at least a full set of genes if transmissions are to occur through the male gametophyte, or broken chromatids with only relatively short terminal deficiencies if transmissions are to occur through the female gametophyte. No deficiencies within the short arm of chromosome 9 which are transmitted through the pollen are known, but deficiencies of terminal segments up to and including one third of the short arm may be transmitted through the female gamete (McCLINTOCK unpublished).

Both methods involve the use of abnormalities in the structural arrangement of chromosome 9 of maize. A photograph of a normal chromosome 9 at the mid-prophase of meiosis is given in figure 2. The genes Yg^2, C, Sh, and Wx are located in the short arm of chromosome 9. Yg normal green plant, yg yellow-green plant; C colored aleurone, c colorless aleurone; Sh normal development of the endosperm, sh shrunken endosperm; Wx normal starch staining blue with iodine, wx waxy starch staining red with iodine). The linear order of these genes is Yg-C-Sh-Wx. Yg is located very near the end of the short arm. *C* is located approximately a quarter of the distance in from the end of the short arm. *Sh* is located very close to it. *Wx* is located at approximately the middle of the short arm, although its exact position has not been determined. It should be noted that the gene *Yg* is a plant character, whereas the genes *C*, *Sh*, and *Wx* are endosperm characters.

[*Editors' Note:* Material has been omitted at this point.]

The method by which crossing over can give rise to broken chromosomes is illustrated in figure 11. In this illustration the plant is considered to be heterozygous for a duplication of the short arm of chromosome 9. In the diagram, the end of the short arm of the normal chromosome 9 terminated in a large knob. No knob has been diagrammed in the duplication chromosome. This constitution has been chosen, since it will apply to all the duplications which will be discussed in sections 4 and 5 of this paper. At meiosis, association of the three homologous segments of the two chromosomes 9 is 2-by-2. Two of these associations are diagrammed (a and b, fig. 11). In a, the duplicated segment is associated with the short arm of the normal chromosome 9. A crossover as indicated would give rise to a dicentric chromatid (c, fig. 11). This dicentric chromatid is the equivalent of two chromosomes 9 fused at the ends of their short arms. In b (fig. 11), the two homologous segments in the duplicated chromosome 9 are associated. A crossover as indicated would result in the same dicentric chromatid (c, fig. 11). Disjunction of homologous centromeres in anaphase I would result in a first division bridge configuration following the crossover in a. Separation of sister centromeres in anaphase II would give rise to a second division bridge configuration following the crossover in b. In the late anaphase or early telophase of these cells, the dicentric chromatid is broken at some position between the two centromeres. If the break occurred at the position of the arrow (c, fig. 11), each of the two broken chromatids would contain a complete set of genes of chromosome 9. If the break occurred at any other position, a deficient chromatid would enter one nucleus and a chromatid with a duplication would enter the sister nucleus. In this way, chromosomes 9 could be produced with various lengths of duplication and deficiencies of segments of the short arm.

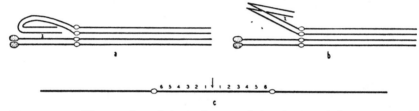

FIGURE 11.—a. Diagram of a meiotic prophase association of a normal chromosome 9 with a large terminal knob and a chromosome 9 with a duplication of the short arm but with no knob. The duplicated segment is homologously associated with the short arm of the normal chromosome 9. A crossover as indicated produces a chromatid composed of two attached chromosomes 9, as shown in c. This will produce a bridge configuration at anaphase I. b. Association of homologous regions of the duplication chromosome 9. A crossover as indicated will give rise to the dicentric chromatid shown in c. A bridge configuration will result in anaphase II.

However, each would possess a newly broken end. Each, therefore, should be capable of inducing variegation in the endosperm tissues to which it is delivered, provided the appropriate genic markers are present to allow detection of variegation. Plants arising from these variegated kernels should show a new series of various types of broken chromosomes 9. All plants with duplications of the type illustrated in figure 9 were heterozygous— that is, contained a duplication chromosome 9 with dominant genes in the duplicated segments and a normal chromosome 9 carrying the recessive alleles. In deriving the constitution of the dicentric chromosome produced following crossing over, it is necessary to insert the knobs and pycnotic regions where they occur in each of the duplicated chromosomes.

[*Editors' Note:* Material has been omitted at this point.]

[*Editors' Note:* Material has been omitted at this point.]

IV. THE CORRELATION OF VARIEGATION AND BROKEN CHROMOSOMES IN THE PROGENY OF A NATURALLY ARISING DUPLICATION IN CHROMOSOME 9

Before the investigations described in section III of this paper had been undertaken, studies were underway on a chromosome 9 possessing a duplication of the short arm. Cultures containing this chromosome gave relatively high percentages of variegated kernels in their progeny. The origin of this duplication was unknown. It was discovered in a genetic strain of maize belonging to DR. L. J. STADLER. The author is indebted to DR. STADLER for use of this duplication in the study now reported. The consti-

tution of the duplicated chromosome 9 is essentially similar to the duplicated chromosome 9 illustrated in type III of figure 9. The duplication
included practically all of the short arm of chromosome 9. No knob was
present. This duplicated chromosome carried the genes I and Wx in the
two homologous segments. I, an inhibitor of aleurone color development, is
placed at the same locus as the gene C (HUTCHISON 1922). The color pattern of variegation induced by broken chromosomes derived from this duplicated chromosome is the reverse of that described in section III, for loss
of the I gene from cells of the aleurone allows color (i) to appear in these
cells.

Since the origin of the broken chromosome at meiosis in plants heterozygous for this duplication is exactly the same as that described in the previ-

TABLE 8

$$\frac{I\ Wx\ \text{duplication chromosome 9}}{i\ wx\ \text{normal chromosome 9}} \times i\ wx\ \text{normal chromosome 9}$$

CROSS	NON-VARIEGATED KERNELS				VARIEGATED KERNELS	
	$I\,Wx$	$I\,wx$	$i\,Wx$	$i\,wx$	$I\text{-}i\,Wx\text{-}wx$	$I\text{-}i\ \ wx$
♀ Parent heterozygous	438	0	23	445	17	0
♂ Parent heterozygous	62	0	29	765	27	3

ous section, the evidence obtained from this duplicated chromosome will be
but briefly reviewed.

In plants heterozygous for the duplicated chromosome 9,—that is, possessing one duplicated chromosome 9 and a normal chromosome 9,—bridge
configurations are seen in both the first and second meiotic mitoses. Of the
193 anaphase I cells recorded, 26 or 13.4 percent showed a bridge configuration. Among 74 dyads in anaphase II, 10 or 13.5 percent showed a bridge
configuration in one of the cells of the dyad.

In these plants, the duplication chromosome 9 carried I and Wx in each
of the duplicated segments. No knobs were present in the duplication
chromosome 9. The normal homologue carried i and wx. Its short arm terminated in a large knob. The chromosome 9 constitution of these plants
was similar to that shown in figure 11. Following crossing over as diagrammed in figure 11, the order of the genes between the two centromeres of
the dicentric chromatid could be $Wx\,I\,i\,wx$, $Wx\,I\,I\,wx$, or $Wx\,I\,I\,Wx$.
Following breakage of the dicentric chromosome at a meiotic anaphase,
broken chromosomes with various genic compositions should arise. Those
possessing $I\,Wx$ or $I\,wx$, either singly or in duplication, should produce
variegation for color if delivered to an endosperm following the cross of this
plant to one possessing two normal chromosomes 9 carrying $i\,wx$.

The results of such a cross are given in table 8. In the first line of the table, the female was the heterozygous parent. In the second line, the male was the heterozygous parent. The low frequency of the *I Wx* non-variegated class in this latter cross is due to the low transmission through the pollen of the chromosome carrying the duplication. Six plants derived from the *I Wx non-variegated* kernels in this latter cross were examined cytologically. Five plants showed the presence of the duplication chromosome 9; one plant possessed two normal chromosomes 9, the short arms of each terminating in a large knob. Since the normal chromosome 9 in all the plant of this study possessed a large terminal knob, the chromosome with *I Wx* in this latter plant obviously arose through a crossover between the normal chromosome 9 and the proximal segment of the duplication chromosome 9. The chromosome 9 constitutions were determined in 26 plants derived from the *I Wx non-variegated* class of the first line of table 8. All 26 possessed the duplicated chromosome 9 and a normal chromosome 9 terminating in a large knob.

The chromosome 9 constitution has been determined for 28 plants derived from the variegated kernels of table 8. A broken chromosome 9 was present in each plant. For illustrative purposes, the type of recovered

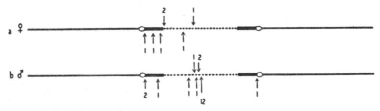

FIGURE 14.—Types of recovered broken chromosomes in plants arising from variegated kernels. The parental chromosome constitution of the heterozygous parent was exactly as diagrammed in figure 11. The dicentric chromatid produced following crossing over is shown in a and b. The constitution of the recovered broken chromosome is that to the right of the arrow in each case. The numbers above or below the arrows indicate the number of plants with this particular broken chromosome. In a, the female parent contributed the broken chromosome. In b, the male parent contributed the broken chromosome.

broken chromosome may be referred to the original dicentric chromatid from which it arose (fig. 14). The broken chromosome in seven of these plants was introduced by the female parent (variegated kernels in line 1 table 8). The broken chromosomes in the remaining 21 plants were introduced by the male parent (variegated kernels in line 2, table 8). Since the proportion of types of recovered broken chromosomes are not comparable in the two crosses, the types received from the female parent are shown in a (fig. 14), those received from the male parent in b (fig. 14). The composition of the broken chromosome is that to the right of the arrow in each case. The numbers above or below each arrow indicate the number of plants with this particular broken chromosome. Totalling both crosses, 14 of the plants possessed a duplicated segment. These varied in length from a single chromomere in one plant to the full short arm in three plants. There was only one plant with a deficiency (b, fig. 14), but this deficiency included all of the short arm. The remaining 13 plants had a complete chromosome 9 with neither a duplication nor a deficiency. Twelve of these were introduced by the male parent. In these plants, the broken chromosome 9 could readily be distinguished from the normal chromosome 9, since the former possessed no knob whereas the short arm of the latter terminated in a large knob.

[*Editors' Note:* Material has been omitted at this point.]

VI. DISCUSSION

(a) *The positions of the breaks in anaphase bridge configurations in successive nuclear divisions*

In the preceding sections of this paper the determination of the chromosome constitution of 186 plants, derived from variegated kernels, has been given. In 180 of these plants, a chromosome with a broken end was found. In each case, the chromosome with the broken end carried the genes associated with the variegation, these genes being located in the arm of the chromosome which possessed the broken end. Of the six exceptional cases, three resulted from hetero-fertilization; consequently, the chromosome constitution of the plant tissues could not be related to that in the endosperm tissues. The remaining three exceptional cases showed an altered chromosome, but the relation of the alteration to the variegation appearing in the endosperm could be determined in only one case (see section III d). There can be little doubt that the variegation in the endosperm tissues, described in this paper, is related to the presence in this tissue of a newly derived broken chromosome. It is necessary that the broken end be newly derived for a broken end heals when introduced into a zygote and is no longer capable of producing variegation either in the resulting sporophytic tissues or in the gametophytic and endosperm tissues of succeeding generations. This has been clearly established.

In all cases mentioned in this paper, the broken chromosome had its origin in a meiotic mitosis. The variegation was confined to the endosperm tissues in the generation immediately following. It made no difference whether the broken chromosome was introduced through the pollen grain or through the embryo sac. In either case variegation resulted in the endosperm tissues. The character of this variegation was clearly of the type which should arise following the breakage-fusion-bridge cycle where successive breaks do not always occur at positions of previous fusions (see fig. 1). Evidence for the initiation of this cycle has been given in a previous publication (McCLINTOCK 1938b). The observations were confined to the first division of the microspore or male gametophyte. This is the first division following the production of the broken end. The evidence for continuation of this cycle in the following gametophyte divisions is obtained inferentially in the study reported here. In several cases the broken chromosome in the endosperm and in the embryo were not alike in chromosome constitution. The difference could readily be accounted for if fusion had occurred between the broken ends of sister chromatids in the generative nucleus followed by an anaphase bridge configuration during the division of this nucleus. If the break in this bridge configuration occurred closer to one centromere than to the other, each sperm would receive a broken chromosome, but the constitution of the broken chromosome in each sperm would differ.

[*Editors' Note:* Material has been omitted at this point.]

(c) *The healing of broken ends of chromosomes*

It has been repeatedly emphasized that the broken end of a chromosome becomes healed in the sporophytic tissues. The breakage-fusion-bridge cycle, which characterizes its behavior in the gametophytic and endosperm tissues, ceases. The healing is permanent. When this chromosome is reintroduced into endosperm tissues in the following generations, no fusions occur at the broken end between the two sister halves. Furthermore, when two such broken chromosomes are brought together after each has passed through a sporophytic generation, no fusions occur between the two broken ends. No obvious explanation is available for the healing of the broken ends in the sporophytic tissues as contrasted with the lack of healing in the gametophytic and endosperm tissues. The question might arise as to whether the broken end ever heals in the gametophyte or endosperm tissues. Although this may occur in some cases, the evidence suggests that it must be rare. It will be recalled that the majority of variegated kernels arising from the rearranged chromosome 9 (see table 1) had the constitution C Sh wx. There were some non-variegated kernels with this constitution. It was assumed that the variegated kernels had broken chromosomes and the non-variegated kernels, non-broken chromosomes. If healing of broken ends occurs, the plants arising from some of the non-variegated C Sh wx kernels should have shown a broken chromosome. However, as table 6 shows, none of the plants arising from the C Sh wx, non-variegated kernels possessed a broken chromosome. Under certain physiological conditions it is possible that the broken end might heal in the endosperm tissues, but at present these conditions are not known.

Realization of the healing of the broken end in the sporophytic tissues came as a surprise. Evidence from the behavior of ring-shaped chromosomes had indicated that fusions of broken ends of *chromosomes* occurred in the sporophytic tissues when the break originated in this tissue. Although extensively looked for, no evidence for healing of these broken ends was obtained in the ring-chromosome material.

[Editors' Note: Material has been omitted at this point.]

SUMMARY

By use of (1) a rearranged chromosome 9, (2) a duplication arising from this rearrangement, (3) a deficiency derived from this rearrangement, (4) a duplication occurring in a genetic strain of maize, and (5) new duplications derived from this latter duplication it was possible to obtain functional gametes carrying a chromosome 9 whose short arm terminated in a broken end.

In all cases, the broken end arose following crossing over at meiosis which produced a dicentric chromatid. Rupture of the dicentric chromatid at a meiotic anaphase produced the broken end.

During the following gametophytic division, fusions occurred at the position of breakage between the two sister halves of the broken chromatid, resulting in an anaphase bridge configuration. Rupture of this bridge at late anaphase or early telophase again introduced a broken chromosome into the sister telophase nuclei. This breakage-fusion-bridge cycle continued in the successive gametophytic divisions.

When such a chromosome, initially broken in the previous meiotic mitosis, is introduced into the endosperm tissues of the following generation through either the male or the female gametophyte, the evidence indicates that the breakage-fusion-bridge cycle continues in each successive division. When this chromosome carries dominant genes in the arm with the broken end and when the normal homologue carries the recessive alleles, variegation for these genes appears in the endosperm tissues. This is caused by non-median breaks in the bridge configurations in many anaphase figures which deletes dominant genes from one telophase nucleus and duplicates them in the sister telophase nucleus.

When such a broken chromosome is introduced into the zygote, the broken end heals, discontinuing the breakage-fusion-bridge cycle. This healing is permanent. The broken end behaves in every respect like a normal end. When a broken chromosome has passed through a sporophytic generation, it no longer is capable of producing variegation in the endosperm of the following generation. When two such chromosomes are brought together after each has passed through a sporophytic generation, no fusions occur between their broken ends.

Thus, the breakage-fusion-bridge cycle occurs only in the nuclear divisions of the gametophytic and endosperm tissues when the broken end is newly derived and has not passed through a sporophytic generation.

Evidence is presented which suggests that following an initial meiotic anaphase break, the breaks in the successive anaphase bridge configurations in the following gametophytic divisions tend to occur at the position of previous fusion, but many breaks occur at other positions.

Once a complete fusion has occurred at the position of breakage between the sister halves of a broken chromosome, the fusion results in a union which is as permanent and strong as that between other parts of the chromosome.

The factors responsible for fusions of broken ends or for the healing of a broken end are not understood but are probably related to the method by which the chromosome becomes broken and to the physiological conditions surrounding the broken end.

REFERENCES

Hutchison, C. B. 1922. The linkage of certain aleurone and endosperm factors in maize, and their relation to other linkage groups. *Cornell Univ. Agric. Exp. Stn. Mem.* **60**:1419–1473.

McClintock, B. 1938b. The fusion of broken ends of sister half-chromatids following chromatid breakage at meiotic anaphases. *Missouri Agric. Exp. Stn. Res. Bull.* **290**:1–48.

_____. 1939. The behavior in successive nuclear divisions of a chromosome broken at meiosis. *Proc. Nat. Acad. Sci. U.S.A.* **25**:405–416.

42

Cytological and Genetical Studies of Important Cereal Species with Special Consideration of the Behavior of the Chromosomes and Sterility in the Hybrids

Hitoshi Kihara

These excerpts were translated expressly for this Benchmark volume by C. R. Burnham, R. L. Phillips, and Patrick Buescher, University of Minnesota, from pp. 3–5, 11, 29, 30, 33, 144–148, 181, 182, 183, 186–187 of "Cytologische und genetische Studien bei wichtigen Getreidearten mit besonderer Rücksicht auf das Verhalten der Chromosomen und die Sterilität in der Bastarden," Kyoto Univ. Coll. Sci. Mem. Ser. B, **1**(1):1–200 (1924) *with the permission of the Faculty of Science, Kyoto University.*

INTRODUCTION

Since the cereal species are the most important cultivated plants, people have been interested in their origin and mode of inheritance for a long time. The exact understanding of the inheritance phenomena, especially of species or genus hybrids (although it in part has already been considerably advanced through genetic studies), can only be satisfactorily solved when one does the cytology of the chromosomes.

In the *Triticum* species, the correct chromosome numbers were first found in 1918 by Sakamura. Until then an incorrect number had been generally accepted as correct, which meant no small hindrance for the progress of cytological studies of this plant. Through the establishment of the correct chromosome number of *Triticum* species, which I soon could confirm as correct, it became clear that the wheats fell cytologically into three different chromosome groups which agrees well with the phylogenetic classification established by Schulz (1913). Further, it is well known that the hybrids between these three groups were more or less fertile. This circumstance caused me to thoroughly investigate cytologically the species hybrids and their progeny, bearing in mind their genetic characteristics as well as sterility.

Moreover, I have worked with investigations concerning the chromosome number relationships in *Avena* species and the abnormal behavior of chromosomes in wheat-rye hybrids through which I have extended my cytological studies considerably.

The previous experimental results have been communicated since 1919 under the title "Cytological studies in some cereal species." (Kihara 1919a, b; 1921). The present work contains a summary of my earlier cytological and genetical investigations of the cereal species mentioned above. The first part is first of all devoted to the establishment of the correct chromosome numbers of important cereal species in order to make possible further studies concerning the behavior of chromosomes in the hybrid progeny, especially of wheat.

By drawing from previously published work of other authors and also from the results of my own experiments, I have established the degree of affinity as well as the behavior of paternal and maternal chromosomes in the reduction-division of hybrid plants and arranged them into groups.

For the X and non-X ploid relationship of chromosomes in the closely related species as well as their changes in number in the course of phylogenetic development, I have also drawn upon a large amount of literature.

In the progeny of 35 chromosome pentaploid wheat hybrids, one must pay special attention to the variation of chromosome number as well as sterility which is brought about by the recombination of the different chromosomes. The inheritance of these progeny is naturally so complicated that the law of simple Mendelian segregation is not applicable. At any rate, the distribution of chromosomes to the progeny is closely correlated with the morphological characteristics of the plants.

Therefore one is not justified to explain the sterility, manner of inheritance, etc., of the progeny of wheat hybrids without considering the variation in the number of chromosomes in the progeny. My morphological studies are communicated in the second part of this work.

*　　*　　*　　*

[Based on the chromosome numbers found by Sakamura (1918) and confirmed by myself (1919a, 1921) and also by Sax (1921)], we are now in a position with these counts to reexamine karyologically the Schulz summary of the hereditary relationships of the Eutriticum species. In Table 2, I give only the haploid numbers.

Table 2. Chromosome numbers and hereditary relationships of Triticum species.

Groups of cultivated forms					
	malformed		14 *T. polonicum*		
			↑		
	normal		14 *T. durum*　14 *T. turgidum*	21 *T. compactum*　21 *T. vulgare*	
			↖　↗	↖　↗	
	Spelt types	7 *T. monococcum*	14 *T. dicoccum*	21 *T. spelta*	
		↑	↑	↑	
	original species	7 *T. aegilopoides*	14 *T. dicoccoides*	unknown	
	Series	Einkorn	Emmer	Spelt (Dinkel)	

(*Naked types* brace groups malformed and normal)

THE REDUCTION DIVISION IN F_1 HYBRIDS BETWEEN PARENTS WITH DIFFERENT CHROMOSOME NUMBERS.

Pentaploid hybrids between Emmer and Dinkel series.

[*Editors' Note:* This section is not translated. See paragraphs 2, 3, 4, and 18 in the summary.]

Triploid hybrids

T. dicoccum × *T. monococcum*
T. aegilopoides × *T. dicoccum*

Both of these hybrids were always completely sterile in my experiments, but according to Tschermak (1914), they should have been almost but not completely sterile. The behavior of the chromosomes in the reduction-division of pollen mother cells in these two hybrids is more or less identical; therefore I would like to concern myself here only with meiosis in *T. dicoccum* × *monococcum* hybrids.

The distribution of chromosomes to the daughter cells in the heterotypic nuclear division is irregular. It differs considerably from that of the above mentioned pentaploid hybrids. The bivalent pairing is loose, and their number is also variable. Already in the diakinesis stage the affinity of the parental chromosomes is somewhat weak. They often associate in the heterotypic metaphase only at their ends. The number of bivalents and univalents varies between 4 and 7, 13 and 7, respectively [i.e., there were cells with 7 bivalents and 7 univalents, 6 and 9, 5 and 11, or 4 and 13, a total of 21 chromosomes in each case]. Sax (1922) had determined in the triploid hybrid (*T. turgidum* × *monococcum*) 7 bivalent and 7 univalent chromosomes.

The division of the 4–7 bivalents follows in a normal manner as in the pentaploid hybrids. The equational division of univalents does not always occur after the halves of the bivalents have arrived at both poles. Some univalents reach the pole unsplit; others split lengthwise, and the longitudinal halves diverge from each other toward the poles as in the pentaploid hybrids. In rare cases I have seen 7 chromosomes in anaphase, all of which remained between the two groups of chromosomes that had arrived at the poles.

In telophase, however, some of the lagging chromosomes after the equational division go toward both poles to participate there in the formation of daughter nuclei. One can, however, often observe chromosomes not reaching either pole that remain isolated in the cytoplasm as a small dark stained clump, exactly as in the pentaploid hybrids.

The homeotypic nuclear plate usually shows 11–12 chromosomes, among which there are dyad and monad chromosomes. With regard to the dyad chromosomes, the second division occurs in the normal manner. The monad chromosomes are delayed naturally once again. Their number amounts usually to 1–4. [*Ed. note:* As in the pentaploid hy-

381

brid, the monad chromosomes are distributed unsplit among the four microspores in the tetrad.] Tetrad formation is usually regular. From one pollen mother cell, usually 4 microspores are formed. However, there are frequently also micronuclei that contain the lagging chromosomes in the homeotypic nuclear division.

The heterotypic nuclear division of these hybrids is characterized by the homologous parental chromosomes showing weak affinity and some of the univalent chromosomes going unsplit to one pole while the others are at the equator and split into two longitudinal halves. According to the literature, the *Pilosella* hybrids (for example, *Hieracium auricula* × *H. aurantiacum*, Rosenberg 1917) and *Papaver atlanticum* × *dubium* (Ljungdahl 1922) should behave similarly. However, the chromosome affinity in our hybrids seems to me to be stronger than in those just mentioned.

[MODE OF ORIGIN OF X-PLOID NUMBERS IN RELATED SPECIES]

Winge (1917) has in his work "The chromosomes, their numbers and general importance" considered this question also of the numbers in related species and how they might have originated. I will give a short account of his view.

If we assume that there are 3 different species (A, B, and C) with the same chromosome number ($x = 9$), then the hybrids among them would change the chromosome number to tetraploidy, hexaploidy in the following manner:

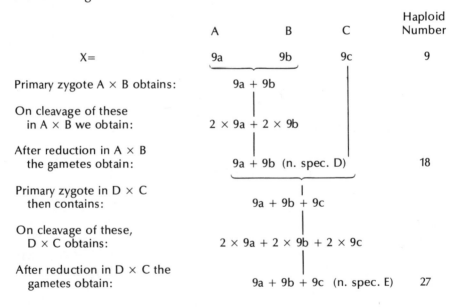

	A	B	C	Haploid Number
X=	9a	9b	9c	9
Primary zygote A × B obtains:		9a + 9b		
On cleavage of these in A × B we obtain:		2 × 9a + 2 × 9b		
After reduction in A × B the gametes obtain:		9a + 9b (n. spec. D)		18
Primary zygote in D × C then contains:		9a + 9b + 9c		
On cleavage of these, D × C obtains:		2 × 9a + 2 × 9b + 2 × 9c		
After reduction in D × C the gametes obtain:		9a + 9b + 9c (n. spec. E)		27

About the behavior of the chromosomes in the zygote of the F_1-hybrid, he said:

How this zygote will behave, must depend upon circumstances; the constitution of the sporophyte may be more or less harmonious. It would be natural, however, that the 9a derived from D at any rate should unite directly with the homologous 9a from A. The 9b must either unite indirectly, i.e., if the sporophyte is to be normally capable of development or remain unpaired, when the sporophyte will be normal. Rosenberg's investigations (1909) with *Drosera* hybrids might be an instance of such a case. *Drosera rotundifolia* has $x = 10$, *D. longifolia* $x = 20$, and the hybrid $2x = 2 \times 10 + 10$, the 10 chromosomes from *D. rotundifolia* uniting with the 10 from *D. longifolia* while the remaining 10 continue unpaired, so that a natural sexual further development cannot take place.

According to this assumption, all hybrids with the *Drosera*-scheme should be sterile. There are, however, many fertile hybrids in this scheme (for example, triploid *Oenothera*, pentaploid *Triticum* hybrids). Although Winge's opinion cited above has a certain significance, it must still not be accepted incontestably.

I will therefore now give the possibilities for the change in number from diploid to tetraploid to hexaploid in *Triticum*. Possibility I is the following:

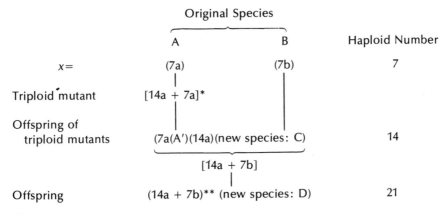

	Original Species		
	A	B	Haploid Number
$x=$	(7a)	(7b)	7
Triploid mutant	[14a + 7a]*		
Offspring of triploid mutants	(7a(A')(14a)(new species: C)		14
	[14a + 7b]		
Offspring	(14a + 7b)** (new species: D)		21

*This doubling of the chromosome number occurs before fertilization.

** If in these cases the chromosome numbers of the offspring of C × B are not (14a + 7b) + (14a + 7b), namely $3x + 3x = 6x$, but $5x + n$ (where $n < x$), then their further offspring will ultimately reach an even 6x.

The opportunity, whereby the tetraploid plant can change into hexaploid, must be very rare because here high sterility and mortality among the zygotes prevail.

Percival (1921) is of the opinion that the plants of the Spelt (Dinkel) series may be none other than the offspring of the hybrid *Aegilops* × Emmer. One can therefore also assume that the hexaploidy might be engendered by hybridization between the 28-chromosome tetraploid plants. If we, for example, represent the chromosome makeup of two tetraploid 28-chromosome species A and B, with 7a + 7c—7b + 7c—then both C-chromosome constellations could form pairs (with each other), whereas the affinity for pair formation is lacking between the a- and b-chromosome constellations. The change of the chromosome number through their hybridization will then be as follows for Possibility II:

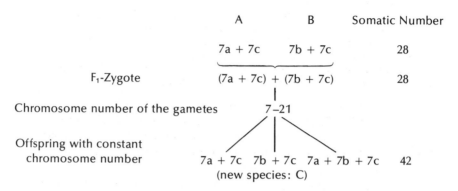

	A	B	Somatic Number
	7a + 7c	7b + 7c	28
F$_1$-Zygote		(7a + 7c) + (7b + 7c)	28
Chromosome number of the gametes		7–21	
Offspring with constant chromosome number	7a + 7c	7b + 7c	7a + 7b + 7c	42
		(new species: C)	

The third possibility derived from the above cited possibilities is the following:

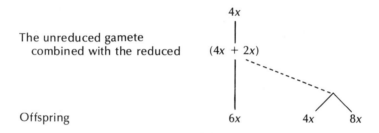

	4x		
The unreduced gamete combined with the reduced	(4x + 2x)		
Offspring	6x	4x	8x

As this sequence shows, the zygote obtained through the union of reduced and unreduced gametes from tetraploid plants is therefore hexaploid. Their chromosome number is, however, generally not constant because again univalent chromosomes segregate. This is shown on the right side of the above diagram.

The possibility is not excluded that the 2–4–8 series might be obtained in this manner inasmuch as the extra chromosomes form no bivalents. It is not impossible that these mutant plants at a certain time in the course of the phylogenetic development had shown, as in diploid plants, pairing of the 2x extra chromosomes and that this affinity of the chromosomes was lost later.

It is also thinkable that soon after the origin of the new tetraploid species from the diploid, the affinity [for pair formation as in the diploid] of the homologous chromosomes was not lost. Then through the union of these gametes (4x + 2x) a constant new hexaploid species with the reduced chromosome number 3x would be formed immediately.

Through possibility III we can better understand the different degrees of relationship between the three *Triticum* series, the 14II and 21II species are more closely related than are the 7II and 14II species.

*   *   *   *

The plants of the Spelt (Dinkel) series are hexaploid. They have 21 haploid chromosomes (3×7), which had been obtained through trebling of the original chromosome composition with 7 chromosomes. If we next assume that a gene for a character (for example, seed color) is located in any one chromosome of every chromosome set, then one can easily understand that the hexaploid wheats possess 3 synonymous Mendelian factors for this character. From that, one also can conclude that tetraploid wheats possess 2 because here one chromosome set (a–g) is missing. We name, for example, the three genes A, B, and D. Then emmer wheats have the formula A_E A_E B_E B_E and Spelt (Dinkel) wheats the A_D A_D B_D B_D D D. The mode of inheritance of the pentaploid hybrids with reference to these characters can be shown in the following:

$$A_E\ A_E\ B_E\ B_E \qquad A_D\ A_D\ B_D\ B_D\ D\ D \qquad P$$
$$(28) \qquad\qquad\qquad (42)$$

$$A_E\ A_D\ B_E\ B_D\ D \qquad\qquad\qquad F_1$$
$$(35)$$

$A_E\ A_E\ B_E\ B_E$	$A_E\ A_E\ B_E\ B_E\ D\ D$	Homozygous offspring with 28 and 42 chromosomes.
$A_E\ A_E\ B_D\ B_D$	$A_E\ A_E\ B_D\ B_D\ D\ D$	
$A_D\ A_D\ B_E\ B_E$	$A_D\ A_D\ B_E\ B_E\ D\ D$	
$A_D\ A_D\ B_D\ B_D$	$A_D\ A_D\ B_D\ B_D\ D\ D$	
(28)	(42)	

This conception of the inheritance is indeed also valid with reference to different characters. Therefore very many new combinations of different genes appear in the offspring of this hybrid.

Because D represents all Dinkel genes (D_1, D_2, D_3, etc.), which the chromosome set (a–g) possesses, it is not incorrect to suppose that they all together give the corresponding characters of *T. vulgare*, *T. spelta*, and *T. compactum*. The manner of inheritance of spike form of *T. durum* × *vulgare* and *T. polonicum* × *Spelta* is a good example. D for spike form is in this case epistatic to A and B for this character. That some 42-chromosome offspring of *T. turgidum* × *compactum* possess *speltoides*-spikes shows a complicated mutual relation which thereby takes place among the genes (A, B, and D). Therefore it is understandable that they [may] retain the characteristics of the Dinkel plants (for example, hollow stems, not keeled glumes). Yet they may not show the typical Dinkel form because they do not have all the D-genes.

385

SUMMARY

1. The chromosome number of the *Triticum-*, *Aegilops-*, *Secale-*, *Hordeum-*, and *Avena*-species has been established (for certain) by the present investigations. Their haploid numbers are the following:

Triticum-species	7	14	21
Aegilops-species		14	
Secale cereale	7 or 8		
Hordeum-species	7		
Avena-species	7	14	21

2. The pentaploid hybrids between the 14-chromosome Emmer and the 21-chromosome Spelt (Dinkel) series have 35 chromosomes, corresponding to the sum of the haploid chromosomes of the parental plants. In the heterotypic nuclear division of the pollen mother cells, 14 Spelt (Dinkel) chromosomes form 14 bivalents with just as many Emmer chromosomes, the 7 surplus Spelt (Dinkel) chromosomes remain as univalents.

3. The heterotypic nuclear division of the pentaploid hybrids is, in relation to the 14 bivalents, a normal reduction division and in relation to the 7 univalents, a longitudinal splitting. The homeotypic nuclear division of these hybrids is in relation to the 14 dyad chromosomes, which come from the 14 bivalents in the heterotypic nuclear division, an equational division. The 7 surplus chromosomes distribute themselves unsplit to the 4 microspores and indeed according to the law of probability. Therefore the resulting pollen grains receive $14 + i$ chromosomes, where $i = 0–7$.

4. Often are seen chromosomes delayed and not arrived at the pole. They do not participate in the future division process (chromatin-diminution). At times they form small pollen.

5. The triploid hybrids between the Einkorn- and the Emmer-series have 21 somatic chromosomes. The number of bivalents of these wheat hybrids varies between 4–7, the number of univalents correspondingly between 13–7. The bivalents show a regular behavior during the entire meiotic nuclear division. A part of the univalents divide lengthwise in the heterotypic nuclear division, while the others go unsplit to any one pole. In the homeotypic nuclear division, the dyad chromosomes divide normally lengthwise, and then the delayed monad chromosomes go to the poles.

18. Constant 40-chromosome [nullisomic] plants were found among the progeny of pentaploid wheat hybrids (21 II × 14 II). The first plant and its progeny were dwarf in growth habit. They lacked one pair of chromosomes. Another constant 40-chromosome plant and its progeny were semidwarf.

REFERENCES

Kihara, H. 1919a. Über cytologische Studien bei einigen Getreidearten. Mit. I. Spezies-Bastard des Weizens und Weizenroggen-Bastard. *Bot. Mag.* **33**:17–38.

———. 1919b. Mit. II. Chromosomenzahl und Verwandtschaftsverhältnisse unter *Avena* Arten. *Bot. Mag.* **33**:94–97.

———. 1921. Mit. III. Über die Schwankungen der Chromosomenzahlen bei den Spezies-bastarden der *Triticum*-Arten. *Bot. Mag.* **35**:19–44.

Ljungdahl, H. 1922. Zur Zytologie der Gattung *Papaver*. *Svensk. Bot. Tidskr.* **16**:103–114.

Percival, J. 1921. *The wheat plant*. A monograph. London.

Rosenberg, O. 1909. Cytologische und morphologische Studien an *Drosera longifolia* × *rotundifolia*. *K. Sven. Vetenskapsakad Handl.* **43**(11):1–65.

———. 1917. Die Reduktionsteilung und ihre Degeneration in *Hieracium*. *Svensk. Bot. Tidskr.* **11**:145–206.

Sakamura, T. 1918. Kurze Mitteilung über die Chromosomenzahlen und die Verwandtschaftsverhältnisse der *Triticum* Arten. *Bot. Mag.* **32**:150–153.

Sax, K. 1921. Sterility in wheat hybrids. 1. Sterility relationships and endosperm development. *Genetics* **6**:399–416.

———. 1922a. II. Chromosome behavior in partially sterile hybrids. *Genetics* **7**:513–558.

Schulz, A. 1913. *Die Geschichte der kultivierten Getreide I*. Halle.

Tschermak, E. v. 1914. Die Verwertung der Bastardierung für phylogenetische Fragen in der Getreidegruppe. *Z. Pflanzenzucht.* **2**:291–312.

Winge, Ø. 1917. The chromosomes. Their number and general importance. *C. R. Trav. Labor. Carlsberg* **13**:131–275.

43

Reprinted from *Genetics* **24**(4):509–523 (1939)

CYTOGENETIC STUDIES WITH POLYPLOID SPECIES OF WHEAT. I. CHROMOSOMAL ABERRATIONS IN THE PROGENY OF A HAPLOID OF TRITICUM VULGARE[1]

E. R. SEARS*

United States Department of Agriculture, Columbia, Missouri[2]

Received February 13, 1939

INTRODUCTION

THE usual methods of genetic analysis lose much of their efficiency when applied to polyploid organisms, where duplicate factors tend to mask recessive gene mutations, and where the large number of chromosomes greatly increases the difficulty of determining linkages. A different way of analyzing polyploids is being tried with *Triticum vulgare*. This method is based on the fact that whole-chromosome deficiencies and duplications are viable in *T. vulgare*, and that these aberrations, as well as reciprocal translocations, can be obtained from haploids of this polyploid species. The origin and possible uses of two reciprocal translocations and a number of whole-chromosome aberrants from a haploid of *T. vulgare* will be discussed in this paper.

PROGENY OF A HAPLOID WHEAT PLANT

Two haploids were found in a field culture of 105 plants grown from seed of *T. vulgare* var. "Chinese Spring" (n = 21) pollinated by *Secale cereale* (n = 7). They were not conspicuously different from diploid Chinese Spring wheat, except for complete male sterility. One of the haploids was completely female sterile, but the other set 14 seeds from less than 300 florets pollinated by normal wheat pollen, and nine seeds from a somewhat larger number of florets pollinated by rye. Successful pollinations involved several different spikes from various parts of the plant, and were made over a period of about two weeks.

From the nine seeds of the haploid pollinated by rye, two mature plants were obtained. These had 27 and 28 chromosomes, respectively, and resembled ordinary wheat-rye hybrids. Neither plant produced seeds.

The 14 seeds of the haploid pollinated by diploid wheat yielded 13 mature plants. Cytological examination at meiosis in these plants, supplemented by similar studies of their progeny, provided the information given in table 1. The aberrant chromosome numbers and pairing relationships of plants 1 to 11 presumably resulted from abnormal chromosome consti-

[1] Journal Series No. 597, Missouri Agricultural Experiment Station.

[2] Cooperation between the Division of Cereal Crops and Diseases and the Department of Field Crops, Missouri Agricultural Experiment Station.

* The cost of the accompanying plate is borne by the GALTON and MENDEL Memorial Fund.

tution of the eggs produced by the haploid. Although no study was made of microsporogenesis in the plants used as pollen parents, several other diploids of Chinese Spring wheat showed regular meiotic behavior. At least two different individuals were involved in the male parentage of the aberrant plants.

TABLE I

Results of cytological analysis of the progeny of a cross,
Triticum vulgare haploid × T. vulgare diploid.

DESIGNATION OF PLANT	SOMATIC NUMBER	BIVALENTS	UNIVALENTS	TRIVALENTS*	RINGS OF FOUR
1	41	20	1		
2	41	20	1		
3	41	20	1		
4	41	20	1		
5	41	20	1		
6	42	19	1	1	
7	40	19	2		
8	41	18	2	1	
9	41	18	2	1	
10	40	17	2		1
11	42	15	2	2	1
12	42	21			
13	42	21			

* Trivalents were frequently replaced by a bivalent and a univalent.

DESCRIPTION OF THE ABERRANT PLANTS AND THEIR PROGENY

Selfed seeds were obtained from all the aberrant plants but one. Some individuals were abundantly fertile under bag, while others had to be pollinated by hand. The poorly self-fertile plants, including the one which had set no seeds, were later back-pollinated successfully by normal Chinese Spring wheat.

Plants with a single univalent

Five individuals were characterized at meiosis by the presence of twenty pairs of chromosomes and one univalent.

In the following descriptions, numbers given to plants correspond to those in table 1.

1. Plant number 1 showed a slight chlorophyll defect. From 20 selfed seeds, 20 plants were obtained, of which two and possibly a third were monosomic, and two were nullosomic (20 pairs of chromosomes). The monosomic plants under field conditions were somewhat smaller and less vigorous than disomic sibs, and were lighter in color. The nullosomics were bushy dwarfs with stiff leaves more nearly normal in color than those of the monosomics. Neither of the nullosomic plants set seeds, but enough

functional pollen was obtained from one of them to produce a viable seed in a cross to *T. durum*.

2. Plant number 2, grown under greenhouse conditions, was nearly normal, differing mainly in the greater thickness and stiffness of its culms. Among its field-grown progeny, however, the only monosomic which occurred was dwarfed and sterile. The 28 plants (from 30 selfed seeds) included no nullosomics.

3. This monosomic was somewhat defective in size and vigor, although normal in color. From 20 selfed seeds, 16 plants were obtained, including three monosomics and no nullosomics. The aberrant plants resembled their parent in size and vigor.

4. This plant did not differ noticeably from normal, but in the field its monosomic offspring were somewhat lighter in color. Of 34 plants which resulted from 35 selfed seeds, three were found by cytological examination to be monosomic, and several which were not studied cytologically had the characteristic color of monosomics. No nullosomics were observed.

5. This was a rather small, weak plant, which did not set selfed seed until late in the season. Two daughter plants were obtained from four backcrossed seeds, and both were monosomic. These were more vigorous than the parent.

Plant with one univalent and one trivalent

6. No differences from the normal could be detected in plant number 6, nor could the monosomic and trisomic plants in its progeny be identified other than cytologically. From 16 selfed seeds, 16 plants resulted. Of the 15 of these which were studied cytologically, only two were normal, six were monosomic (figure 1), five had one univalent and one trivalent (figure 4), one had a univalent and a quadrivalent (figure 5), and one was nullosomic. The nullosomic plant, although small and lacking in vigor, had normal pollen fertility and set a number of selfed seeds. The monosomic-tetrasomic plant was normal, or nearly so, in appearance and fertility.

Plant with two univalents

7. This plant had narrow leaves and slender spikes and culms. From seven selfed seeds, six plants were obtained, of which five were examined cytologically. Four were of the parental constitution (figure 2), although phenotypically nearer normal, while the fifth had 20 bivalents and one univalent.

Plants with two univalents and one trivalent

8. This was the smallest and weakest of the 11 plants. It had a very short spike and shed no pollen. The only plant obtained from the three

backcrossed seeds had one univalent and one trivalent, and was much more vigorous than its parent.

9. This individual had a glossy appearance, and produced a short tapering spike with an extra spikelet at each of several central nodes. Two selfed seeds were obtained, and these produced one plant of the parental constitution (figure 3) and one with a single univalent. A single backcrossed seed gave an individual with one univalent and one trivalent.

Plant with two univalents and a ring of four

10. This plant was characterized by a dark green striping of its leaves. From 22 selfed seeds 21 plants were obtained, of which all but six were studied cytologically. Three were of the parental constitution (but failed to show the stripe), seven had one univalent and a ring of four, one had one univalent and a chain of four, three had two univalents, and one had one univalent. Figure 6 shows the appearance of the ring at first metaphase.

Plant with two univalents, two trivalents, and a ring of four

11. This plant was not conspicuously abnormal, its chief distinctive characteristic being a glossy appearance. Three plants were obtained from three selfed seeds, and two of these were examined cytologically. One, which died without setting seed, had two univalents, one trivalent, and a ring of four. The other plant had a trivalent and a chain of four. From 11 backcrossed seeds, 10 plants resulted. Two of these had two univalents and two trivalents; one had two univalents, one trivalent, and a ring of four; two had two univalents and one trivalent; one had one univalent, one trivalent, and a ring of four; one had one univalent, one trivalent, and a chain of four; two had one univalent and one trivalent; and one was normal.

EXPLANATION OF PLATE

Photographs from aceto-carmine smears, showing first metaphases (except figure 3) in aberrant plants. ×850.

FIGURE 1.—A monosome with a conspicuous, sub-terminal constriction, from an offspring of plant 6.

FIGURE 2.—Two monosomes, one considerably longer than the other, from an offspring of plant 7.

FIGURE 3.—Two monosomes at first anaphase, showing the chromatids passing to the poles. The upper chromosome has a median centromere, the lower a sub-terminal centromere. From an individual in the progeny of plant 9.

FIGURE 4.—A trisome, from an offspring of plant 6.

FIGURE 5.—A tetrasome (at left), from an offspring of plant 6.

FIGURE 6.—A ring of four, from an offspring of plant 10.

FIGURE 7.—Thirteen bivalents and eight univalents in a 34-chromosome hybrid of *T. durum* with a monosomic offspring of plant 3. u = univalent.

FIGURE 8.—Fourteen bivalents and six univalents in a 34-chromosome hybrid of *T. durum* with a monosomic offspring of plant 10.

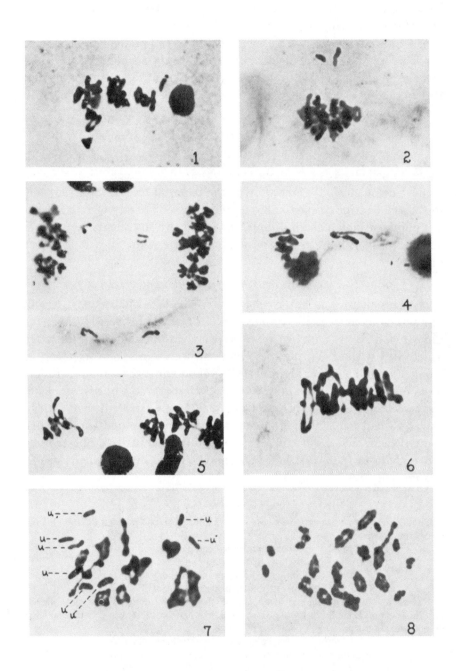

GENERAL CONSIDERATION OF THE ABERRATIONS

Monosomes

In the chromosome constitutions of the 11 aberrant plants in the immediate progeny of the haploid, 16 monosomes were involved, of which two were lost in the succeeding generation. Where two of these occurred in the same plant, as they did in five instances, those two necessarily involved non-homologous chromosomes; and other differences between the monosomes have been indicated by differences in appearance of monosomic plants and by differences in the size and morphology of the chromosomes concerned (figures 1, 2, 3). The chance of all 14 monosomes being different is small, however, even though they may have occurred at random.

The effort involved in determining homologies amongst the 14 monosomes would be much less if there were reliable cytological means of identifying the chromosomes. Recent work of BHATIA (1938) indicates that cytological distinctions do exist among most of the chromosomes.

Since the 14 chromosomes of emmer wheats pair regularly with 14 of the 21 *vulgare* chromosomes (SAX 1922), a 34-chromosome hybrid of emmer with a monosomic plant of *T. vulgare* will show at meiosis whether or not the monosome concerned is a chromosome homologous with one of emmer. Of eight monosomes thus far tested, five (figure 7) have proved to have a homolog in the emmer complement, while three (figure 8) are among the extra seven chromosomes of *T. vulgare*. *T. durum* was the emmer wheat used.

Deficiencies of whole chromosomes or parts of chromosomes have been observed several times in 21-chromosome wheats. KIHARA (1924) obtained in the F_4 of a cross between *T. polonicum* and *T. spelta* two plants which had only 20 pairs of chromosomes and which bred true. One was a dwarf with a somewhat lowered seed set, while the other was semi-dwarf, with nearly normal fertility. NISHIYAMA (1928) studied the monosomic plants which were obtained from these two lines through crosses with normal plants. He found that such monosomics produced functioning 20-chromosome gametes, though the proportion was much lower among male than among female germ cells.

HUSKINS (1928, 1933) and HUSKINS and SPIER (1934) have shown that certain types of speltoids are due to deficiency of a whole chromosome or a part of a chromosome. Where a whole chromosome is missing in the speltoid, homozygotes (that is, nullosomics) are rarely obtained, although selfing gives 5 to 20 times as many heterozygotes as normals. Where only a part of a chromosome is missing, homozygous speltoids are more frequently obtained, giving typically ratios near 1:2:1, and sometimes approaching 1:1:1.

UCHIKAWA (1937) also has found 41 chromosomes in a heterozygous speltoid. The same chromosome number was observed in a short, but otherwise normal, plant.

LOVE (1938) showed that white-chaff off-types in a golden chaff wheat were homozygous for a deficiency of one chromosome arm. LOVE (1939) also reports in F_5 to F_7 of *vulgare-durum* crosses *vulgare*-like plants deficient for as many as four chromosomes.

KATTERMANN (1932) isolated 40- and 41-chromosome plants from an F_6 of a wheat-rye cross.

Trisomes

Not as many instances of trisomes in *T. vulgare* have been observed as of monosomes. HUSKINS (1928, 1933) believed that certain speltoids were due to the presence of an extra chromosome or pair of chromosomes. UCHIKAWA (1937) reported that a semi-*compactum* wheat had 43 chromosomes, and its short-*compactum* derivative 44 chromosomes. In *vulgare*-like derivatives of *vulgare-durum* crosses, LOVE (1939) found numerous trivalents, some of which probably represented trisomes.

Thus far, two of the five trisomes obtained from the haploid of *T. vulgare* have been separated from the monosomes which accompanied them in the original offspring of the haploid. Neither differs conspicuously from normal in phenotype. From one trisomic plant both pollen and ovules of $n+1$ constitution have functioned; a daughter plant of a selfed trisomic individual (number 6) was tetrasomic (figure 5).

Reciprocal translocations

The two associations of four[3] observed were presumably reciprocal-translocation configurations rather than tetrasomes. Rings of four occurred with greater regularity than would be expected for tetrasomes, which frequently form a chain of four, or two pairs. Also, since the male parent presumably contributed only one each of the 21 kinds of chromosomes, a tetrasome would necessarily have involved three homologs from the maternal parent. The occurrence of gametes from a haploid with three homologs is difficult to explain.

Additional evidence is available that the ring of four found in plant 10 (figure 6) was due to a reciprocal translocation. The 17 bivalents and two univalents account for only 19 paternal chromosomes, leaving two to be involved in the ring. Furthermore, if the association of four had been a tetrasome, the plant would have been deficient for one pair of chromosomes, in which case it would probably not have been as vigorous as it was.

---

[3] There is some evidence from a later generation that a third reciprocal translocation was present in the immediate progeny of the haploid (in plant 7).

The interpretation of the ring of four in plant 11 as a reciprocal-translocation association is substantiated by the pairing relationships in one of its offspring. This daughter plant (from a selfing) had 43 chromosomes, 36 of which always formed 18 pairs. The other seven were observed to form a trivalent and a chain of four; a bivalent, a univalent, and a chain

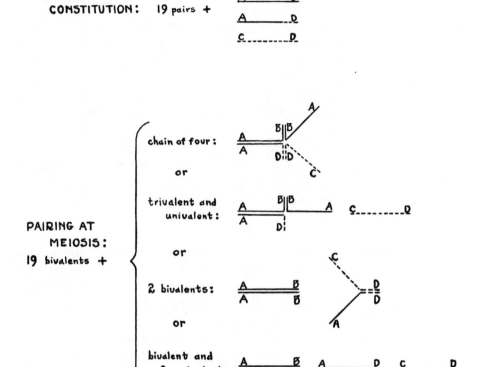

FIGURE 9.—The chromosome constitution and meiotic behavior of a plant which would occur in the progeny of an individual with a ring of four if one of the parental gametes received adjacent chromosomes from the ring and therefore had a duplication and a deficiency.

of four; two trivalents and a univalent; a trivalent, a bivalent, and two univalents; or two bivalents and three univalents. Part of this variability can be accounted for by the presence of a trisome, which sometimes formed a trivalent and sometimes a bivalent plus a univalent. The remaining four chromosomes presumably represented an unstable chain of four, which frequently broke up (figure 9) into a trivalent and a univalent, into two bivalents, or into a bivalent and two univalents. Such a chain of four would be expected in the progeny of a plant with a reciprocal-translocation ring of four, as a result of the functioning of a duplication-deficiency gamete.

Although neither of the two reciprocal translocations has yet been obtained in a plant which was free from monosomes, it has been possible to determine that they bring about no great reduction in male or female fertility. It is yet to be learned whether this lack of sterility is due mainly to directed segregation of the members of the chromosome ring, as shown by THOMPSON and THOMPSON (1937) in *T. durum* and by these authors (1937) and by SMITH (1939) in *T. monococcum*, or mainly to the ability of deficiency-duplication gametes to survive. In the one instance just discussed the functioning of a deficiency-duplication gamete has been indicated.

A reciprocal translocation in *T. vulgare* has been discovered by KOSTOFF (1937) in a 42-chromosome segregate from a cross of *T. vulgare* × *T. monococcum*. KATTERMANN (1934, 1935a, b) backcrossed a wheat-rye hybrid to wheat and obtained plants with multivalent associations of as many as six chromosomes. Some of these associations probably were due to reciprocal translocations. LOVE (1939) found numerous quadrivalent associations in *vulgare*-like derivatives of *vulgare-durum* crosses.

ORIGIN OF THE ABERRATIONS

Since no cytological study was made of the haploid which produced the aberrant plants, there is no direct evidence as to the mode of origin of the abnormalities. However, the observations of others, particularly of GAINES and AASE (1926) on a haploid of *T. compactum*, suggest several ways in which the aberrations could have arisen.

GAINES and AASE observed usually 21 univalents at first metaphase in their hybrid, although "occasionally two and rarely more" paired. The behavior of the chromosomes at first anaphase varied considerably. Sometimes the univalents were distributed to the two poles more or less at random; sometimes each univalent split and the halves went to opposite poles; and sometimes some of the univalents went as halves to opposite poles, while the halves of the rest failed to disjoin and went together to one or the other pole. In still other cells the entire chromosome group cohered in an irregular mass in the center of the cell. Micronuclei occurred frequently, as a result of chromosomes failing to be included in the telophase nuclei. The second division was observed to proceed more or less regularly if the chromosomes went to the poles as wholes at first anaphase, although supernumerary micronuclei frequently occurred. The behavior at second division was not determined for cells where all the univalents split at the first division nor where restitution nuclei were formed.

Origin of reciprocal translocations

KIHARA and NISHIYAMA (1937), KOSTOFF (1937), and HOWARD (1938) have pointed out that pairing between partially homologous chromosomes in hybrids may lead to the occurrence of segmental-interchange rings in the progeny of the hybrids. This explanation could apply also to haploids of *T. vulgare*, where a small amount of pairing occurs. GAINES and AASE (1926) reported occasionally one and rarely more pairs in their haploid, and RAW (1937) observed 0–5 bivalents and an occasional trivalent in *T. vulgare* haploids. Crossing over between a pair of non-homologous or partially homologous chromosomes (figure 10) would constitute a reciprocal translocation, and would result in a ring of four in backcrossed progeny if the two interchanged chromosomes were included in the same

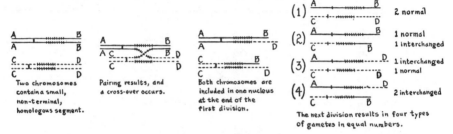

FIGURE 10.—Probable method of origin of reciprocal translocations from a haploid of *Triticum vulgare*. A gamete of type 4, if fertilized by a normal germ cell, will produce a plant with a ring of four at meiosis, while types 2 and 3 will give a plant with a chain of four.

gamete. The formation of a single restitution nucleus such as was sometimes observed by GAINES and AASE at the end of the first division would permit both interchanged chromosomes to go to the same gamete. Or, if all the univalent chromosomes split and disjoined at the first division, the two chromosomes which were paired might disjoin tardily and both be included in the same telophase nucleus.

As emphasized by HOWARD (1938) and illustrated in figure 10, through random segregation of the four chromatids of a bivalent in which crossing over had occurred, only one-fourth of the gametes formed from spore-mother cells in which an interchange had taken place would have two interchanged chromosomes, one-fourth would be normal, and one-half would contain one normal and one interchanged chromosome. An egg of this last type would have a duplication and a deficiency, and if fertilized by a normal sperm, would produce a plant with a chain of four, as in figure 9. The fact that no such plant was discovered in the immediate progeny of the haploid may have been due to the difficulty of detecting the infrequent chains of four and trivalents which would have occurred if the deficient-duplicated segment was small.

Origin of monosomes and trisomes

These two abnormalities will be discussed together because every mode of origin to be suggested for one also applies to the other. Where only

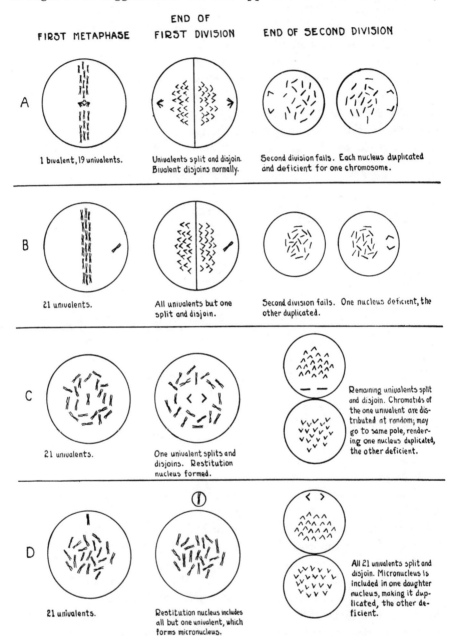

FIGURE 11.—Possible methods of origin of deficient and duplicated gametes from a haploid of *Triticum vulgare*.

monosomes occurred, as in plants 1 to 5 and in plant 7, it is possible that these arose through random segregation of 19 or 20 chromosomes to one pole at the first division in the haploid, a possibility which does not provide for the occurrence of trisomes. However, the chance of such an unequal distribution as 20:1 is so small that its occurrence even once among so few flowers as were pollinated would be highly improbable.

One way in which monosomes and trisomes could originate has already been suggested by von BERG (1935), as a result of his observations on meiosis in a 21-chromosome generic hybrid, *T. turgidum* × *Haynaldia villosa*. He found, as had GAINES and AASE in their wheat haploid, that frequently all the univalents split and disjoined at the first division, and he observed that at the second division, restitution nuclei were formed which included all 21 chromatids. He pointed out (figure 11, A) that the presence and normal disjunction of a bivalent in a cell where all the univalents split and go to opposite poles at the first division would render the resulting gametes deficient for one member of the bivalent and duplicated for the other member. It should be pointed out that if an interchange occurred between the two paired chromosomes, the deficiency and duplication would be equal only to the non-interchanged parts of the chromatids, and a gamete with such a deficiency-duplication for a part of a chromosome, when fertilized by a normal germ cell, would produce a plant with a chromosome constitution like that in figure 9. Pairing in such a plant would depend on the size of the interchanged segments, and might never involve a chain of four if the segments were small.

Another possible explanation of the origin of monosomes and trisomes (figure 11, B) is based on the observation of GAINES and AASE that both halves of one or more univalents sometimes fail to disjoin at the first division and are distributed together to one or the other pole, while the other univalents split and go to opposite poles. A daughter nucleus which received both halves of a univalent would be disomic for that chromosome, and the other nucleus would be nullosomic.

Another type of division in the *T. compactum* haploid of GAINES and AASE which suggests the origin of monosomes and trisomes is the inclusion of all the chromosomes in a single, restitution nucleus following failure of the first division. If one of the univalents were already divided when the restitution nucleus was formed (figure 11, C), the two chromatids might be distributed independently at the next division (which presumably would be equational for the rest of the univalents) and frequently pass to the same daughter nucleus, making that nucleus duplicated and the other deficient.

A fourth method of origin (figure 11, D) was suggested by KIHARA and NISHIYAMA (1937) for disomic and nullosomic gametes produced by the

21-chromosome hybrid *T. polonicum × Haynaldia villosa*. Cytological observation showed that about 30 percent of pollen mother cells formed restitution nuclei at the end of the first division and that over a fourth of these restitution nuclei were accompanied by one or more micronuclei. Each micronucleus contained a single chromosome, and this chromosome presumably divided within its nucleus at the second division. The inclusion of a micronucleus in one of the telophase chromosome groups of the dividing restitution nucleus was suggested as an explanation for disomic gametes, while the omission of micronuclei explained the origin of deficient gametes.

Finally, there must be mentioned the possibility of an interchange occurring (as in figure 10) which was large enough that a gamete of type 2 or 3 (containing only one interchanged chromatid), upon fertilization by a normal gamete, would produce a plant in which a trivalent and a univalent were regularly formed, with never a chain of four. It is doubtful, however, that such a large interchange occurs other than very rarely. Most of the chromosomes of *T. vulgare* have median or near-median centromeres, so that interchanges on both sides of the centromere, each involving the same chromatid, would be necessary to make a chromatid much more than half interchanged. From the observable looseness of the association at metaphase in wheat haploids, it is unlikely that interchanges often occur on both sides of the centromere.

DISCUSSION

Several possibilities present themselves for the use of chromosome aberrants in a broad genetic analysis of *T. vulgare*. One possible study is of the immediate effects on the plant of deficiencies and duplications of chromosomes and parts of chromosomes. Another line of attack is to use monosomes and trisomes to locate on specific chromosomes the genes of common wheat.

Hybrids of the various aberrants with wheats of the emmer group will show at meiosis whether any of the chromosomes concerned are homologous with emmer chromosomes. As previously noted, five monosomes have thus far proved to have a homolog in the emmer complement. The hybrids which involve these monosomes can presumably be used to introduce, through backcrossing, single, intact emmer chromosomes into *T. vulgare*. Observation of the effects of these emmer chromosomes may shed light on the phylogeny of *T. vulgare*, and will have obvious practical applications. By backcrossing these same hybrids to the emmer parent, it should be possible to obtain monosomics in emmer, if they are viable, and to compare the effects of the deficiencies there with their effect in *T. vulgare*, where an additional set of seven chromosomes is present.

Homologies among the 21 chromosomes of *T. vulgare* are indicated by the formation of bivalents in the haploid. Little is known about these homologies. They may consist of duplicated segments scattered among all 21 chromosomes, or they may be duplications confined to a few chromosomes. Up to five meiotic pairs have been observed in a haploid, but there is no knowledge whether these five represent the total amount of homology, or whether they are merely the random result of homologies which involve more than ten chromosomes but which are too slight ever to result in more than five pairs in a single cell. Reciprocal translocations obtained from haploids of *T. vulgare*, if due to crossing over between homologous regions, furnish a means of determining the homology of definite chromosomes and regions. If association in the haploid is restricted to certain, partially homologous chromosomes, then the translocations which result should always involve the same chromosome combinations. These could be identified by crosses among a number of plants containing translocations of different origin, and by crosses with specific nullosomics or monosomics.

It is possible that many other supposedly polyploid plants will prove amenable to the same sort of analysis as that under way on *T. vulgare*. The results of STADLER (1931), KATAYAMA (1934), YEFEIKIN and VASILYEV (1936), GERASSIMOWA (1936), and IVANOV (1938), indicate that X-rays may be effective in inducing haploidy, and SINGLETON (1938) noted the occurrence of several haploids after ultra-violet treatment of maize pollen. Twin seedlings constitute another source of haploids (NAMIKAWA and KAWAKAMI 1934; HARLAND 1936; KIHARA 1936; YAMAMOTO 1936; MÜNTZING 1937, 1938; WEBBER 1938; KASPARYAN 1938). Whether haploids of other polyploid species than *T. vulgare* will give rise to the same types of chromosome aberrations remains to be seen.

Hybrids with little chromosome pairing might be studied with profit in the foregoing fashion. If an amphidiploid could be produced and backcrossed successfully to the hybrid, the resulting plants might then contain segmental-interchange associations and whole-chromosome deficiencies and duplications.

ACKNOWLEDGMENTS

The author is indebted to DR. L. J. STADLER for encouragement during the investigation, and to him and DR. BARBARA McCLINTOCK for suggestions in the preparation of the manuscript.

SUMMARY

1. From a haploid of *Triticum vulgare*, 13 viable seeds were obtained by application of pollen from diploids.
2. Eleven of the resulting plants showed abnormal meiotic associations.

A total of 16 univalents, five trivalents, and two rings of four were present in the 11 individuals, the most aberrant of which had two univalents, two trivalents, and a ring of four.

3. Nearly all of the aberrant plants differed from normal in morphology, color, size, or vigor. Selfed seeds were obtained on all but one, and that one was fertile to pollen from a normal plant.

4. Nullosomic plants were obtained in the progeny of two monosomics. One of these was dwarfed and sterile, while the other was only semi-dwarf and was fertile.

5. Crosses of *T. durum* with eight monosomic plants of different origin showed that five of the eight involved a chromosome homologous to one of those of *T. durum*.

6. A tetrasomic plant, phenotypically indistinguishable from normal, was found in the progeny of a trisomic.

7. Rings of four were presumably the result of segmental interchanges which occurred in the haploid as crossovers between paired, partially homologous chromosomes.

8. From a plant with a ring of four, an individual with an unstable chain of four was obtained, supposedly as a result of the functioning of a deficiency-duplication gamete.

9. The various chromosome aberrations apparently resulted from some regulatory process in the haploid which favored the production of spores with near the somatic number of chromosomes. The use of these aberrations provides a new method for the genetic anlysis of *T. vulgare*—a method which may be applicable to polyploid plants in other genera.

LITERATURE CITED

BHATIA, G. S., 1938 Cytology and genetics of some Indian wheats. II. The cytology of some Indian wheats. Ann. Bot., N. S. **2**: 335–371.

GAINES, E. F., and AASE, H. C., 1926 A haploid wheat plant. Amer. J. Bot. **13**: 373–385.

GERASSIMOWA, H., 1936 Experimentell erhaltene haploide Pflanze von *Crepis tectorum* L. Planta **25**: 696–702.

HARLAND, S. C., 1936 Haploids in polyembryonic seeds of sea island cotton. J. Hered. **27**: 229–231.

HOWARD, H. W., 1938 The fertility of amphidiploids from the cross *Raphanus sativus*✕*Brassica oleracea*. J. Genet. **36**: 239–273.

HUSKINS, C. L., 1928 On the cytology of speltoid wheats in relation to their origin and genetic behavior. J. Genet. **20**: 103–122.
    1933 The origin and significance of fatuoids, speltoids, and other aberrant forms of oats and wheat. Proc. World's Grain Exhib. and Confer., Regina, Saskatchewan **2**: 45–50.

HUSKINS, C. L., and SPIER, J. D., 1934 The segregation of heteromorphic homologous chromosomes in pollen-mother-cells of *Triticum vulgare*. Cytologia **5**: 269–277.

IVANOV, M. A., 1938 Experimental production of haploids in *Nicotiana rustica* L. Genetica **20**: 295–397.

KASPARYAN, A. S., 1936 Haploids and haplo-diploids among hybrid twin seedlings in wheat. C. R. (Doklady) Acad. Sci. URSS **20**: 53–56.

KATAYAMA, Y., 1934 Haploid formation by X-rays in *Triticum monococcum*. Cytologia **5**: 235–237.

KATTERMANN, G., 1932 Genetische Beobachtungen und zytologische Untersuchungen an der Nachkommenschaft einer Gattungskreuzung. II. Zytologische Untersuchungen. Z. i. A. V. **60:** 395–466.

1934 Die zytologischen Verhältnisse einiger Weizen-Roggen-Bastarden und ihrer Nachkommenshaft ("F₂"). Züchter **6:** 97–107.

1935a Die Chromosomenverhältnisse bei Weizenroggenbastarden der zweiten Generation mit besonderer Berücksichtigung der Homologiebeziehungen. Z. i. A. V. **70:** 265–308.

1935b Die Paarungsintensität der Chromosomen bei Weizen-Roggenbastarden zweiter Generation im Vergleich zum Weizenelter. Planta **24:** 66–77.

KIHARA, H., 1924 Cytologische und genetische Studien bei wichtigen Getreidearten mit besonderer Rücksicht auf das Verhalten der Chromosomen und die Sterilität in den Bastarden. Mem. Coll. Sci., Kyoto Imp. Univ. **B:** 1–200.

1936 Ein diplo-haploides Zwillingspaar bei *Triticum durum*. Agric. and Hort. (Tokyo) **11:** 1425–1433.

KIHARA, H., and NISHIYAMA, I., 1937 Possibility of crossing over between semihomologous chromosomes from two different genoms. Cytologia, Fujii Jubil. Vol.: 654–666.

KOSTOFF, D., 1937 Formation of a quadrivalent group in a hybrid between *Triticum vulgare* and a *Tr. vulgare* extracted derivative. Curr. Sci. **5:** 537.

LOVE, R. M., 1938 A cytogenetic study of white chaff off-types occurring spontaneously in Dawson's Golden Chaff winter wheat. Genetics **23:** 157.

1939 Cytogenetics of *vulgare*-like derivatives of pentaploid wheat crosses. Genetics. **24:** 92.

MÜNTZING, A., 1937 Polyploidy from twin seedlings. Cytologia, Fujii Jubil. Vol.: 211–227.

1938 Note on heteroploid twin plants from eleven genera. Hereditas **24:** 487–491.

NAMIKAWA, S., and KAWAKAMI, J., 1934 On the occurrence of haploid, triploid and tetraploid plants in twin seedlings of common wheat. Proc. Imp. Acad. Tokyo **10:** 668–671.

NISHIYAMA, I., 1928 On hybrids between *Triticum spelta* and two dwarf wheat plants with 40 somatic chromosomes. Bot. Mag. (Tokyo) **42:** 154–177.

RAW, A. R., 1937 Genetical studies with wheat—haploids of *Triticum vulgare*. J. Dept. Agri. Victoria **35:** 300–306.

SAX, K., 1922 Sterility in wheat hybrids. II. Chromosome behavior in partially sterile hybrids. Genetics **7:** 513–552.

SINGLETON, W. R., 1938 Cytological observations on deficiencies produced by treating maize pollen with ultra violet light. Collecting Net **13:** 158.

SMITH, L., 1939 Reciprocal translocations in *Triticum monococcum*. Genetics **24:** 86.

STADLER, L. J., 1931 The experimental modification of heredity in crop plants. I. Induced chromosomal irregularities. Sci. Agr. **11:** 557–572.

THOMPSON, W. P., and Thompson, M. G., 1937 Reciprocal chromosome translocations without semi-sterility. Cytologia, Fujii Jubil. Vol: 336–342.

UCHIKAWA, I., 1937 Cytogenetic studies on compactoid wheat. Jap. J. Genet. **13:** 9–15.

VON BERG, K., 1935 Cytologische Untersuchungen an den Bastarden des *Triticum turgidovillosum* und an einer F₁ *Triticum turgidum×villosum*. III. Weitere Studien am fertilen konstanten Artbastard *Triticum turgidovillosum* und seinen Verwandten. Z. i. A. V. **68:** 94–126.

WEBBER, J. M., 1938 Cytology of twin cotton plants. J. Agr. Res. **57:** 155–160.

YAMAMOTO, Y., 1936 Ein haplo-diploides Zwillingspaar bei *Triticum vulgare* Vill. Bot. Mag. (Tokyo) **50:** 573–581.

YEFEIKIN, A K., and VASILYEV, B. I., 1936 Artificial induction of haploid Durum wheats by pollination with X-rayed pollen. Bull. Appl. Bot. Genet., Pl. Br., ser. II, no. **9:** 39–45.

44

Reprinted by permission of the University of California Press from *Univ. Calif. Publ. Bot.* **11**(13):245–256 (1928)

INTERSPECIFIC HYBRIDIZATION IN NICOTIANA

VIII. THE SYLVESTRIS–TOMENTOSA–TABACUM HYBRID TRIANGLE AND ITS BEARING ON THE ORIGIN OF TABACUM

BY

T. H. GOODSPEED AND R. E. CLAUSEN

In studies of interspecific hybridization in *Nicotiana* we have endeavored to determine whether any consistent cytological behavior may be demonstrated in groups of hybrids. The simplest condition for such a demonstration should be provided by a study of the three possible hybrids which may be obtained between three given species. Thus with the species, *sylvestris, tomentosa,* and *tabacum,* the three possible hybrids, F_1 *sylvestris-tabacum,* F_1 *tomentosa-tabacum,* and F_1 *sylvestris-tomentosa,* have been obtained and studied. We call such a series a hybrid triangle. Despite the fact that some forty interspecific hybrids have been obtained in *Nicotiana,* very few complete hybrid triangles are represented among them. The one dealt with in this paper, however, is of special interest because of the significance it may have for the origin of *tabacum.*

F_1 MORPHOLOGY

F_1 *sylvestris-tabacum* is characterized by a remarkable resemblance to its *tabacum* parent, as is shown by the fact that a wide range of highly distinct *tabacum* varieties give F_1 hybrids with *sylvestris* which exhibit with great fidelity the characteristic features of their particular *tabacum* parents on an enlarged scale. Despite the recent contentions of Brieger (1928) as to the characteristic features of these hybrids, we believe that the evidence set forth in our own account (Goodspeed and Clausen, 1917) is adequate and convincing; moreover, our subsequent observations are in harmony with it.

F₁ *tomentosa-tabacum,* on the other hand, clearly represents a synthesis of the characteristic features of the parental species, as may be appreciated by examination of plate 8, figure 2, where it is shown beside a *tabacum* haplont. Its habit approaches that of *tomentosa,* although it apparently never grows so large nor so tall, and it is more

Fig. 1.   Flowers of (*a*) F₁ *sylvestris-tomentosa,* (*b*) F₁ *tomentosa-tabacum,* and (*c*) *N. tomentosa.*   × %₁₀.

or less strictly annual as contrasted with the perennial character of that species. The flowers are somewhat larger than those of *tomentosa,* but they are very much like them in form as is manifested by their bilabiate tendency, reflexed corolla lobes, and exserted style and stamens (fig. 1*b*). The color is more pronounced than that of *tomentosa,* but it is a rusty shade of red, not the clear purple red of *tabacum* var. *purpurea* (U. C. B. G. 25/06). The leaves are intermediate in shape and texture, and they lack the distinctive constriction at the base so characteristic of the *tabacum* parent.

F_1 *sylvestris-tomentosa,* the third member of the series, is illustrated in plate 8, figure 1; its parents in plate 9. The *tomentosa* plant has been cut back and is therefore not wholly typical of the species. Flowers of *tomentosa,* and F_1 *sylvestris-tomentosa* are illustrated in figure 1. Flowers of the F_1 hybrid are pinkish in color, intermediate between the white of *sylvestris* and the salmon red of *tomentosa.* The F_1 hybrid is exceedingly vigorous, possibly even surpassing *tomentosa* in this respect; and its characters are mainly a synthesis of those of the two parents.

For the sake of comparison, measurements of flowers of the pertinent forms in the *sylvestris-tomentosa-tabacum* series are presented in table 1.

TABLE 1

FLOWER SIZE MEASUREMENTS IN THE *sylvestris-tomentosa-tabacum* SERIES

Type	Spread in mm.	Length in mm.
tabacum var. *purpurea*	36.0	49.3
sylvestris	42.5	85.3
tomentosa	30.0	31.0
tomentosa (including anthers)	30.0	52.0
tabacum var. *purpurea* haplont	25.0	40.0
F_1 *sylvestris-tomentosa*	35.0	56.0
F_1 *tomentosa-tabacum*	27.0	34.0
F_1 *sylvestris-tabacum*	44.1	59.0

F_1 CYTOLOGY

Chromosomal behavior in F_1 *sylvestris-tabacum* has already been described in some detail (Goodspeed and Clausen, 1927*b*). The hybrid uniformly exhibits the Drosera scheme of conjugation, $12_{II} + 12_{I}$, in diaphase and I–M. The conjugants separate normally; the univalents are distributed for the most part at random in I and divide in II. The twelve bivalents of this hybrid are taken to represent conjugation of *sylvestris* with *tabacum* homologues in view of the prevailing absence of conjugation in *tabacum* haplont (Chipman and Goodspeed, 1927) and the chromosomal situation in backcross progenies of F_1 *sylvestris-tabacum* $\times$ *sylvestris,* which are found to conform to the formula $12_{II} + i_{I}$, i exhibiting values from 0 to 12, with, however, a marked concentration of values in the lower and upper portions of the range (Goodspeed and Clausen, 1927*b*). The F_1 hybrid has not given progeny from self-fertilization but from ten to twenty seeds per capsule are obtained by backcrossing it to its parental species, *sylvestris* and *tabacum*.

F_1 *tomentosa-tabacum* also exhibits the Drosera scheme, $12_{II} + 12_I$, in diaphase and I–M, with clear distinction between bivalents and univalents (figs. 2–4). The subsequent behavior, however, is somewhat different from that characteristic of F_1 *sylvestris-tabacum*. Separation of the conjugants is effected normally, but I–A (fig. 3) usually exhibits a number of univalents in advanced stages of division.

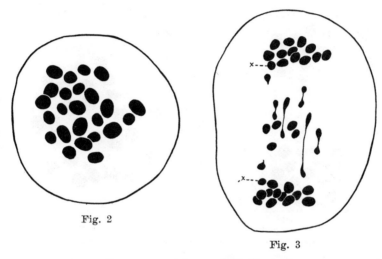

Fig. 2

Fig. 3

Fig. 2. F_1 *tomentosa-tabacum*, I–M, $12_{II} + 12_I$.

Fig. 3. F_1 *tomentosa-tabacum*, I–A, illustrating division of univalents, *x, x*, halves of univalents.

In F_1 *sylvestris-tabacum* (Goodspeed and Clausen, 1927) and F_1 *paniculata-rustica* (Goodspeed, Clausen and Chipman, 1926) similar figures are also seen but in them it was shown that II–M plates rarely exhibit a total in excess of 36 chromosomes, which indicates that separation of the halves rarely is completed. In F_1 *tomentosa-tabacum*, however, II–M counts usually exhibit a total of more than 36 chromosomes. In 18 PMC in which both plates were countable, only three gave a total of 36 chromosomes and the average total per PMC was 40.4. Counts of 112 single II–M plates gave an average of 20.0 chromosomes per plate, which is substantially in agreement with the value obtained from PMC in which both plates were countable.

The subsequent behavior does not appear different from that described for F_1 *sylvestris-tabacum* and F_1 *paniculata-rustica*. Lagging in II–A was not conspicuous, but tetrad counts indicated the occurrence of a considerable amount of elimination and also the production of a small percentage of dyads. Two separate counts are included in table 2.

The F_1 *tomentosa-tabacum* hybrid exhibits a slight degree of fertility when *tabacum* pollen is applied to it, apparently approximately equal to that shown by F_1 *sylvestris-tabacum*. Trials with *tomentosa* have not been made.

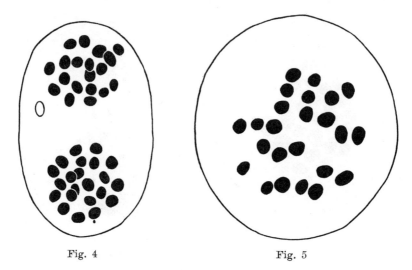

Fig. 4   Fig. 5

Fig. 4. F_1 *tomentosa-tabacum*, II–M, 22 chromosomes in left plate, 19 in right plate, and one in the plasma.

Fig. 5. F_1 *sylvestris-tomentosa*, ''I-M,'' with 24 unpaired chromosomes.

TABLE 2

Tetrad Counts in the *sylvestris-tomentosa-tabacum* Series

Hybrid	Types of pollen groups									Anomalous
	2_0	2_1	2_2	2_3	4_0	4_1	4_2	4_3	4_4	
F_1 *sylvestris-tabacum*	1	....	2	....	89	76	27	4	....	3_3
F_1 *tomentosa-tabacum*	....	....	....	....	82	67	42	8	1	
F_1 *tomentosa-tabacum*	3	....	1	....	55	62	33	14	1	3_0, 3_0
F_1 *sylvestris-tomentosa*	45	10	....	....	283	4	1	....	....	3_0
tabacum haplont	8	9	6	4	111	27	7	9	1	$3_0(11)$, $3_1(3)$, $3_2(4)$

F_1 *sylvestris-tomentosa* usually exhibits 24_I at diaphase and I–M (fig. 5) although a small proportion of PMC in I–M and I–A may exhibit a few "bivalents" as in the *tabacum* haplont (Chipman and Goodspeed, 1927). In figure 7, I–A shows 7_I approaching one pole, 9_I near the other, 6_I in the equatorial region and a "bivalent" disjoining. Sometimes as in figure 6 there is evidence of division of univalents. In this PMC, an extreme example, there were 11½ near one pole, 6½ near the other, and in the center three undivided univalents together with three which were undergoing division. Instances in which both I–M plates were countable, however, indicate that division of univalents in I is not characteristic of this hybrid.

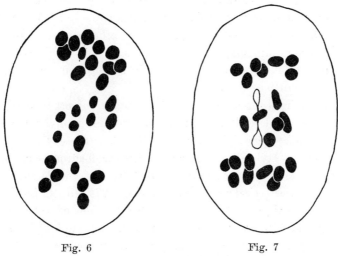

Fig. 6   Fig. 7

Fig. 6. F_1 *sylvestris-tomentosa*, I–A, 11½ at one pole, 6½ at the other, $3\frac{6}{2}$ in the equatorial region, illustrating division of univalents.

Fig. 7. F_1 *sylvestris-tomentosa*, I–A, 7 at one pole, 9 at the other, 6 in the equatorial region and a "bivalent" (in outline) disjoining.

The counts of pollen groups contained in table 2 disclose the occurrence of a relatively large percentage of "restitution" divisions in this hybrid. The $2_0 + 2_1$ pollen groups amount to 16.0 per cent of the total number. Meiotic behavior leading to the production of dyads was not studied but a single record of a 0/24 distribution was presumably of this type. Dyad production is a not uncommon feature of *Nicotiana* hybrids, and while it appears to be most characteristic of hybrids in which the chromosomes fail to conjugate, it is also observed in those which exhibit the Drosera type of conjugation.

F_1 *sylvestris-tomentosa*, despite its conspicuous production of dyads, has as yet produced no seed either from selfing or from back-crossing to the parental species.

DISCUSSION

According to the data reported above and in previous papers, the F_1 *sylvestris-tabacum* and *tomentosa-tabacum* hybrids both exhibit $12_{II} + 12_I$ chromosomes in meiosis. In view of the fact that *sylvestris* and *tomentosa* are so distinctly different morphologically, and that *tabacum* haplont (Chipman and Goodspeed, 1927) and F_1 *sylvestris-tomentosa* typically exhibit no conjugation of chromosomes, it seems reasonable to assume that *tabacum* possesses two sets of chromosomes; one homologous with that of *sylvestris*, the other with that of *tomentosa*. Further data may be necessary to establish this conclusion: but at present it appears to be the only assumption consistent with all the facts outlined above.

Tabacum hybridizes readily with a number of other species. In addition to those described above we have studied its hybrids with the 12-chromosome species, *glauca* and *glutinosa;* but as yet our evidence is incomplete. Hybrids of *tabacum* with the 9-chromosome species, *alata,* the 12-chromosome species, *solanifolia,* and the 24-chromosome species, *Bigelovii* (Goodspeed and Clausen, 1927), all exhibit complete absence of pairing in F_1 meiosis. The hybrids of *tabacum* with *sylvestris* and *tomentosa* appear, therefore, to be unique as respects meiotic phenomena; and in addition, they are the only partially fertile ones in the series. Brieger (1928) has, however, reported a hybrid of *tabacum* with the 12-chromosome species, *Rusbyi.* He finds that this hybrid exhibits the Drosera scheme of conjugation, and also that it produces some offspring when backcrossed to *tabacum.* We have *Rusbyi* in our collection, and find that it is clearly a species closely allied to *tomentosa.* It is not surprising, therefore, that it exhibits parallel behavior in crosses with *tabacum.*

Since *tabacum* possesses chromosome sets homologous with *sylvestris* and *tomentosa* it may be assumed that its progenitor arose through hybridization of two species, progenitors or close allies of *sylvestris* and *tabacum,* followed by doubling of chromosome number after the manner described for the origin of *digluta* (Clausen and Goodspeed, 1925; Clausen, 1928) by doubling of the chromosome number in a *glutinosa-tabacum* hybrid. If this assumption is correct, F_1 *sylvestris-*

tomentosa should bear a general morphological resemblance to *tabacum*. To a certain extent this is the case, but a considerable enlargement of floral and vegetative organs and certain structural differences preclude an appeal to these species as they now exist as the immediate ancestors of *tabacum*. Moreover, our studies of inheritance in the *sylvestris-tabacum* hybrid indicate the existence of extensive genetic differences between homologous chromosomes of these two species. It is probable that *tabacum* has undergone evolutionary changes since establishment of its chromosome number, and that its putative ancestors have also been modified; but the cytological evidence presented above indicates that these subsequent changes have not been sufficient to destroy the affinity between their chromosomes. It seems that further studies by the method here outlined may lead eventually to production, as Jorgensen (1928) suggests, of an existing species from hybridization of two other species.

LITERATURE CITED

BRIEGER, F.
   1928. Über Artkreuzungen in der Gattung Nicotiana. Verh. v. Internat. Kong., Vererb. Berlin, 1927, vol. 1, pp. 485–495.

CHIPMAN, RUTH H., and GOODSPEED, T. H.
   1927. Inheritance in *Nicotiana tabacum*. VIII. Cytological features of *purpurea* haploid. Univ. Calif. Publ. Bot., vol. 11, pp. 141–158, pls. 4–6.

CLAUSEN, R. E.
   1928. Interspecific hybridization in Nicotiana. VII. The cytology of hybrids of the synthetic species, *digluta*, with its parents, *glutinosa* and *tabacum*. *Ibid.*, vol. 11, pp. 177–211, 43 figs. in text.

CLAUSEN, R. E., and GOODSPEED, T. H.
   1925. Interspecific hybridization in Nicotiana. II. A tetraploid *glutinosa-tabacum* hybrid, an experimental verification of Winge's hypothesis. Genetics, vol. 10, pp. 278–284.

GOODSPEED, T. H., and CLAUSEN, R. E.
   1917. The nature of the F₁ species hybrids between *Nicotiana sylvestris* and varieties of *N. tabacum*. Univ. Calif. Publ. Bot., vol. 5, pp. 301–346, pls. 37–48.
   1927*a*. Interspecific hybridization in Nicotiana. V. Cytological features of two F₁ hybrids made with *Nicotiana Bigelovii* as a parent. *Ibid.*, vol. 11, pp. 117–125, 8 figs. in text.
   1927*b*. Interspecific hybridization in Nicotiana. VI. Cytological features of *sylvestris-tabacum* hybrids. *Ibid.*, vol. 11, pp. 127–140, 9 figs. in text.

GOODSPEED, T. H., CLAUSEN, R. E., and CHIPMAN, RUTH H.
   1926. Interspecific hybridization in Nicotiana. IV. Some cytological features of the *paniculata-rustica* hybrid and its derivatives. *Ibid.*, vol. 11, pp. 103–115, 6 figs. in text.

JÖRGENSEN, C. A.
   1928. The experimental formation of heteroploid plants in the genus Solanum. Jour. Genetics, vol. 19, pp. 133–211.

EXPLANATION OF PLATES

PLATE 8

Fig. 1. F₁ *sylvestris-tomentosa.*

Fig. 2. On the left, a plant of F₁ *tomentosa-tabacum;* on the right a haplont *tabacum* which occurred in the same F₁ population.

PLATE 9

Fig. 1. *N. sylvestris.*

Fig. 2. *N. tomentosa,* three years old, somewhat cut back.

Fig. 2

Fig. 1

Fig. 2

Fig. 1

Part VII

GENETICAL AND CYTOLOGICAL MAPPING

HISTORICAL PERSPECTIVES

Sex-linked Inheritance

The first case of sex-linked inheritance showed that the female was heterogametic, producing two kinds of gametes (Doncaster and Raynor 1906, in the currant moth *Abraxas*). This evidence was contrary to previous cytological observations on several other species that the male was heterogametic. Morgan's 1910 paper (Paper 45) is the first report of sex linkage in *Drosophila* that demonstrated that in this species the male was heterogametic.

Morgan reported breeding results starting with a male with white eyes that appeared in a pedigree culture of *Drosophila* being grown in the laboratory at Woods Hole, Massachusetts. When mated with red-eyed sisters, essentially all the offspring were red-eyed. When these were crossed, there was a ratio of about 2 red-eyed females to 1 red-eyed male and 1 white-eyed male, but no white-eyed females. When these red-eyed females were crossed with white-eyed male sibs, there were about equal numbers of the four classes, i.e., red-eyed females and males and white-eyed females and males. To explain the results, Morgan assumed that all the spermatozoa of the white-eyed male carried the factor for white eyes, but only half carried the sex factor X; and that each egg of the female carried an X, half of them carrying the factor for red eyes (R) and half the factor for white eyes (W). Among the progeny, when white-eyed females were mated with red-eyed male sibs, the female offspring were red-eyed, the male white-eyed.

Hence the red-eyed males bred as if they were heterozygous for the white-eye factor. At the time the male was believed to have a single sex chromosome, the female two. It was not until Bridges' studies of nondisjunction of the sex chromosomes (beginning in 1913) that the male was found to have an X and Y-chromosome, the latter without alleles of most of the genes in the X. Thus any gene showing sex linkage is expected to be in the X-chromosome. Morgan pointed out that in *Abraxas*, the color mutant (lacticolor) appeared in nature only in females. The explanation is that in *Abraxas* the females produce two types of eggs, just the opposite of *Drosophila*.

Bridges' 1916 paper (Paper 46) on nondisjunction, published in two sections, is a classic. The first section which was the first paper published in *Genetics*, the new journal, dealt with the genetic effects of nondisjunction. The second portion that appeared in the second issue of the journal dealt with the cytological proofs of nondisjunction and its correlation with the genetic results.

In the mating of vermilion-eyed females and red-eyed (wild-type) males, which produced mainly wild-type females and vermilion males, there was an occasional exceptional vermilion female and wild-type male. These are the results expected if nondisjunction of the two X-chromosomes of the female had occurred to produce XX and no-X eggs. When fertilized by X-sperm, two classes are expected-XXX (dies) and XO (sterile male); when fertilized by Y-sperm, two classes are expected-XXY (exceptional female) and YO (dies). These exceptional females, mated to normal males, again produced some exceptional offspring. These were said to be the result of *secondary nondisjunction*. The occasional XX nondisjunction that occurred in normal females was termed *primary nondisjunction*. Breeding tests showed that in XXY females X-Y versus X-X synapsis was not random, X-Y synapsis occurred with a frequency of only 16.5 percent (66.7 percent is expected if at random). In XYY males the different kinds of synapsis were approximately random, i.e., 2X-Y:1Y-Y synapsis. In XXY females, the XX eggs come from X-Y synapsis. Tests using genetic markers in the X showed that the X-chromosomes in these XX eggs were noncrossovers; tests also showed that the first meiotic division produces this XX combination. Cytological examination showed that exceptional females were always XXY. The evidence from cytology and the genetic tests showed conclusively that the X-chromosomes carry the genes for the sex-linked characters. The distribution to the offspring of sex-linked characters carried by the exceptional females parallels the distribution of the X-chromosomes from the female and male parents to the offspring.

Aneuploids

The trisomics in *Datura* were probably the first to be used in plants to identify a linkage group with a specific chromosome (Blakeslee, Belling, and Farnham, Paper 29). In maize trisomic plants among the progeny of a triploid crossed with the normal diploid were used by McClintock in cytological studies to identify the extra chromosome and also in crosses with genetic markers to determine which particular linkage group was carried by each chromosome. The paper by McClintock and Hill (1931) summarized those tests for chromosome 10, the shortest one. The same method was used to associate all but a few of the other linkage groups in maize with their chromosomes.

By using aneuploid individuals in which the extra chromosome was a telocentric (one arm of the chromosome missing) or an isochromosome (both arms the same), it was possible to determine the portion of the linkage group carried by each arm. An example of this is Rhoades' report (1936) using a telocentric for the short arm of chromosome 5 in maize. For two very closely linked genes, brittle endosperm-1 (*bt*) and brown midrib-1 (*bm*) with less than 1 percent recombination, *bt* was shown to be in the long arm and *bm* in the short arm.

In allopolyploids such as common bread wheat (*Triticum vulgare*) and *Nicotiana*, the most common method of locating genes to chromosomes is through the use of monosomics ($2n - 1$) and nullisomics ($2n - $ one pair) (Sears 1944, for wheat and Clausen and Cameron 1944, for *Nicotiana*). Often genes can be located to chromosome by utilizing only F_1 data.

Deficiencies

By using irradiated pollen from plants homozygous for the dominant alleles on plants homozygous for the recessives, any deficiency for a dominant allele would allow the recessive character to be expressed in the plant grown from that seed. In species having analyzable pachytene chromosomes, this deficiency may result in a buckle (loop) in the normal chromosome at or near the deficient region. The position of the buckle may vary. Several genes in maize were located in the physical chromosome by this method (McClintock 1931a). Mackensen (Paper 47), using a method similar in principle in *Drosophila*, analyzed the salivary-gland chromosomes of individuals heterozygous for deficiencies of known genes. The position of the buckle relative to the banding pattern served to cytologically place the gene. When two or more linked genes were located in this manner, it was usually possible to determine the way the linkage map is oriented in relation to the physical chromosome.

Translocations

If the pachytene chromosomes can be analyzed, the translocation breakpoints can be placed in the physical chromosome through direct cytological examination of the heterozygotes. If those heterozygotes are partially sterile, tests for linkage between the sterility and genetic markers can be used to place the breakpoints in the linkage map. Doing this for more than one translocation involving a particular chromosome makes it possible to orient the linkage group in the chromosome and also to obtain information on the positions of the genes. In a species such as *Datura* with directed segregation in the chromosome rings, such linkage tests are possible only for those translocations with interstitial segments long enough to permit sufficient crossing-over with resultant sterility to identify the translocation heterozygotes.

Progeny of a translocation heterozygote occasionally carry an extra chromosome, as the result of disjunction of three of the chromosomes in the ring to one pole and one to the other pole. McClintock (1931b) used such a derived 21-chromosome (tertiary trisomic) maize plant that carried one interchange chromosome in addition to the normal set in order to orient the genes in linkage group 9, the first one for which this was accomplished. Those genes in linkage group 9 carried by the interchanged chromosome showed abnormal (trisomic) ratios. This fact coupled with the known order of the genes in the linkage map and the known position of the interchange breakpoint in the chromosome allowed her to orient the linkage group in the physical chromosome.

To determine which chromosomes were translocated in *Drosophila*, males heterozygous for the second chromosome dominant gene Bristle (*Bl*) and the third chromosome dominant gene Dichaete (*D*) were X-rayed and crossed to untreated females with an attached-X chromosome homozygous for recessive yellow on the X and eyeless-2 on chromosome 4. By selecting the *Bl D* males and crossing them to homozygous eyeless females, all four chromosomes were marked (sex, *Bl, D* and ey^2). Since there is complete linkage in the male, by testing single males, a translocation between any two chromosomes in an individual culture would show complete linkage between the markers on those two chromosomes (Dobzhansky 1929).

Analysis of the salivary-gland chromosomes in translocation heterozygotes to determine the position of the cross-shaped configuration placed the breakpoint position in the chromosomes and also with respect to known genes that had been cytologically located.

In maize the use of interchanges between the B and normal chromosomes for locating genes in the chromosomes was described earlier (Roman, Paper 40). See Beckett (1975) for a list of those interchanges.

Also in maize a series of interchanges between one chromosome with a closely linked genetic marker and the other chromosomes has been used to map genes (Anderson 1943). Another extensive series of interchanges marks all or most of the chromosome arms with two breakpoints (Burnham 1966, 1968).

In *Neurospora* ($n = 7$) a stock with three interchanges, three circles of four in the heterozygote, each circle with a closely linked mutant marker, is being used in locating new mutants to chromosome (Perkins, et al. 1969).

New Mapping Techniques

Several advances have been made recently in genetical and cytological mapping. The technique of DNA/ribosomal RNA hybridization on filters has been used to localize DNA complementary to rRNA to the nucleolus organizer region (Ritossa and Spiegelman, Paper 48, for animals; and Phillips, Kleese, and Wang 1971 for plants). The degree of redundancy in genes coding for rRNA apparently renders quite difficult the mapping of these genes by conventional means. A mutation in a single copy of the ribosomal gene complex is not likely to be expressed so as to allow conventional mapping (unless the mutation occurs in a supposed master gene). By utilizing stocks that differed in rRNA gene dosages, Ritossa and Spiegelman nicely applied the DNA/rRNA hybridization technique to mapping.

Subsequently Gall and Pardue (Paper 49) developed the technique of *in situ* DNA/rRNA hybridization in which chromosomes of a standard cytological preparation are directly hybridized (after denaturation) with labelled rRNA, or other nucleic acids such as satellite DNA (Pardue and Gall 1970), and the hybridization visualized by autoradiography. The salient feature of this technique as opposed to the filter technique is that special stocks such as duplications or special mutants are not necessary. The cytological location of the DNA complementary to the hybridizing nucleic acid can be observed directly on normal chromosomes; that is, genes can be mapped without alternative forms or special cytogenetic stocks being available.

Electron microscopy of DNA has contributed greatly to mapping in prokaryotes. Inman (1966) first recognized A-T rich sequences by coupling electron microscopy and thermal denaturation; A-T rich regions are the first to denature thus separating the double helix into its constituent polynucleotide strands. Following this method the technique of heteroduplex mapping was developed for the cytological placement and measurement of various aberrations in the DNA such as deletions and insertions (Davis and Davidson 1968; Westmoreland, Szybalski, and Ris 1969).

The methods of mapping genes in man also have undergone significant expansion in recent years. Weiss and Green (1967) demonstrated the significance of human-mouse cell hybridization for gene mapping in humans. Hybrid cell lines isolated in a selective medium gradually lose human chromosomes. Weiss and Green studied a gene, thymidine kinase, furnished by the human cells and necessary for survival on the selective medium. The chromosome that carried the gene could be identified by determining the human chromosome that allowed the hybrid cells to survive on the selective medium. Thus the gene could be mapped to its physical chromosome. The use of somatic cell genetics in mapping human genes has been rapidly expanding since this discovery. Methods are now being developed for mapping the gene within its linkage group by hybridizing specialized cell lines that carry certain chromosomal aberrations such as a translocation.

Editors' Comments
on Papers 45 Through 49

Paper 45 was the first report of sex linkage that agreed with previous cytological studies that had shown the male to be heterogametic. From results of crosses starting with a single white-eyed mutant male that appeared in pedigree culture, Morgan hypothesized that the male was heterozygous for sex but carried the eye color with the sex factor and that all the eggs of the female carried the eye-color factor. This is followed by the results of four experiments that demonstrated that the hypothesis was correct.

The excerpts (Paper 46) from the original paper on nondisjunction that appeared in two separate issues in the new journal *Genetics* furnish only a small sample of Bridges' experiments that demonstrated the "identity of distribution between specific genes and a specific chromosome." The original article includes 86 tables of data from the various experiments. The relationship between genetic segregation and chromosome distribution, the very basis of cytogenetics, was emphatically illustrated in this paper.

Paper 47 gives numerous examples of cytological analysis of salivary-gland chromosomes in *Drosophila* larvae heterozygous for deletions and other chromosome changes from crosses of X-rayed males to females with genetic markers in the X-chromosome. The breadth of the application of the salivary-gland chromosome technique illustrated in this paper makes it benchmark.

In this paper, Ritossa and Spiegelman (Paper 48) describe the first experiment in which DNA extracted from special cytological stocks of *Drosophila* was used for molecular hybridization with labeled ribosomal RNA. The authors were able to demonstrate that the nucleolus organizer in the X and the Y-chromosome is the site of the DNA complementary to rRNA.

Gall and Pardue (Paper 49) describe the first experiment in which molecular hybridization was accomplished between nucleic acid in solution and chromosomes of intact conventional smear preparations. The authors illustrate the technique by demonstrating the cytological localization of rDNA, and they discuss the possibilities of localizing other types of DNA in the chromosomes of higher organisms. This technique has allowed the mapping of redundant gene systems and of DNA that likely does not produce a gene product. Both are difficult to map by conventional means since alternative forms of the genes may not be available. Thus the technique described in this paper opened the door to a new realm of cytogenetic investigations.

REFERENCES

Anderson, E. G. 1943. Utilization of translocations with endosperm markers in the study of economic traits. *Maize Genet. Coop. News Letter* **17**:4–5.

Beckett, J. B. 1975. Genetic breakpoints of the B-A translocations of maize. *Maize Genet. Coop. News Letter* **49**:130–134.

Burnham, C. R. 1966. Cytogenetics in plant improvement. Chapt. 4:139–187. *Plant Breeding Symposium*, Iowa State University Press, Ames, Iowa.

———. 1968. "All-arms" interchange tester set. *Maize Genet. Coop. News Letter* **42**:122–123.

Clausen, R. E., and D. R. Cameron. 1944. Inheritance in *Nicotiana tabacum*. XVIII. Monosomic analysis. *Genetics* **29**:447–477.

Davis, R. W., and N. Davidson, 1968. Electron-microscopic visualization of deletion mutations. *Proc. Nat. Acad. Sci. U.S.A.* **60**:243–250.

Dobzhansky, T. 1929. Genetical and cytological proof of translocations involving the third and fourth chromosomes of *Drosophila melanogaster*. *Biol. Zentralbl.* **49**:408–419.

Doncaster, L., and G. H. Raynor. 1906. On breeding experiments with Lepidoptera. *Proc. Zool. Soc. London* **1**:125–133.

Inman, R. B. 1966. A denaturation map of the λ phage DNA molecule determined by electron microscopy. *J. Mol. Biol.* **18**:464–476.

McClintock, B. 1931a. Cytological observations of deficiencies involving known genes, translocations and an inversion in *Zea mays. Missouri Agric. Exp. Stn. Res. Bull.* **163**:1–30.

———. 1931b. The order of the genes *C, Sh* and *Wx* in *Zea mays* with reference to a cytologically known point in the chromosome. *Proc. Nat. Acad. Sci. U.S.A.* **17**:485–491.

——— and H. E. Hill. 1931. The cytological identification of the chromosome associated with the *R-G* linkage group in *Zea mays. Genetics* **16**:175–190.

Pardue, M. L., and J. G. Gall. 1970. Chromosomal localization of mouse satellite DNA. *Science* **168**:1356–1358.

Perkins, D. D., D. Newmeyer, C. W. Taylor, and D. C. Bennett. 1969. New markers and map sequences in *Neurospora crassa*, with a description of mapping by duplication coverage, and of multiple translocation stocks for testing linkage. *Genetica* **40**:247–278.

Phillips, R. L., R. A. Kleese, and S. S. Wang. 1971. The nucleolus organizer region of maize (*Zea mays* L.): chromosomal site of DNA complementary to ribosomal RNA. *Chromosoma* **36**:79–88.

Rhoades, M. M. 1936. A cytogenetic study of a chromosome fragment in maize. *Genetics* **21**:491–502.

Sears, E. R. 1944. Cytogenetic studies with polyploid species of wheat. II. Additional chromosomal aberrations in *Triticum vulgare. Genetics* **29**:232–246.

Weiss, M. C., and H. Green. 1967. Human-mouse hybrid cell lines containing partial complements of human chromosomes and functioning human genes. *Proc. Nat. Acad. Sci. U.S.A.* **58**:1104–1111.

Westmoreland, B. C., W. Szybalski, and H. Ris. 1969. Mapping of deletions and substitutions in heteroduplex DNA molecules of bacteriophage lambda by electron microscopy. *Science* **163**:1343–1348.

SEX LIMITED INHERITANCE IN DROSOPHILA

T. H. Morgan

In a pedigree culture of *Drosophila* which had been running for nearly a year through a considerable number of generations, a male appeared with white eyes. The normal flies have brilliant red eyes.

The white-eyed male, bred to his red-eyed sisters, produced 1,237 red-eyed offspring, (F_1), and 3 white-eyed males. The occurrence of these three white-eyed males (F_1) (due evidently to further sporting) will, in the present communication, be ignored.

The F_1 hybrids, inbred, produced:

> 2,459 red-eyed females,
> 1,011 red-eyed males,
> 782 white-eyed males.

No white-eyed females appeared. The new character showed itself therefore to be sex limited in the sense that it was transmitted only to the grandsons. But that the character is not incompatible with femaleness is shown by the following experiment.

The white-eyed male (mutant) was later crossed with some of his daughters (F_1), and produced:

> 129 red-eyed females,
> 132 red-eyed males,
> 88 white-eyed females,
> 86 white-eyed males.

The results show that the new character, white eyes, can be carried over to the females by a suitable cross, and is in consequence in this sense not limited to one sex. It will be noted that the four classes of individuals occur in approximately equal numbers (25 per cent.).

An Hypothesis to Account for the Results. —The results just described can be accounted for by the following hypothesis. Assume that all of the spermatozoa of the white-eyed male carry the "factor" for white eyes "W"; that half of the spermatozoa carry a sex factor "X" the other half lack it, *i. e.*, the male is heterozygous for sex. Thus the symbol for the male is "WWX," and for his two kinds of spermatozoa WX—W.

Assume that all of the eggs of the red-eyed female carry the red-eyed "factor" R; and that all of the eggs (after reduction) carry one X, each, the symbol for the red-eyed female will be therefore RRXX and that for her eggs will be RX—RX.

When the white-eyed male (sport) is crossed with his red-eyed sisters, the following combinations result:

WX — W (male)

RX — RX (female)

RWXX (50 %) — RWX (50 %)

Red female   Red male

When these F_1 individuals are mated, the following table shows the expected combinations that result:

RX — WX (F_1 female)

RX — W (F_1 male)

RRXX—RWXX — RWX—WWX

(25 %)   (25 %)   (25 %)   (25 %)

Red   Red   Red   White

female   female   male   male

It will be seen from the last formulæ that the outcome is Mendelian in the sense that there are three reds to one white. But it is also apparent that all of the whites are confined to the male sex.

It will also be noted that there are two classes of red females—one pure RRXX and one hybrid RWXX—but only one class of red males (RWX). This point will be taken up later. In order to obtain these results it is necessary to assume, as in the last scheme, that, when the two classes of the spermatozoa are formed in the F_1 red male (RWX), R and X go together—otherwise the results will not follow (with the symbolism here used). This all-important point can not be fully discussed in this communication.

The hypothesis just utilized to explain these results first obtained can be tested in several ways.

Verification of Hypothesis

First Verification.—If the symbol for the white male is WWX, and for the white female WWXX, the germ cells will be WX—W (male) and WX—WX (female), respectively. Mated, these individuals should give

WX — W (male)

WX — WX (female)

WWXX (50 %) — WWX (50 %)

White female   White male

All of the offspring should be white, and male and female in equal numbers; this in fact is the case.

Second Verification.—As stated, there should be two classes of females in the F_2 generation, namely, RRXX and RWXX. This can be tested by pairing individual females with white males. In the one instance (RRXX) all the offspring should be red—

RX — RX (female)

WX — W (male)

RWXX — RWX

and in the other instance (RWXX) there should be four classes of individuals in equal numbers, thus:

RX — WX (female)

WX — W (male)

RWXX—WWXX — RWX—WWX

Tests of the F_2 red females show in fact that these two classes exist.

Third Verification.—The red F_1 females should all be RWXX, and should give with any white male the four combinations last described. Such in fact is found to be the case.

Fourth Verification.—The red F_1 males (RWX) should also be heterozygous. Crossed with white females (WWXX) all the female offspring should be red-eyed, and all the male offspring white-eyed, thus:

RX — W (red male)

WX — WX (white female)

RWXX — WWX

Here again the anticipation was verified, for all of the females were red-eyed and all of the males were white-eyed.

Crossing the New Type with Wild Males and Females

A most surprising fact appeared when a white-eyed female was paired to a wild, red-eyed male, *i. e.*, to an individual of an unrelated stock. The anticipation was that wild males and females alike carry the factor for red eyes, but the experiments showed that all wild males are heterozygous for red eyes, and that all the wild females are homozygous. Thus when the white-eyed female is crossed with a wild red-eyed male, all of the female offspring are red-eyed, and all of the male offspring white-eyed. The results can be accounted for on the assumption that the wild male is RWX. Thus:

RX — W (red male)
WX — WX (white female)
_____
RWXX (50 %) — WWX (50 %)

The converse cross between a white-eyed male RWX and a wild, red-eyed female shows that the wild female is homozygous both for X and for red eyes. Thus:

WX — W (white male)
RX — RX (red female)
_____
RWXX (50 %) — RWX (50 %)

The results give, in fact, only red males and females in equal numbers.

General Conclusions

The most important consideration from these results is that in every point they furnish the converse evidence from that given by Abraxas as worked out by Punnett and Raynor. The two cases supplement each other in every way, and it is significant to note in this connection that in nature only females of the sport *Abraxas lacticolor* occur, while in *Drosophila* I have obtained only the male sport. Significant, too, is the fact that analysis of the result shows that the wild female *Abraxas grossulariata* is heterozygous for color and sex, while in *Drosophila* it is the male that is heterozygous for these two characters.

Since the wild males (RWX) are heterozygous for red eyes, and the female (RXRX) homozygous, it seems probable that the sport arose from a change in a single egg of such a sort that instead of being RX (after reduction) the red factor dropped out, so that RX became WX or simply OX. If this view is correct it follows that the mutation took place in the egg of a female from which a male was produced by combination with the sperm carrying no X, no R (or W in our formulæ).

In other words, if the formula for the eggs of the normal female is RX—RX, then the formula for the particular egg that sported will be WX; *i. e.,* one R dropped out of the egg leaving it WX (or no R and one X), which may be written OX. This egg we assume was fertilized by a male-producing sperm. The formula for the two classes of spermatozoa is RX—O. The latter, O, is the male-producing sperm, which combining with the egg OX (see above) gives OOX (or WWX), which is the formula for the white-eyed male mutant.

The transfer of the new character (white eyes) to the female (by crossing a white-eyed male, OOX to a heterozygous female (F_1)) can therefore be expressed as follows:

OX — O (white male)
RX — OX (F_1 female)
_____
RXOX—RXO — OOXX—OOX
Red   Red   White   White
female male   female   male

It now becomes evident why we found it necessary to assume a coupling of R and X in one of the spermatozoa of the red-eyed F_1 hybrid (RXO). The fact is that this R and X are combined, and have never existed apart.

It has been assumed that the white-eyed mutant arose by a male-producing sperm (O) fertilizing an egg (OX) that had mutated. It may be asked what would have been the result if a female-producing sperm (RX) had fertilized this egg (OX)? Evidently a heterozygous female RXOX would arise, which, fertilized later by any normal male (RX—O) would produce in the next generation pure red females RRXX, red heterozygous females RXOX, red males RXO, and white males OOX (25 per cent.). As yet I have found no evidence that white-eyed sports occur in such numbers. Selective fertilization may be involved in the answer to this question.

46

Reprinted from pp. 1–4, 9–14, 107–109, 111, 115–116, 134–136, 159–160, 161, 162 of *Genetics* **1**(1–2): 1–52, 107–163 (1916)

NON-DISJUNCTION AS PROOF OF THE CHROMOSOME THEORY OF HEREDITY[1]

CALVIN B. BRIDGES

Columbia University, New York City

[Received October 21, 1915]

I

INTRODUCTION

There has been a long series of observations and experiments which has led more and more definitely to the conclusion that the chromosomes are the bearers of the hereditary materials.[2] It was observed that male and female contribute equally to the inheritance of the offspring, and yet the contribution of the male consists of little more than a nucleus. That

---

[1] Contribution from the Zoological Laboratory of Columbia University.

[2] For a fuller discussion of the steps in the accumulation of this evidence see "The mechanism of Mendelian heredity" by MORGAN, STURTEVANT, MULLER, and BRIDGES.

inheritance is a function of the nucleus rather than of the cytoplasm is shown by many embryological and cytological facts. Attention was next narrowed to the chromatin and chromosomes. Embryological experiments showed that the chromosomes are qualitatively different and that a full complement is essential to normal development. The increasing cytological knowledge of mitotic division and of gametogenesis made it clear that the chromosomes were qualified to serve as the material basis of heredity.

The next advance was the result of the exact knowledge of heredity which Mendelian analysis furnished, and of the coupling of experimental genetics with cytological investigation. It was shown that the genes for characters and the chromosomes have the same *method* of distribution. More recently cases have arisen in which genes and chromosomes have the *same* distribution. The final step has been to demonstrate the *identity of distribution between specific genes and specific chromosomes* in such a way that the argument of "the-cell-as-a-whole" cannot be applied, and in such a way that the chromosomes must be regarded as the means and not the consequence of the inheritance of characters and of sex. The experimental and cytological evidence in the case of non-disjunction furnishes such a demonstration.

An account of the discovery of non-disjunction and of its effect upon sex-linked inheritance was published (BRIDGES 1913 b) in the JOURNAL OF EXPERIMENTAL ZOOLOGY for November 1913. A further account in the form of a summary was given in SCIENCE, July 17, 1914 (BRIDGES 1914).

The work on non-disjunction started from exceptions in certain experiments which I was carrying out with Prof. T. H. MORGAN (MORGAN and BRIDGES 1913). The work was continued while I was assisting Dr. MORGAN, and my most sincere thanks are due for the opportunity and the encouragement which he offered. The frequent consultations and the constant association with Dr. A. H. STURTEVANT, Dr. H. J. MULLER, and other workers in the laboratory have brought out possibilities that would otherwise have been overlooked.

A brief statement concerning sex determination, normal sex-linked inheritance, etc., which form the background of these experiments may first be made.

THE SEX CHROMOSOMES AND SEX

The female of *Drosophila ampelophila* has a pair of sex chromosomes (X chromosomes) and three pairs of autosomes. The medium sized straight chromosomes shown by the diagram to the left in figure 1 are identified as the pair of X chromosomes.

FEMALE   **MALE**

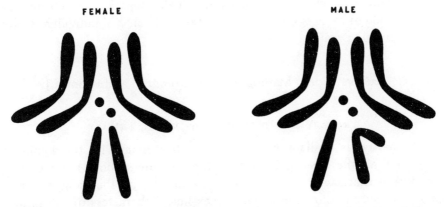

FIGURE 1.—Diagram showing the characteristic pairing, size relations, and shapes of the chromosomes of *Drosophila ampelophila*. In the male an X and a Y chromosome correspond to the X pair of the female. The relative lengths of the chromosomes of this diagram are based upon averages of the measurements of the figures in plate 1; on the basis of X = 100 the length of each long autosome is 159, of each small autosome 12, of the whole Y 112, of the long arm of the Y 71, and of the short arm 41.

The chromosomes of the male have presented serious difficulties in spite of the very great amount of study devoted to them. Miss STEVENS (1908 a) made a prolonged study involving the dissection and examination of some two thousand individuals. She described the three pairs of autosomes of the male as the same as those in the female, and the sex chromosomes as *"a clearly unequal pair"*.[3] The cytological work which I have done on males and on females having an extra Y shows that X is somewhat shorter than Y, that it is typically straight, and attached to the spindle fiber by its end. On the other hand, the Y chromosome (see figure 1 to the right) is somewhat longer than the X, is attached not by its end, but sub-terminally, and typically has the shape of a J or of a V with one of the arms shorter than the other. Drosophila is therefore a member of the group of forms in which the male produces two kinds of sperm, half with an X and half with a Y chromosome. All

[3] Miss STEVENS supposed the longer of these two chromosomes to be the chromosome present in duplex in the female, and that a distinct "X" piece constituted the middle of this long heterochromosome, so that at the time at which the first paper on non-disjunction was written (BRIDGES 1913 b), it had become the general belief that the constitution of the male was XO and of the female XX. The X chromosome was supposed to be attached to an autosome, as in Ascaris.

the eggs of the female are alike, each carrying an X after the polar bodies have been given off. The fertilization of an X egg by an X sperm results in an XX individual which develops into a female; the fertilization of an X egg by a Y sperm results in an XY individual which develops into a male.

[*Editors' Note:* Material has been omitted at this point.]

PRIMARY NON-DISJUNCTION IN THE FEMALE

Ordinarily, as in diagram 2, in a cross to a male with the dominant character all the sons and none of the daughters show the recessive sex-linked characters of the mother. Similarly, all the daughters and none of the sons show the dominant sex-linked characters of the father. The peculiarity of non-disjunction is that sometimes a female transcends these rules and produces a daughter like herself (a matroclinous daughter) or a son like the father (a patroclinous son), while the rest of the offspring are perfectly regular, showing the expected criss-cross inheritance. Such exceptions, produced by a normal XX female, may be called primary.

The production of primary exceptions by a normal XX female may be supposed to result from an aberrant reduction division at which *the two X chromosomes fail to disjoin from each other. In consequence both remain in the egg or both pass out into the polar body.* In the former case the egg will be left with two X chromosomes and in the latter case with no X.

If the genes for sex-linked characters are carried by the X chromosomes, then each of the X chromosomes of the XX egg of a vermilion female will carry the gene for vermilion. The fertilization of such XX and zero eggs by the X and by the Y spermatozoa of a wild male will result in four new types of zygotes, as shown in figure 5.

(1) The XX egg fertilized by the X sperm gives an XXX zygote which might be expected to develop into a female. No females of this class have been found, and it is certain that they die.

(2) The fertilization of the XX egg by the Y sperm gives rise to a female having an extra Y chromosome (XXY). Since both of the X chromosomes came from the vermilion-eyed mother, this daughter must be a vermilion *matroclinous exception.*

(3) The fertilization of the zero egg by the X sperm gives rise to a male which has no Y chromosome (XO), and whose X coming from his red-eyed father brings in the red gene which makes the son a *patroclinous exception.* These XO males are viable but are completely sterile.

(4) The zero egg by the Y sperm gives a zygote (OY) which is not viable.

Perhaps the cause of the initial aberrant reduction which constitutes primary non-disjunction is a mechanical entanglement (an incomplete untwisting from a strepsinema stage) of the two X chromosomes, resulting in a delayed reduction. In such cases the formation of the cell boundaries would catch the lagging X's and include them in one or the other cell, and perhaps very often (as in certain nematodes) would pre-

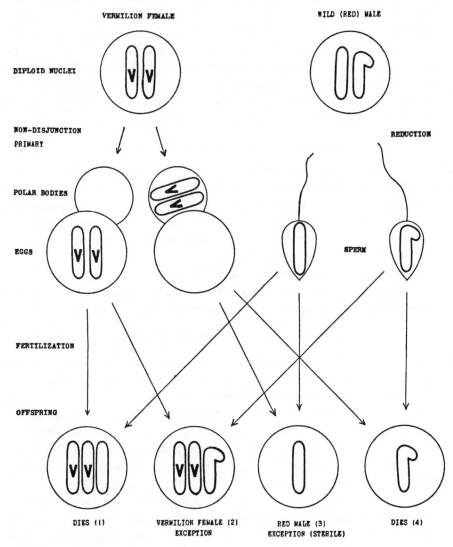

FIGURE 5.—Diagram of the production by a vermilion-eyed female of XX and zero eggs through *primary* non-disjunction, and the progeny resulting from the fertilization of such eggs by the sperm of a red-eyed male.

vent their leaving the middle of the spindle to join either daughter nucleus. If such an occurrence were common there should be more zero than XX eggs and consequently more primary exceptions should be males than females.

In studying primary non-disjunction we are dependent on what material chance offers, since we know of no means of controlling the process. It is equally as likely that an egg produced by primary non-disjunction will become a non-viable zygote (XXX and OY) as that it shall be viable (XXY and XO). For this reason it is impossible to detect half of those cases of primary non-disjunction which really occur. The XO male is viable and should offer an interesting field for further work, but—he is sterile. The direct opening offered for further work is through the matroclinous XXY daughter, which is perfectly fertile and which produces further exceptions which we may call secondary.

PRIMARY NON-DISJUNCTION IN THE MALE

If primary non-disjunction occurred in the male, XY and zero sperm would be formed, but the zygotes from them would not differ in their sex-linked characters from regular offspring, so that such an occurrence could not be detected immediately. However, the XY sperm would give rise to XXY daughters, and these in turn would produce secondary exceptions which could be observed.

Primary non-disjunction has been actually seen to occur in the male of Metapodius. WILSON found three spermatocytes in which X followed Y to one pole at the reduction division (WILSON 1909).

SECONDARY NON-DISJUNCTION IN THE FEMALE

It has been shown that matroclinous daughters of the constitution XXY may arise as the result of primary non-disjunction. The results from the outcrossing of several matroclinous daughters to males having other sex-linked characters were given in 1913. Of unusual interest was the appearance in F_1 of about four[5] percent of further exceptions (secondary exceptions). That is, about four percent of the daughters were like the mother and four percent of the sons were like the father. The remaining sons and daughters were of the kinds expected.

The explanation given at first for the fact that exceptional daughters inherit from their mother the power of producing exceptions, was that each X of the exceptional female carried a gene which caused these

---

[5] In previous papers the percentage of exceptions has been given roughly as five. The mean of all data now on hand is 4.3 percent of exceptions (see page 16).

chromosomes to undergo reduction abnormally in a small percentage of cases. Since these chromosomes descend directly to their exceptional daughters, they would transmit to those daughters the same gene and consequently the same power of producing exceptions.

Later work has provided data which can not be explained by appealing to the action of a gene in the X chromosome, and which prove that these secondary exceptions are due to the presence of the extra Y. In an XXY female there are three homologous sex chromosomes, between any two of which synapsis may occur, that is, synapsis may be of the XX or the XY type (homo- and heterosynapsis). In only about sixteen percent of cases (see page 17) does heterosynapsis occur, while about 84 percent of cases are homosynaptic and the Y remains unsynapsed. At the reduction division the two chromosomes that have synapsed, disjoin, one going to each pole, and the free chromosome goes to one pole as often with the one as with the other of the disjoined chromosomes. Thus, after heterosynapsis the reduction divisions are of two kinds, the XX-Y and the X-XY types. Half the eggs that come from the XX-Y type of reduction are XX and the other half are Y. For the X-XY type the eggs are·X and XY, as many of one kind as of the other. After homosynapsis all the reductions are X-XY. As a result of reduction of these two types there are four classes of eggs—two of which, X and XY, are composite and large (46 percent), and two of which, XX and Y are of single origin and small (4 percent). If these eggs are produced by a vermilion-eyed female, both of whose X chromosomes carry the recessive gene for vermilion, then the eight classes of zygotes shown in figure 6 will result upon fertilization by a wild male, which produces X and Y sperm.

The XX eggs fertilized by X sperm give XXX individuals (figure 6, 1) which are unable to live.

The XX eggs fertilized by the Y sperm give individuals (5) which are exact duplicates of their mother in their sex chromosomes, and like her are females each containing an extra Y chromosome. Since the gene for vermilion is carried by the X chromosome, these females have vermilion eyes and hence are matroclinous exceptions. Since they have not received an X from their father, they can neither show nor transmit his sex-linked characters. If in the mother the presence of the extra Y led to the production of secondary exceptions, then these XXY daughters should also give exceptions, and this is in fact the case.

The Y eggs fertilized by the X sperm give males (2). These males have received their X from their father and they show his sex-linked

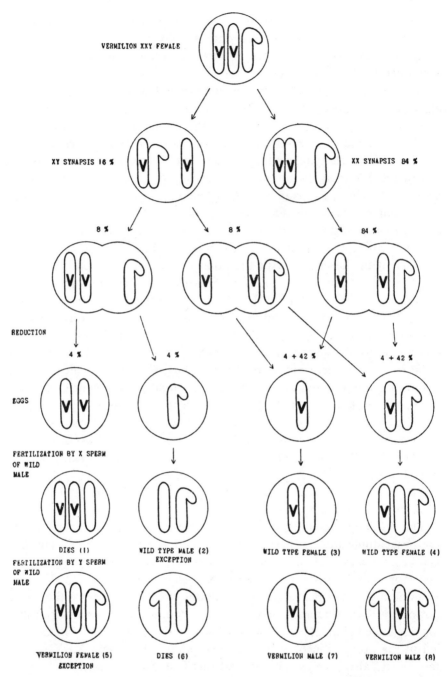

VERMILION XXY FEMALE

XY SYNAPSIS 16 %

XX SYNAPSIS 84 %

8 %

8 %

84 %

REDUCTION

4 %

4 %

4 + 42 %

4 + 42 %

EGGS

FERTILIZATION BY X SPERM
OF WILD
MALE

DIES (1)

WILD TYPE MALE (2)
EXCEPTION

WILD TYPE FEMALE (3)

WILD TYPE FEMALE (4)

FERTILIZATION BY Y SPERM
OF WILD
MALE

VERMILION FEMALE (5)
EXCEPTION

DIES (6)

VERMILION MALE (7)

VERMILION MALE (8)

FIGURE 6.—Secondary non-disjunction in the female. Diagram showing the constitution of an exceptional vermilion female, the two types of synapsis, reduction, and the four classes of eggs produced. Each kind of egg may be fertilized by either of the two (X and Y) kinds of sperm of the wild male, giving the eight classes of zygotes shown.

characters, that is, they are patroclinous exceptions. Since in chromosome constitution (XY) these males are not different from ordinary males, they should have no power of producing exceptions. This has been shown to be the case.

The Y eggs fertilized by Y sperm give YY individuals (6) which are unable to live.

The X eggs by X sperm give regular XX females (3), and by Y sperm give regular XY males (7). Neither of these two classes is able to produce secondary exceptions or to transmit non-disjunction.

The XY eggs by X sperm give XXY females (4) *which, because of the extra Y, possess the power of producing secondary exceptions, though they themselves are not exceptions.*

The XY eggs by Y sperm give XYY males (8). These males do not give rise to genetic exceptions in F_1, but they endow some of their daughters with an extra Y (XY sperm and X egg) which enables these daughters to produce secondary exceptions.

By breeding in each generation the exceptional daughters with the exceptional sons a line was maintained in which the entire set of sex-linked genes of the mother was handed down to the exceptional daughters and the entire set of the father to the exceptional sons.

That the XX eggs which developed into matroclinous exceptions had really been fertilized by normal sperm of the father was proved by the introduction into such daughters of autosomal genes from the father. The inheritance was uniparental with respect to the sex-linked genes, but biparental and wholly normal with respect to the autosomal genes. The fact that exceptional offspring inherit sex-linked characters from only one parent, but at the same time inherit the autosomal characters from both parents is explained if the sex chromosomes are the only chromosomes which have undergone non-disjunction, the ordinary chromosomes disjoining normally.

[*Editors' Note:* Material has been omitted at this point.]

NON-DISJUNCTION AS PROOF OF THE CHROMOSOME
THEORY OF HEREDITY (concluded)

CYTOLOGICAL PROOF OF THE OCCURRENCE OF XXY FEMALES

The breeding work presented in the last two sections has furnished data which show that the cause of secondary non-disjunction cannot be a gene carried by the X chromosome, while the same data are consistent with the assumption that a female which produces secondary exceptions does so because of the presence of an extra Y chromosome. Likewise the data from the tests of the constitution of the regular daughters, the regular sons, the exceptional daughters, and the exceptional sons, all lead to this same conclusion.

Accordingly, the prediction was made that cytological examination of the daughters of an exceptional female would demonstrate the presence of an extra chromosome in half of the daughters while the other half would show only normal figures. This prediction has been completely verified.

The ovaries at the mid-pupal stage of development offer the best material for examination. Nearly a hundred pairs of ovaries were dissected from pupae of cultures the mother of which was an exceptional female. Eighteen of these gave oögonial metaphases which were clear enough to give a reliable count of the chromosomes.

Nine of the females showed figures like those published by Miss STEVENS (1908 a) in which the X chromosomes are a pair of straight rods. (See plate 1, figures 1-3, as well as the generalized group in figure 1, p. 3 of the first instalment).

The other nine females showed these two X chromosomes and in addition a chromosome which differed from both the X's in that it had the shape of a V with one arm shorter than the other. This chromosome was identified as the Y from the following considerations.

No figures which showed this extra chromosome were found by METZ (1914) when he examined the chromosomes of several wild stocks of *Drosophila ampelophila*. From the work of STEVENS (1908 a), METZ (1914) and myself (plate 1, figs. 1-3) there can be no doubt that the normal condition of the female is that shown in figure 1 (p. 3, first instalment.) The new type of figure which I found differs from the normal only by the addition of this chromosome. Fortunately, there are several good figures in each of four or five of the XXY females, and all the figures in any one female show the same condition.

The figures given by STEVENS, METZ, and myself, show that homologous chromosomes usually lie together as actual pairs. In the figures showing the extra chromosome, this chromosome is usually found in company with the two straight chromosomes so that it behaves as a homologue to them.

Very recently I have found several excellent spermatogonial figures in the testes of larval males; these show beyond question that the identification of the Y has been correct, for the Y has in these males the same characteristics as the supernumerary chromosome of the XXY females.

Breeding tests with sex-linked characters have shown that half the regular daughters produce exceptions to the inheritance of sex and sex-linked characters; parallel with this is the fact that half of the regular daughters possess this extra Y chromosome. Normal females do not possess this chromosome and do not produce exceptions, so that the exceptions must be produced by the daughters with the extra chromosome.

Recently over forty freshly hatched females which were first classi-

[*Editors' Note:* Plate I is not reproduced here.]

fied as exceptions were dissected and a cytological examination made of their chromosomes. In over a dozen of these individuals sufficiently clear figures were found to be sure of the number and character of the chromosomes. *In every case the exceptional female was found to be XXY.* This direct examination of the exceptions gives entirely conclusive proof that the cause of the production of secondary exceptions is the presence of the Y.

Miss STEVENS's (1908 a) work upon the male showed a pair of unequal chromosomes in place of the pair of equal straight rods of the female The longer of these two chromosomes seems to have the shape of a J in some of STEVENS's figures. The perpetuation of this longer chromosome in the male line can only be explained if these chromosomes have a causal connection with the differentiation of sex.

[*Editors' Note:* Material has been omitted at this point.]

THE XX EGGS OF XXY FEMALES

Thus far in our analysis, the XX eggs of an XXY female have been considered mainly in their relation to the production of matroclinous daughters, but the method of origin of these eggs is interesting, if, as so far assumed, they are preceded by synapsis between an X chromosome and a supernumerary Y. In an ordinary XX female, synapsis takes place between X and X, and in an XXY female synapsis must be supposed to follow this same female type in 83.5 percent of cases. During or after synapsis of X with X, crossing over would have an opportunity to occur, and does occur, as will be shown in another section. While the frequency of this crossing over between the two synapsed X's of an XXY female is a little higher than normal the process must be essentially the same as that in an XX female.

But if synapsis in the XXY female should take the course that it follows in the male, where X has Y for its mate, *there would now be no chance of crossing over between the two X's, for one of them is in synapsis with another chromosome, the Y*. At the reduction division, Y and the X which conjugated with it, would pass to opposite poles, and the free X would go either with the Y or with the disjoined X. From those cases where the free X went with the disjoined X, XX eggs and Y eggs would result. The XX eggs should therefore always be *non-crossovers*, and this has been shown to be true.

It was noticed that when an XXY female carried eosin in one of her X chromosomes and vermilion in the other, the exceptional daughters were always like the mother in that they still carried eosin in one X and vermilion in the other. Linkage experiments have shown that about a third of the X eggs of an ordinary XX female are crossovers, that is, they carry both eosin and vermilion in the single X or conversely they carry neither. An experiment was made to find what percentage of XX eggs of an XXY female are crossovers, and the conclusion was reached that none of the XX eggs are crossovers.

[*Editors' Note:* Material has been omitted at this point.]

SYNAPSIS IN XXY FEMALES

Females of the constitution XXY have been produced in the following three ways, namely, XX egg by Y sperm, XY egg by X sperm, and X egg by XY sperm. Irrespective of their origin, these XXY females have given the same percent of secondary exceptions. This means that the method of synapsis is not influenced by the origin of the chromosomes. Two chromosomes from the *same* parent synapse with each other as readily as though they were from different parents. Thus when both X's come from the mother, they synapse with each other in about 84 percent of cases, and crossing over takes place in about the normal amount. This is of interest in connection with the case of the tetraploid Primula, for as MULLER (1914) has shown, the data of GREGORY (1914) are in accord with the view that the four homologous chromosomes pair with each other, two by two, irrespective of whether they are from the same or from opposite parents.

That synapsis in an XXY female does not involve all three chromosomes at once, but is between two of them with the third chromosome left unsynapsed, is proved by the fact that the chromosomes of XX eggs are never crossovers, while the X's of the X and XY eggs are crossovers in about the usual percent. A difference between the paths followed by these two kinds of chromosomes originated before the stage at which crossing over became possible. If all three chromosomes synapsed together, there should not be this difference, for the X's which enter the XX egg should be crossovers in the same percentage as those which enter the X and XY eggs. Furthermore, it is difficult to see how two X's in synapsis with each other and at the same time with Y can cross over with each other without involving crossing over between the Y and one of the X's. There is no evidence that there is crossing over between X and Y in an XXY female. This leads us to suppose that in the male the lack of crossing over between X and Y is a property of Y and not entirely due to the general fact that there is no crossing over in the male of Drosophila between any chromosomes.

[*Editors' Note:* Material has been omitted at this point.]

There have been eighteen instances of equational exceptions. In thirteen of these cases, crossing over is known to have occurred, and this is *in every case where evidence upon this point was obtainable.* Again, of these thirteen cases which involved crossing over, *twelve were cross-overs in only one chromosome and were non-crossovers in the other.* The thirteenth case was a double crossover in one chromosome and at least a single crossover in the other. Although fifteen of the instances occurred in XXY females, three occurred in females known to be simply XX, and it is thus evident that equational non-disjunction is not caused by the presence of a Y. The descendants of equational exceptions seem to have a greater tendency to produce further equational exceptions, though no basis for this tendency has been found.

As previously explained, all these exceptions are easily accounted for in the following way: XX synapsis took place; each X split so that a four strand stage occurred; crossing over took place between two only of these strands, one from each X; the reduction division separated the paternal X from the maternal X, each cell receiving a non-crossover and at the same time a crossover strand; at the next division these two strands ordinarily enter different cells, but by an occasional non-disjunction these two strands do not separate from each other at the equational division and consequently enter the same nucleus. In the case of an XX female the presence of the Y might favor the process by entering the other cell so that one cell receives two X chromosomes and the other two Y's. Equational non-disjunction thus enables us to examine at leisure the products of reduction.

It is impossible that a non-crossover and a crossover chromosome come from a cell in which only two strands are present. Therefore the proof that these exceptions arise from XX (or XXY) oöcytes would at the same time prove that crossing over took place in this manner at a four strand stage.

If, however, the oöcyte contained three X chromosomes, synapsis and crossing over might occur between two of them and the equational split appear after the crossing over had been accomplished. At the reduction division the two synapsed X's (which would also be crossovers in a certain percent of cases) would disjoin and pass to separate poles, while the free X, which would always be a non-crossover, would go with one or the other according to chance. In those cases in which crossing over has occurred and a crossover and the free non-crossover X have remained in the egg after reduction, the equational division would make no quantitative change. These eggs fertilized by Y sperm would give exceptions having all the characteristics of those just described.

But none of these equational exceptions came from XXX mothers; for it is quite certain that XXX females do not live. It is possible, however, that a small group of XXX cells in an XX female might live if such cells were produced. Let us suppose that in a germ-tract division the two daughter chromosomes from one X were late in separating from each other, while the other X divided normally. Then both these X's might become caught in one cell as in other cases of primary non-disjunction. One cell from such a division would be X and the other X X' X' in composition, having two maternal (or paternal) X's and one of the contrary type. This X X' X' cell might give rise to a group of such cells in the germ-tract and these might cause the production of several equational exceptions in a single culture, as in culture 1217, table 77.

The twelve cases in which one of the X's was a non-crossover and the other a crossover chromosome are equally well explained on either the chiasmatype or the X X' X' view. The thirteenth case in which one X was a double crossover and the other X probably also a double crossover could be explained on either view as a case of primary non-disjunction at the second maturation division.

Yet if an equational exception should arise such that one of its X's were a double crossover and the other a single crossover between two of the same genes involved in the double crossover, then the X X' X' view could not be applied, for one strand cannot cross over at the same point with each of two other strands. Of the two exceptions which might have answered this question, one did not give enough offspring to settle its nature and the other could have arisen by either method. The evidence in hand thus allows us no conclusion as to whether crossing over takes place at a two or four strand stage; but equational non-disjunction offers the possibility of answering this question definitely.[9]

SOMATIC NON-DISJUNCTION AND GYNANDROMORPHISM

If the same sort of primary non-disjunction which has been assumed to give rise to X X′ X′ cells in an XX female, should take place at a cleavage stage, gynandromorphs and mosaics would result. One might expect at an early cleavage division, particularly the first, a relatively large number of X—X X′ X′ divisions, for the greatly condensed chromosomes introduced by the sperm do not for some time attain the state or the appearance of those of the egg nucleus. If the paternal X of a female were slower than the maternal in preparing for division, it might lag upon the spindle so that both daughter X's would become included in the same cell. The portion of the fly which comes from the X cell should be male and should show the sex-linked characters of the mother. Such a process may be the explanation of the large number of lateral gynandromorphs of Drosophila. When the X—X X′ X′ division occurs at a later cleavage stage we may have mosaics, as for example, a red female with a patch of white facets in the eye.

HIGH NON-DISJUNCTION

The mean of the percentages of exceptions given by all the XXY cultures is 4.3 percent and the mode is at about 2.3 percent. Occasionally rather high percentages occur. For example, one of three regular white-eosin daughters from culture 800 gave nearly 14 percent of exceptions, while a sister gave 8 percent (table 73).

[9] The conclusive evidence that crossing over does take place at a four strand stage according to the chiasmatype hypothesis came Dec. 16, 1915, with the discovery of an equational female one of whose X chromosomes had undergone double crossing over while the other had undergone single crossing over at the same level (between the same two genes) at which the first crossover of the double had occurred. The mother of this exception was an XXY female of the constitution $\dfrac{v \quad f}{sg}$; the two recessive characters vermilion and forked were in one X, and the two recessive characters, sable and garnet, were in the other (for garnet see p. 151). The equational exception was of the constitution $\dfrac{v \quad sg}{}$; the chromosomes represented by the space below the line (unmutated genes only) is a double crossover, and the chromosome represented above the line is a single crossover which had taken place between vermilion and sable, that is, in the same region in which the first single of the double had occurred.

[*Editors' Note:* Table 73 is not reproduced here.]

[*Editors' Note:* Material has been omitted at this point.]

IV

SUMMARY AND CONCLUSIONS

Evidence has been presented which proves that the occasional (1 in 1700) matroclinous daughter or patroclinous son produced by females known to be XX in composition is due to primary non-disjunction, that is, the X's fail to disjoin and are both included in the egg or both extruded to the polar cell.

The fertilization of the zero egg by an X sperm of a normal male results in a patroclinous XO son. He is entirely unaltered in somatic appearance, both as to sex-linked characters and as to sexual characters, but he is absolutely sterile. This difference between XO and XY males proves that the Y has some normal function in Drosophila.

The fertilization of an XX egg by a Y sperm of a normal male gives rise to a matroclinous daughter of the constitution XXY. The constitution of a matroclinous female as XXY has been proved by direct cytological examination and by conclusive genetic tests.

Matroclinous females always produce further exceptions which we may call secondary, to the extent of 4.3 percent. The cause of this production is the fact that the presence of the extra Y forces both X's to enter the same cell in a certain percent of reductions.

In an XXY female the sex chromosomes do not synapse as a triad, but two synapse, leaving the third unsynapsed. Synapses are not at random, but are highly preferential; in 16.5 percent of cases Y synapses with one or the other X (heterosynapsis) and the remaining X is unsynapsed; in 83.5 percent of cases synapsis is between X and X (homosynapsis) and the Y is unsynapsed.

At the reduction division the two synapsed chromosomes disjoin and pass to opposite poles. The free chromosome goes with one or the other at random.

Reductions are not preferential; the polar spindle delivers two chromosomes to the polar cell as often as to the egg.

After XX synapsis the amount of crossing over is slightly increased (13.5 percent) in some manner by the presence of the extra Y.

After XY synapsis there is no crossing over—either between the X and Y or between the synapsed X and the free X.

After XY synapses the eggs are XX and Y, and X and XY. These four classes of eggs are in equal numbers and are non-crossovers.

The XX eggs by Y sperm give matroclinous daughters which are exact reproductions of the mother in all respects.

The Y eggs by X sperm give patroclinous sons which can give non-disjunctional effects neither in F_1 nor in F_2.

The X and XY eggs from XY synapses are indistinguishable from the non-crossover classes of the X and XY eggs which are from XX synapses. As a result the linkage values must be corrected.

The XY egg by X sperm gives an XXY regular daughter which nevertheless gives 4.3 percent of secondary exceptions by virtue of the extra Y.

YO, YY, and XXX zygotes are unable to live.

The XY egg by Y sperm gives XYY males. These males produce no exceptions in F_1 but produce XY sperm which, fertilizing X eggs, give rise to XXY daughters, and these produce secondary exceptions.

Synapses in an XYY male are probably at random.

The source of chromosomes, whether maternal or paternal, is without effect upon their subsequent behavior at synapsis and reduction.

The predominant type of non-disjunction has been shown to be preceded by XY synapsis and to take place at the reduction division.   A rare type of non-disjunction takes place at an equational division.   Equational non-disjunction is apparently always preceded by XX synapsis and crossing over.   Equational non-disjunction offers the possibility of determining whether crossing over in Drosophila takes place by the chiasmatype (four strand stage) method or at a two strand stage.

[*Editors' Note:* Material has been omitted at this point.]

The genetic and cytological evidence in the case of non-disjunction leaves no escape from the conclusion that the X chromosomes are the carriers of the genes for the sex-linked characters.   The distribution of sex-linked genes (as tested by experimental breeding methods) has been demonstrated to be identical, through all the details of a unique process, with the distribution of the X chromosomes (as tested by direct cytological examination).   The argument that the cell as a whole possesses the tendency to develop certain characters is completely nullified by the fact that in these cases the cells that produce exceptions are of exactly the same parentage as those which do not produce exceptions, the only difference being the parentage of a particular chromosome, the X.   Those eggs which have lost nothing but the X chromosome have completely lost therewith the ability to produce any of the maternal sex-linked characters, and with the introduction of an X from the father these eggs have developed all of the sex-linked characters of the father.   Conversely, those eggs which have retained both X's of the mother and have received no X from the father show all of the sex-linked characters of the mother and none from the father.

[*Editors' Note:* Material has been omitted at this point.]

Experimental proof is given that particular chromosomes, the X chromosomes, are the differentiators of sex; the X chromosome constitution of an individual is the cause of the development by that individual of a particular sex, and is not the result of sex already determined by some other agent. The sex is not determined in the egg or the sperm as such, but is determined at the moment of fertilization; for the X sperm of a male gives rise to a female when it fertilizes an egg containing an X, but to a male if it fertilizes an egg containing a Y or no sex chromosome at all. Likewise the Y sperm of a male gives rise to a female when fertilizing an XX egg and to a male when fertilizing an X egg. These facts in connection with the fact that an X egg of a female produces a male if fertilized by an X sperm prove that the segregation of the X chromosomes is the segregation of the sex-differentiators. The presence of two X chromosomes determines that an individual shall be a female, the presence of one X that the individual shall be a male. The origin of these chromosomes whether maternal or paternal is without significance in the production of sex.

The Y chromosome is without effect upon the sex or the characters of the individual, for males may have one Y, two Y's, or may lack Y entirely (males lacking Y are sterile) ; and females may have one or two supernumerary Y's with no change in appearance in any case.

REFERENCES

Bridges, C. B. 1913b. Non-disjunction of the sex chromosomes of *Drosophila. J. Exp. Zool.* **15**:587–606.

———. 1914. Direct proof through non-disjunction that the sex-linked genes of *Drosophila* are borne by the X-chromosome. *Science* **40**:107–109.

Gregory, R. P. 1914. On the genetics of tetraploid plants in *Primula sinensis. Proc. Roy. Soc. B.* **87**:484–492.

Metz, C. W. 1914. Chromosome studies in the Diptera I. *J. Exp. Zool.* **17**:45–59.

Morgan, T. H., and C. B. Bridges. 1913. Dilution effects and bicolorism in certain eye colors of *Drosophila. J. Exp. Zool.* **15**:429–466.

Muller, H. J. 1914. A new mode of segregation in Gregory's tetraploid *Primulas. Am. Nat.* **48**:508–512.

Stevens, N. M. 1908a. A study of the germ cells of certain Diptera with reference to the heterochromosomes and the phenomena of synapsis. *J. Exp. Zool.* **5**:359–374.

Wilson, E. B. 1909. Studies on chromosomes. V. The chromosomes of *Metapodius*. A contribution to the hypothesis of the genetic continuity of chromosomes. *J. Exp. Zool.* **6**:147–205.

47

Copyright © 1935 by the American Genetic Association

Reprinted from *J. Hered.* **26**(4):163–174 (1935)

LOCATING GENES ON SALIVARY CHROMOSOMES

Cyto-Genetic Methods Demonstrated in Determining Position of Genes on the X Chromosome of *Drosophila melanogaster*

OTTO MACKENSEN

University of Texas

IN a recent article in this journal Painter[7] gave an account of how salivary gland chromosomes came to be studied at this laboratory, and briefly discussed some of the major problems which this new method has allowed us to attack. Foremost among these is the question of the morphological position of the genes along the chromosomes. As he has pointed out, there are several different ways in which this can be determined. We can study breaks (accompanied by translocations) which have occurred between two known gene loci, and having found where the chromosome is actually broken we can say that genes *a* and *b*, for example, lie to the right and left of a given morphological point. This is the method which Painter has used in making new cytological maps of the X[4], the second[6], and third[5] chromosomes, largely because this was the main type of aberration available. It was recognized at the very beginning of the work that simple translocations would only occasionally allow us to localize genes to very small areas along the chromosomes. At the same time that translocations are produced by irradiation, however, Dr. J. T. Patterson[8] has shown that very often small sections are deleted from a single chromosome and this seemed to offer a more exact and a much simpler method of determining the position of gene loci. Knowing what genes are absent and what part of the chromosome is missing, we can say that these gene loci lie within the deleted section. It was therefore suggested to the writer that he use this method and apply it to the loca-

tion of gene loci along the X chromosome. A preliminary statement of the problem, the methods employed, and some of the results obtained has already been published (Mackensen[1]).

In the present article it is proposed to present the evidence which has made it possible to localize eight genes along the X chromosome to very small areas, in two instances to one "band" or a part of a band. At this point the author wishes to express his appreciation for the assistance given him by Dr. T. S. Painter from the cytological side, and Dr. J. T. Patterson and Mr. Wilson Stone from the genetic standpoint.

Material and Methods

We had at this laboratory among our stocks a few deletions obtained by Patterson[8], but most of the cases described below were obtained by the writer by irradiating normal male flies and then mating them to females which carried a series of recessive genes along the X chromosomes. Any F$_1$ female which showed one of these recessives must have received from the father either a chromosome in which the normal gene had mutated or a chromosome from which the normal gene at this locus had been deleted (giving "pseudo-dominance"). Subsequently, these aberrant females were tested and individuals carrying deleted chromosomes were analyzed cytologically by a study of the salivary gland chromosomes. For this purpose larvae heterozygous for the deletion were used.

In all of the cases of deletions to be described, an actual absence of a part of the chromosome could be demonstrated but it may be well to mention the genetic criteria used for the separation of deletions. These are recessive lethal effects, pseudo-dominance as explained above, and exaggeration. Mohr[2] has shown that most genes when heterozygous with a deletion (acting in a single dose) show a more extreme (exaggerated) phenotype than when homozygous.

After deletions had been picked up, their genetic limits were tested by matching the deleted chromosome with all available genes in the vicinity of the original locus. All genes showing pseudo-dominance and exaggeration were assumed to be deleted. This test may not always be dependable, however, because, as Mohr has shown, an extreme allelomorph at a given locus (e.g., white) may have the same effect as an actual absence of this locus. It is very important to keep this in mind when making genetic tests, because even though a deletion is demonstrable in the salivary chromosomes it is still possible that an extreme mutation may have occurred just outside the deleted area. The points of breakage of long deletions and translocations were determined by examining hyperploid males and females, and hypoploid females in which the abnormal chromosome, or piece, covered certain recessives. It may be pointed out that as a result of any of these tests we are only able to say that a break lies somewhere in the region between two gene loci. Crossing over tests are of little value in making closer determinations for either translocations or deletions on account of the reduction in crossing over near breakage points.

A number of complex cases have been worked out and have proved to be very valuable but, in general, simple deletions and the simpler types of translocations have been the most useful. Before taking up some of the cytological advantages and difficulties encountered in the use of these, the reader must understand that the relative ease with which we can determine the exact point (or points) at which

a chromosome has been broken (or deleted) is due to the fact that in the salivary gland homologous chromosomes undergo somatic synapsis. (In the following work we shall call the cross striations of the chromosomes "bands" or "lines" without prejudice as to their real nature.) For example, in the case of a deletion in the X chromosome, the normal X will synapse band for band with the abnormal chromosome to each side of the point of the reattachment, causing a buckle in the normal chromosome which corresponds to the lost region, so that we can read right off just what bands are missing (Figure 15B). In doing this, however, certain points must be kept in mind. When both breaks of a deletion are within clear areas then the limiting bands are complete and will synapse with ease, so that a mistake as to true homology will not be easily made. However, there are two possible conditions under which one can confuse non-homologous with homologous parts. These are: (a) when one of the breaks has occurred in a clear area and the other within a band, resulting in the formation of a half band, and (b) when each break has occurred within a band, resulting in the formation of a composite band. In these two cases the stretching, compression, and twisting about each other, which is caused by the two chromosomes of different lengths trying to synapse completely, will often cause non-homologous parts to lie side by side and appear to be synapsed. If the above points are kept in mind very accurate determinations of loci are possible by the use of overlapping deletions.

Cytologically there are certain advantages in studying translocations. Since in them no part of the chromosome is lost (at least in non-lethal mutuals), a complete synapsis of all parts is possible. This is a very distinct advantage in making exact determinations, because the possibility of non-homologous parts becoming closely associated is largely eliminated. At least, this is true when all parts are synapsed to form a cross figure. If in a translocation the break has occurred within a band, the normal band will at times synapse with the half which lies in the left-hand piece, and sometimes with the half which lies in the right-hand piece. Such behavior can then be taken to mean that the break has taken place within the band.

The analysis of one deletion at a particular locus may not place this locus within very narrow limits, but when two slightly overlapping deletions remove the same locus, then a very accurate determination is possible. The same can be said of translocations. If the points of breakage of two translocations practically coincide and genetically a given locus lies to the left of the break in one case and to the right in the other, then the locus is plotted as accurately as we can hope to place it. This implies, of course, that the locus in each case lies very close to the point of breakage. It is on this account that the sort of aberration which has been called "simultaneous breakage and mutation" promises to be of considerable value in mapping chromosomes. The exact rate of occurrence of this type of rearrangement has not been determined, but if the results obtained at the forked locus represent a random sample, then it occurs as frequently as actual deletions. Of a total of nine changes at this locus two proved to be simultaneous translocations and mutations, two true deletions, two mutations accompanied by recessive lethal effects, and three mutations. In the two translocations the forked locus has been shown to be very close to the point of breakage, but in each case to the right of the break. If in one of these the locus had been to the right and in the other to the left of the break then, since the two breaks practically coincide, the locus would have been placed very accurately. It may later be shown that the points of breakage in such cases can be taken as the exact locus of the mutated gene with a fair degree of accuracy and certainty.

Description of Cases and the Limiting of Loci

Before taking up each case separately a few points concerning the figures will be mentioned. A camera lucida drawing is given of each case analyzed. The course of the normal chromosome is indicated by arrows, and the points of reattachment in the abnormal elements by lines running up to their margins. If a deletion has occurred this line is labeled "Del." (Figure 15B). If a translocation has occurred the chromosome parts, involved are given as in Figure 15E, for example 4L-1R means that at this point the left piece of the IV chromosome is attached on to the right piece of the X. In order that the breaks involved in certain cases may be easily compared they are diagrammatically reproduced upon sections of the complete map (example, Figure 15A). On these the points of breakage of translocations are indicated by lines, deleted sections by brackets below the chromosome, and the narrowest limits to which certain loci or groups of loci have been localized are indicated above the chromosome. All these gene localizations are reproduced on the complete morphological map (taken from Painter[7]) shown in Figure 17J. This figure also gives all the available data on gene localizations in the X chromosome obtained to date, at this laboratory.

The prune locus

T 1-2, 50b A. This translocation consists of a long deletion in the X chromosome with the deleted section inserted into the right arm of the II chromosome. The left break genetically, occurred between broad (br) and prune (pn), and the right break between rudimentary (r) and forked (f). Cytologically the left break occurred as is shown diagrammatically in Figure 15.1. This case will be described in more detail in connection with the forked locus.

SYNAPSIS OF NORMAL AND DELETED CHROMOSOMES

Figure 15

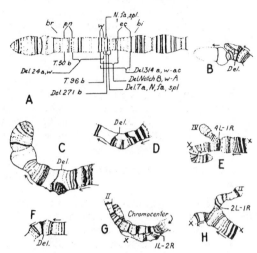

A—This is a section of the *X* chromosome map on which the aberrations illustrated below (*B-H*) are shown diagrammatically. Below the chromosome, translocation breaks are indicated by single lines and the limits of deletions by brackets. Indicated above the chromosome are the narrowest limits to which the genes studied have been localized. *B-H* are camera lucida drawings showing how the various deleted or aberrant chromosomes synapse with a normal *X*. The course of the normal *X* is indicated by arrows. *B* is taken from a female larva heterozygous for deletion 24a. In like manner *C* is taken from Notch 8, *D* from deletion 314a, *E* from T 1-3-4, 96b, and *F* from deletion 7a. *G* and *H* are taken from deletion 271b. The synapsis of the 1L-2R element with the normal II is shown in *G* and in *H* we can see how the 2L-1R element unites with the unbroken *X*.

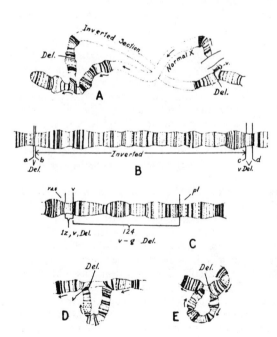

THE 1z AND THE T 1-2, 124 ABERRATIONS

Figure 16

The aberration known as 1z, resulted in the inversion of a part of the *X* chromosome which was accompanied by a deletion at each end of the inverted section. *A* shows the synapsis of a normal with the aberrant *X*. The points at which parts of the chromosome were removed are indicated by *Del.* The inverted section lies between these two labels. *B* is a section of the normal map of the *X* showing diagrammatically just what occurred in the 1z aberration. *C* is a section of the normal *X* showing schematically how the vermilion locus is restricted to part of a compound band. In aberration T 1-2, 124 a section of the *X* was deleted and inserted into the second chromosome. *D* shows the synapsis of this aberration with a normal *X*, complete for the left hand side of the break while *E* shows synapsis to the right of the break.

Del. 24a A, genetically removes white but not prune, split (*spl*) or facet (*fa*). Cytologically the deletion (Figure 15*B*) begins three bands to the right of the 50b break and extends through the first two of the five bands which have been termed the "facet" bands by Painter.[4] Figure 15*A* shows diagrammatically just what portion of the *X* was deleted. The evidence from these two cases (50b and 24a) limits the prune locus to the area of the three bands indicated in Figure 15*A*.

The loci of white, white-mottled and echinus

Mohr's Notch 8 genetically removes white (*w*), facet and Abnormal abdomen (*A*), but not prune and echinus (*ec*) (Mohr[2]). The split locus has also been tested for and found to be absent. Cytologically (Figure 15*C*) this deletion removes the right four of the five "facet" bands and a region to the right including two sharply defined dark bands. This is shown diagrammatically in Figure 15*A*. The evidence from this deletion with the foregoing (24a) limits the white locus to the neighborhood of the second of the five facet bands.

Del. 314a A, of Patterson[8], genetically removes white to echinus inclusive, but not prune and bifid (*bi*). Morphologically the left break occurred between the first and second of the facet bands (Figure 15*D*). In this figure the first and second facet bands of the normal chromosome are almost completely fused to form one wide band, and it will be noticed that only the left part of this wide band is present in the abnormal chromosome. Bands 3 and 4 of the facet area are also fused. Such a fusion is often characteristic in this region. The right break occurred just to the left of the first of the three heavily staining bands which have been termed the "three banded area" by Painter. The limits of this deletion compared with those of Notch 8, places echinus within fairly narrow limits (Figure 15*A*). The deleted piece of 314a is inserted in the

left arm of the third chromosome and gives a white-mottled effect when present in a female homozygous for white. This places white-mottled within the limits of this deletion.

T 1-3-4, 96b A. This is a complicated translocation in which the *X*, III, and IV chromosomes are involved. Although this case is not yet completely worked out it is known that the break is between white-mottled and facet, that white-mottled and gray are linked with the third chromosome and that most of the IV chromosome is attached to the proximal piece of the *X*. Figure 15*E* shows the 4L-1R element synapsed with the normal *X*. Since only the right three of the facet bands are synapsed the break must have occurred between the second and third facet bands. This is shown diagrammatically in Figure 15*A*. The evidence from this case and the foregoing (314a) places white-mottled on the same band with white, and thus we have independent confirmation as to the locus of white since white-mottled is an allelomorph of this gene.

The loci of Notch, facet and split

Del. 7a A, genetically shows Notch (*N*), and removes facet and split, but not white and echinus. Abnormal abdomen was not tested. Morphologically the right three of the five facet bands are removed (Figure 15*F*). The loci of Notch, facet and split must therefore lie within these three bands as is shown diagrammatically in Figure 15*A*.

Del. 271b A, of Patterson,[8] is a mutual translocation between the *X* and left arm of II with a deletion at the point of breakage in the *X*. It shows Notch, and when crossed to facet and split these two genes are exaggerated. The deletion has not been demonstrated morphologically, but the break and reattachment occurred between the third and fourth facet bands (diagrammatically shown in Figure 15*A*). Figure 15*G* shows the 1L-2R element synapsed with the normal second. Here the left three of the facet bands can be clearly seen to the left of the point of reattachment. In Figure 15*H*, where the 2L-1R chromosome is shown synapsed with the normal *X* we see the fourth and fifth facet bands synapsed with the normal *X*. A part of the clear area between bands three and four of the facet area must be deleted and it is very possible that some part of one or both of the limiting bands has been lost. The exact limits of such a deletion are extremely difficult to demonstrate. We can only say at most that the loci involved lie within a region including parts of the third and fourth facet bands (Figure 15*A*).

The cut locus

T 1-4, 12*A*. This is a complicated case found and analyzed genetically by Stone,[10] and cytologically by Painter.[4] The cut (*ct*) locus has been deleted from the *X* and inserted into the left arm of the II chromosome. Genetic tests show that this piece does not include singed (*sn*), ocelliless (*oc*) or carmine (*cm*). The morphological limits of this insertion have been revised as shown in Figure 17*J*.

The vermilion locus

Del. 1z A, genetically removes vermilion (*v*), but not raspberry (*ras*) or lozenge (*lz*). When this case was examined cytologically it was found that it involved an inversion with a short deletion at each end of the inverted section. The right break is within a wide heavily staining band labeled *v* in the figures, which has been called the "vermilion" band. In Figure 16*A* three components are visible within this band on the normal chromosome, a heavily staining one in the middle and a lighter one to each side. The right and the middle ones are apparently synapsed, but the left one is clearly missing on the abnormal chromosome. In Figure 16*B* the four breaks are indicated schematically on a section of the map, and using this as a guide the reader can follow the patterns in Figure 16*A*. The regions from *a* to *b* and from *c* to *d* are deleted and the section between *b* and *c* is inverted. This case then shows that vermilion must

lie within the deleted area indicated diagrammatically in Figure 16*B*.

T 1-2, 124 A. This case involves a deletion in the *X* chromosome with the deleted section inserted into the right arm of the second chromosome. Genetic tests made by Bedichek (see Patterson et al.[9]) show that vermilion and garnet (*g*) have been deleted, but cut and pleated (*pl*) have not. The fact that the II chromosome carries the normal allelomorphs of vermilion and garnet excludes the possibility of a mutation to vermilion at the left point of breakage. The vermilion locus must therefore be to the right of the left point of breakage of this deletion, shown diagrammatically in Figure 16*C*. In Figure 16*D* we see the left end of the deleted X completely synapsed with the normal up to the point of reattachment, and in Figure 16*E* a complete synapsis of the right end. The point of reattachment in the abnormal chromosome is indicated by "Del." In both figures this line separates a lighter band to the left from a darker one to the right. The lighter band has been interpreted as the left-hand part of the vermilion band. The left break of 124 and the right break of 1z practically coincide, as is shown diagrammatically in Figure 16*C*, and since in 124 the vermilion locus is to the right and in 1z to the left of the break, the vermilion locus probably lies in the left part of the vermilion band. (The word "probably" is used here because it is possible, though not probable, that in Del. 1z a mutation might have occurred near the point of breakage which would be indistinguishable from a deficiency, because vermilion does not display the exaggeration phenomenon).

The forked locus

Del. 60b A, and 14z A, both remove forked (*f*), but not rudimentary (*r*) or fused (*fu*). They are both simple deletions which are sufficiently explained by the Figures: 60b in Figure 17*A*, and 14z in Figure 17*B*. Morphologically both remove a heavily staining band near their left limits which will for convenience be called the "Bar" band provisionally. 60b takes out the first and second bands to the left, and 14z only the first band to the left of the "Bar" band. The limits of these deletions are shown diagrammatically in Figure 17C.

T 1-4, 4 A. This translocation was found by Stone and analyzed genetically by him. The X is broken between the loci of forked and Bar (B), the left end being translocated to the IV chromosome. A preliminary cytological analysis was made by Painter[4] by a study of males. Further study of this case in females has resulted in placing this break within the heavily staining "Bar" band as is indicated schematically in Figure 17C Evidence that the break is within this band and not to one or the other side of it is given below.

Figure 17D was taken from a female larva hyperploid for the 4L-1R element. The 1R piece is completely synapsed with the two normal X chromosomes from the spindle fiber end through the Bar band. The unsynapsed nipple to the left of the Bar band represents the tip of the IV chromosome. In this figure the break seems to have taken place just to the left of the Bar band, but in reality it has occurred within this band as is revealed by a study of Figure 17E. This figure (17E) shows the 1L-4R chromosome synapsed with the normal X and the normal IV chromosome. The tip of the normal IV passes underneath the X at the point of reattachment so that it is not visible in the figure. The important point here is that the Bar band of the normal chromosome synapses with what must be a part of the Bar band in the 1L-4R element. Furthermore in nuclei in which the 1L-4R element is not synapsed a dark band is seen at the point of reattachment, which is very obviously not a part of the normal pattern of the fourth (Figure 17F). Since at times the normal Bar band synapses with a band to the left of the break and at other times with a band

to the right of the break we must conclude that the Bar band has been split by the break. The forked locus therefore lies to the left, and the Bar locus to the right of a point within the Bar band. Since there is only three-tenths of one per cent. crossing over between these two loci, their position must be very close morphologically to the point of separation.

T 1-2 61b, A. This is a mutual exchange with an inverted left arm of the II chromosome. The break is near forked and accompanied by a mutation to forked. That this is a true mutation and not a deficiency is shown first, by the fact that it does not cause forked to be exaggerated when crossed to this gene, as the true deficiencies do at this locus, and second, by its behavior in hyperploid females, which is described below. To determine whether the forked gene is in the left or right piece, hyperploid tests were carried out. These tests are based upon the fact that most forked allelomorphs are hypomorphic according to Muller's[3] classification, that is, their action is toward normal so that a diploid forked female shows a less extreme phenotype than a forked male, and a triple X forked female is less forked than the diploid forked female. That this behavior is a function of the gene itself and not of modifying factors in other parts of the X has been shown by the following considerations. That the region to the right of forked has no effect is indicated by the fact that yellow, vermilion, forked attached X females hyperploid for the right end of translocation 1-4, 4 broken between forked and Bar, are no less forked (only two forked genes present) than the plain yellow, vermilion, forked attached X females. The evidence that the region to the left of forked has no effect comes from 61b itself. Hyperploid females containing two forked chromosomes plus the left-hand piece of this translocation are just as forked (only two forked genes present) as diploid forked females. That the forked gene lies in the right piece in 61b is

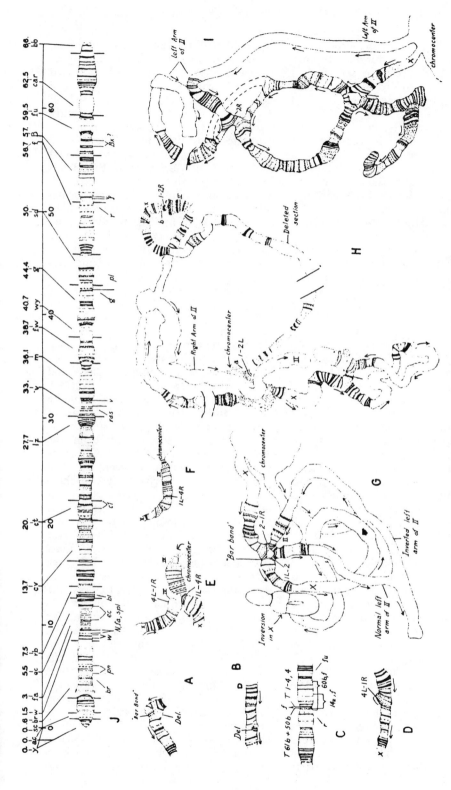

DELETIONS AND TRANSLOCATIONS OF THE X CHROMOSOME
Figure 17

A shows deletion 60b synapsed with a normal element; and *B* deletion 14z. In *C* we have a section of the normal *X* showing diagrammatically the breaks near the forked locus. *D*, *E*, and *F* are drawings of the 1-4, 4 translocation, for explanation see text. *G* shows the aberration known as T 1-2, 61b partly synapsed with a normal *X* and second chromosome, which is explained in the body of this paper. *H* and *I* are complex aberrations known as T 1-2, 50b and T 1-2, 106b. *J*, which is taken from Painter, is a map of the *X* chromosome showing the position or morphological limits of all gene loci which have been placed to date. Single lines on each side of the *X* show approximately where a translocation has broken this element, and lines from the crossover map above the figure of the *X* indicate the limiting gene loci. Below the *X* are brackets or lines showing the position of the genes discussed in the present study.

shown by the fact that hyperploid females containing two forked chromosomes plus this piece are less forked than the diploid forked females. This means that the hyperploids contain three forked genes.

That the break is very near the forked locus is shown by the fact that it occurred within the region included in the short forked deletion 14z (Figure 17C). Figure 17G shows the normal X chromosome practically completely synapsed with the 1L-2, and the 2-1R chromosomes. Here we see the proximal region of the X synapsed from the spindle fiber end up to and including the heavily staining Bar band. Just distal to the Bar band of the normal X are two lighter staining bands (almost completely fused in Figure 17G). The most distal of these is clearly synapsed, but the proximal one which for convenience in description will be called the "forked" band, appears to be only partly synapsed. The unsynapsed proximal portion has been interpreted to be homologous with the fine band which is seen just distal to the Bar band in the 2-1R chromosome. A more complete synapsis than is shown in this figure has never been seen and it would probably be useless to continue to search for better synapsis on account of the complexity of the case. According to the above interpretation the break occurred within the band which has been called the "forked" band, and the forked locus lies to the right of this break as is shown diagrammatically in Figure 17C.

T 1-2, 50b A, consists of a deleted X chromosome with the deleted section inserted into the right arm of the second. The left break is between broad and prune, and has already been described (Figure 15A). The right break is just to the left of forked and is accompanied by a mutation to forked. That this is a true mutation and not a deficiency, and that the forked gene lies to the right of this break was demonstrated by tests similar to those carried out with 61b. The complete configuration of this abnormality, partly

synapsed with the normal chromosomes, is shown in Figure 17H. In the lower left-hand corner the deleted X is shown synapsed with the normal X. The right end of the deleted section, which is entirely unsynapsed, apparently arises from the chromocenter at a and continues to the upper right-hand corner of the figure where the left end of this section attaches on to the proximal end of the right arm of the second, which soon synapses with the normal second. Beginning at the spindle fiber end of the X we find the proximal portion synapsed almost to the heavily staining Bar band. In other preparations good synapsis of this band has been seen. In the deleted X there is a fine lightly staining band just distal to the Bar band, which has been interpreted as the proximal portion of the forked band. There are three reasons for this interpretation. First, no such lightly staining band has ever been found associated with the Bar band even when this region is stretched. Second, it is lighter and finer than the normal forked band. Third, it cannot be a part of the most distal band of the deleted section indicated at b in the upper right corner of the figure, because a clear space exists between this last band and the pattern of the II chromosome. It should be added that near the attachment of this arm of the normal II chromosome to the chromocenter no areas as clear as this are ever seen. Therefore the left break must have taken place within this clear area of the normal chromosome as indicated diagrammatically in Figure 15A. At the other end (proximal end) of the deleted section which arises from the chromocenter, a pair of bands is seen which has been interpreted as homologous to the pair just distal to the Bar band. The most proximal of these is the "forked" band which is not quite complete if the interpretation given is correct. The evidence gathered from these three cases, translocations 1-4 4, 61b and 50b, places the forked locus between a point within the "forked" band and a point within the "Bar" band as indicated diagrammatically in Figure 17C.

T 1-2, 106b, A. This is a complicated case in which a long deletion has occurred in the X chromosome and the deleted section has been inserted into the left arm of the second. In Figure 17I, we see the deleted X chromosome completely synapsed. The circle in the center of the figure represents the deleted portion almost completely synapsed with the normal X. Beginning at the chromocenter the left arm of the second is synapsed up to about the middle of this arm. Here the deleted section of the X attaches on to the 2R piece, and as we follow the deleted section around the loop we come first to the point where the loop was closed when the deletion was formed, and then to the point where this loop should attach on to the distal half of the left arm of the second. This connection has been reconstructed with dash lines for the sake of completeness. This case is a very good demonstration of the manner of occurrence of long deletions. The deleted loop would of course have been lost if it had not been inserted into the second. Short deletions might occur in the same manner. It will be noticed that the parts of this insertion are not in normal sequence. This figure also illustrates how an insertion may occur such as that in 50b, where all parts of the whole deleted-inserted section are apparently in normal sequence. For example if the translocation of 106b had occurred very close to the point of breakage and reattachment of the deleted loop of the X, the sequence would still be abnormal but one of the sections would be too small to be recognized. The result would be an insertion with all of the inserted portion apparently in normal sequence.

This stock was originally picked up as a Beadex (Bx) mutation. When the salivary gland analysis revealed a break and reattachment near the expected neighborhood of the Beadex locus, it was assumed that this was another case of breakage accompanied by mutation, such as those at the forked locus, and the Beadex locus was tentatively placed in the region of this break (Figure 17J).

Discussion

Among the many problems which the use of salivary glands have allowed us to attack none is more important than the exact localization of gene loci along the chromosomes. Since this problem will be one of the foremost in the field of cyto-genetics until most of the genes have been accurately located in *Drosophila melanogaster*, it is pertinent to discuss the relative merits of the translocation and deletion methods.

Translocations occur frequently and are easily detected, but the disadvantage with these is the difficulty of determining exactly, genetically, the point of breakage. Unless hyperploid flies will live, and the experience of this laboratory is that autosomal hyperploids usually do not survive, then laborious crossover counts must be made and unusually large numbers are required. In addition there is some evidence that the breaks tend to occur in definite regions, at least in 2-4 and 3-4 translocations, as if there was a rather definite orientation of the elements in the sperm head at the time of irradiation, so that many breaks may be bunched within narrow morphological limits (see Patterson et al.[9] and Painter[5,6] for examples). These factors make translocations a less desirable method than the use of deletions, but when the former are available and have been analyzed genetically, they offer a quick method for determining the morphological limits within which a given gene must lie.

Deletions, like translocations, occur frequently as a result of irradiation and they offer a more direct method of locating gene loci. They are easily picked up, can be tested genetically very readily, and with the proper genetic set-up limited regions of a chromosome can be investigated. There are, however, certain pitfalls to be avoided when deletions are studied, i.e., possible confusion concerning homology when one or both breaks fall within bands and the possible occurrence of an extreme mutation (i.e. "amorph" by Muller's[3] terminology) having the same phenotypic effect as a true deletion of this region near the points of breakage. Both of these sources of error have been fully discussed in the body of this paper.

This brings us to the discussion of a type of abnormality which has been found to be very useful and which seems to occur quite frequently, namely, simultaneous mutation and chromosome breakage. The usefulness of this type hinges upon the fact that in the cases analyzed up to the present, the break occurred very near the locus of the mutation, and that therefore breaks near a given locus can be picked up by the same direct method used to isolate deletions and at the same time that deletions are being isolated.

We come now to the question of where the genes are placed. Do they lie within the bands or are they distributed along the matrix irrespective of these chromatic differentiations? This question has not been definitely answered, but the evidence to date is in favor of the genes lying within bands in so far as the maximum limits of each localization include at least part of a band, and up to the present no gene has been limited to within a relatively clear area. It must be pointed out, however, that in each case the inclusion of at least part of a clear area is possible, either to each side of a band or within the chromosome if these "bands" are in the nature of rings rather than disks. Even in the case of vermilion which has been limited to part of a band, this is true, since the wide vermilion band is often seen to be made up of component bands with small clear areas between. If in time many loci are definitely limited to bands and none to clear areas then we may assume with some degree of safety that the chromatic bands represent the gene bearing material.

We have called the chromatic cross striations of the salivary gland chromosomes "bands," realizing from the beginning that they are composite structures, that the wider bands are often seen to be aggregates of several smaller bands, that the finer bands often show a particulate structure similar to a string of beads running around the chromosome, and that these particles are often fused to form a solid band. These finer details of structure are best seen in lightly stained preparations and since it was necessary in the present study to stain deeply in order to bring out faint lines, most of the figures do not show the particulate nature of the bands. It has been felt that the first step in the localization of the gene was the accurate placing of the loci with relation to the visible structures. In his recent paper in this JOURNAL Painter[7] reviewed the various ideas which have been expressed about the nature of these "bands."

The possibility that all lethals are due to small deletions has been suggested and there can be no doubt that many of the abnormalities that have been classed as lethals are in reality short deletions which remove no known loci. Up to the present we have not been able to distinguish between a lethal caused by a small deletion and one caused by a point mutation. The salivary glands of a number of lethal stocks have been examined in which no deletion could be demonstrated. This of course does not exclude the possibility of an undetectable one being present. Dominant Notch stocks have been examined in which no absence could be demonstrated. Here we are confronted with the problem of distinguishing between an actual absence and a possible dominant point mutation, accompanied by a recessive lethal effect, which has the same effect as an actual absence. So far all demonstrable deletions have recessive lethal effects. Therefore, as far as the evidence goes, a lethal may still be either a point mutation or a deletion (visible or invisible).

It has long been pointed out that a large number of translocations are accompanied by recessive lethals. (For a discussion of simultaneous effects resulting from irradiation see Muller.[3]) Just how many of these lethals are due to actual losses of chromatin at the point of breakage, has not been determined. In the one case of translocation accompanied by a forked mutation and a recessive lethal (50b) no deletion was demonstrable. In one case (271b) a deletion probably exists because this stock has a dominant Notch and a recessive lethal effect, and genetic tests show that facet and split are deleted (show pseudo-dominance and exaggeration when heterozygous with the abnormal chromosome).

Since the preparation of this article the work of Bridges on the detailed normal morphology of salivary chromosomes has appeared (*Journal of Heredity*, Vol. 26, 60-64). For the reader's convenience the following readings are given (according to Bridges' system) by which the reader may place the limitations for genes determined in the present paper upon the map prepared by Bridges:

prune	2D6 - 2F2	inclusive
white	3C2	
facet, split, Notch	3C3 - 3C6	inclusive
echinus	3E2 - 3E4	"
cut	7A3 - 7B4	"
vermilion	10A1	"
forked	15E1 - 15F1	"
Beadex	17C1 - 17D1	"

Literature Cited

1. MACKENSEN, OTTO. A Cytological Study of Short Deficiencies in the *X* Chromosome of Drosophila melanogaster. *Amer. Nat.* 67: 76, 1934.

2. MOHR, OTTO L. A genetic and Cytological Analysis of a Section Deficiency Involving Four Units of the *X* Chromosome in Drosophila melanogaster. *Zeits. Ind. Abst. u. Vererb.* 32: 108-132, 1923.

3. MULLER, H. J. Further Studies on the Nature and Causes of Gene Mutations. *Proc. of the Sixth Int. Cong. of Genetics,* 1: 213-255, 1932.

4. PAINTER, T. S. The Morphology of the *X* Chromosome in Salivary Glands of Drosophila melanogaster and a new Type of Chromosome Map for this Element. *Genetics* 19: 448-469, 1934.

5. ——————. The Morphology of the Third Chromosome in the Salivary Gland of Drosophila melanogaster and a New Cytological Map of this Element. In Press.

6. ——————. The Morphology of the Second Chromosome in the Salivary Gland of Drosophila melanogaster and a New Cytological Map for this Element. Ready for press.

7. ——————. Salivary Chromosomes and the Attack on the Gene. *Journal of Heredity,* 25: 465-476, 1934.

8. PATTERSON, J. T. Lethal Mutations and Deficiencies Produced in the *X* Chromosome of Drosophila melanogaster by X-Radiation. *Amer. Nat.* 66: 193-206, 1932.

9. PATTERSON, J. T. Wilson Stone, Sarah Bedichek and Meta Suche, The Production of Translocations in Drosophila. *Amer. Nat.* 68: 359-369, 1934.

10. STONE, W. S. Linkage between the *X* and IV Chromosomes in Drosophila melanogaster. In Press.

48

Reprinted from *Natl. Acad. Sci. (U.S.A.) Proc.* **53**(4):737–745 (1965)

LOCALIZATION OF DNA COMPLEMENTARY TO RIBOSOMAL RNA IN THE NUCLEOLUS ORGANIZER REGION OF DROSOPHILA MELANOGASTER*

By F. M. Ritossa† and S. Spiegelman

DEPARTMENT OF MICROBIOLOGY, UNIVERSITY OF ILLINOIS, URBANA

Communicated by T. M. Sonneborn, January 27, 1965

The use of hybrid formation between DNA and isotopically labeled RNA[1] combined with RNAase treatment to eliminate unpaired RNA permitted the detection[2-4] in bacteria of sequences in DNA complementary to the two (16S and 23S) homologous ribosomal RNA components. The methods developed with microorganisms sufficed to establish that a similar situation exists in higher plants,[5] mammals,[6, 7] and insects.[8]

The proportion (0.3%) of the total genome involved was constant in the bacteria examined and indicated[3, 4, 9] a multiplicity of sites for each of the two ribosomal components. The densities of the DNA-RNA hybrids suggested[3] that the multiple sites were clustered rather than scattered throughout the genome. However, the bacteria were not convenient material for a more detailed attempt at illuminating the relation of these cistrons to each other and to the rest of the genome.

It seemed likely that higher organisms would furnish a better opportunity by permitting the correlation of cytogenetic and cytochemical information with data derived from molecular hybridization. Thus, diverse observations implicate the nucleolus with protein synthesis,[10-12] ribosomes[13-16] and ribosomal RNA formation,[17-19] the most striking being the absence of ribosomal RNA synthesis in a lethal anucleolate mutant of the aquatic toad, *Xenopus laevis*.[20]

462

The available facts are consistent with the hypothesis that identifies the nucleolus as the site of ribosomal RNA synthesis. They do not, however, eliminate the alternative possibilities that it is a repository for RNA synthesized elsewhere or that the nucleolus serves some other indirect function in the assembly of ribosomes. Nevertheless, if we adopt it as a working concept, the notion of a nucleolar location for ribosomal RNA formation leads to some interesting and experimentally testable predictions.

The nucleolus characteristically occupies a specific position (the nucleolar organizer, or "NO," segment) in the chromosome complement. This invariant relation between the nucleolus and a particular chromosomal locus suggests the obvious possibility that the DNA complements which generate the ribosomal RNA strands are confined to the nucleolar organizer region. Confirmation would resolve the question of nucleolar function as well as decide between scattered versus clustered distributions of the multiple ribosomal RNA cistrons.

A direct attack requires a reliable method for isolating pure nucleoli still attached to their chromosomal organizer segments and uncontaminated by other chromatin fragments. Comparative hybridizations could then be carried out between ribosomal RNA and "nucleolar DNA" versus "nonnucleolar DNA." Two attempts[5, 6] to carry out such experiments have yielded contradictory results and it is apparent from both investigations that neither had available "nucleolar chromatin" of sufficient purity to permit a truly decisive experiment. Under the circumstances, it seemed worth while to consider other experimental approaches which bypass the limitations of physical separations by employing biological devices to achieve the desired end result.

Mutants of *Drosophila melanogaster* are known[21] which contain inversions involving[22-24] the "NO" region on the X-chromosomes and others which contain useful linked markers.[25, 26] Stocks can be derived from these possessing duplications or deletions of the "NO" region, making it possible to construct[27] by suitable crosses strains which have 1–4 doses of the nucleolar organizer region. Hybridization experiments between the DNA derived from these stocks and isotopically labeled ribosomal RNA should, in principle, provide a precise answer to the following question: Are the DNA complements of the ribosomal RNA confined to the "nucleolar organizer" region? An affirmative outcome would be indicated if the amount of RNA hybridizable per unit of DNA is directly proportional to the dosage of "NO" per genome. The absence of proportionality would indicate that the ribosomal RNA cistrons are not localized in the chromatin region contained in the deleted and duplicated regions.

These experiments have been performed and it is the primary purpose of the present paper to present the results. The data indicate that the DNA sequences complementary to ribosomal RNA are confined to the segment contained in the deletion employed. We conclude that the "nucleolar organizer" region of the chromatin contains the cluster of cistrons for each of the two ribosomal RNA components.

Materials and Methods.—(a) *Growth medium:* The various stocks were raised on a medium made by adding the following in grams to 1 liter of H_2O; corn meal (100); sucrose (100), or 100 ml of molasses (Brer Rabbit, Gold Label); fresh baker's yeast (100); agar (10); methyl-p-hydroxy benzoate (2.7). The flies were collected at the adult stage and stored at $-14°C$ for use as a source of DNA.

(b) *Isotopic labeling:* To achieve a high level of labeling, the standard medium was modified to contain 0.5 gm of yeast per 10 ml, an amount calculated to be the minimum necessary to support the growth of 2–3 gm of larvae, corresponding to approximately 2–3 mg of RNA. To 10 ml of the modified medium 7 mc of H^3-uridine (21 C/mM, Nuclear-Chicago) were added. In labeling experiments the eggs were collected with a brush, washed, and loaded on wet strips of blotting paper which were laid on the surface of the labeled medium. After 7–8 days of growth, larvae were harvested by Mead's[28] method for extraction of the RNA. Final preparations of pure ribosomal RNA had a specific activity of 77,300 cpm/µg. All counting was done in a liquid scintillation counter on nitrocellulose membranes[1] permitting assay of P^{32} and H^3 in the same samples.

(c) *RNA extraction and purification:* After repeated washing with water, the larvae were suspended in 10 vol (w/v) of an extraction fluid which contained NaCl (0.1 M), 1% sodium dodecyl sulphate (SDS), and heparin (0.001%). The suspension was homogenized in a glass homogenizer kept at 0°C, following which 1 vol of phenol saturated with 0.1 M NaCl and containing heparin (0.001%) was added and the mixture shaken for 20 min at 0–4°C. This was then followed by a centrifugation and removal of the aqueous layer. The phenol treatment was repeated three more times, the first for 20 min and the others for 10 min each. The phenol was then removed from the aqueous layer with two ether extractions and the ether eliminated by bubbling air through the solution. The RNA was then precipitated by the addition of 2 vol of a cold 80% ethanol containing Na acetate (0.02 M), taken up in a buffer (0.1 M NaCl; 0.05 M phosphate, pH 6.8) and passed through a methylated albumin column as detailed by Yankofsky and Spiegelman.[4] The two ribosomal RNA components do not separate, the 18S appearing as a shoulder on the leading edge of the 28S peak at about 0.7 M NaCl. The ribosomal fractions were pooled and dialyzed 24 hr at 4°C against 100 vol of 2 × SSC (SSC is 0.15 M NaCl, 0.015 M Na citrate pH 7.4). For analysis in sucrose gradients, the RNA was dissolved in a buffer[29] at pH 5.1 containing NaCl (0.05 M), Na acetate (0.01 M), and Mg acetate (0.0001 M). It was dialyzed for 5 hr against the same buffer and layered on a 5–20% sucrose (dissolved in the same buffer) gradient and spun for 12 hr at 25,000 rpm at 4°C in a Spinco SW 25 rotor. The two ribosomal components (28S and 18S) are readily identified and separated.

(d) *Preparation of DNA:* The DNA employed was extracted exclusively from adults suspended in 10 vol (w/v) of the following buffer at pH 7.6, Tris (0.05 M), KCl (0.025 M), Mg acetate (0.005 M), sucrose (0.35 M). Homogenization was carried out in a mortar and the homogenate filtered through 8 layers of gauze. The resulting filtrate was centrifuged for 10 min at 1,000 rpm in the International centrifuge (rotor 253). The pellet was resuspended in 0.15 M NaCl, 2% SDS, and 0.1 M EDTA, and adjusted to pH 8. For each 10 gm of flies, about 35 ml of this medium were used and the resulting suspension was shaken for 10 min at 60°. All subsequent steps followed the procedure detailed by Marmur,[30] with the exception that at the first precipitation 1 vol (rather than 2) of cold ethanol was added. This avoided interference by salt precipitation and increased the yield. Fibers were collected on a glass rod, and any flocculent DNA precipitate remaining behind was centrifuged, resuspended in SSC to which 1 vol of ethanol was added yielding fibrous precipitates. The fibrous DNA, dissolved in SSC, was subjected to a 4-hr digestion at 37°C with heated (100°C for 10 min) RNAase (Sigma 5 × crystallized) at 150 γ/ml. At this stage there was consistent contamination with a polysaccharide detected by model E Schlieren optics in CsCl equilibrium gradients as a hyperfine band with no absorption at 260 mµ. To remove this material, the preparation was digested for 45 min at 37°C with 250 µg/ml of alpha-amylase (2 × crystallized, Worthington). This step effectively avoided any subsequent difficulties in the final purification of DNA. The enzymes were removed by adding SDS (1%) and 2 phenol treatments at room temperature followed by 2 successive deproteinizations with chloroform-isoamyl alcohol for 10 min each. The DNA was precipitated by the addition of 2 vol of ethyl alcohol, and the fibers were collected and dissolved in 1/100 SSC. Purity of the final DNA preparations was monitored by optical density measurements at 230, 260, 280 mµ, and analysis for ribose and deoxyribose. It yielded a single peak in a CsCl gradient at a position corresponding to 38% GC. Alkaline denaturation of the DNA was carried out in 1/100 × SSC at a concentration of 150 µg/ml. The pH was adjusted to 12.2 with freshly prepared 1 N NaOH, the solution left to stand at room temperature for 10 min, followed by readjusting the pH to 7. Denaturation was complete as measured by either hyperchromicity or adherence to nitrocellulose membrane filters.

(e) *Hybridization and detection of hybrid structures:* The method of Gillespie and Spiegelman[31] was used and it involves three steps. *Step 1—Irreversible fixation of DNA to filters:* The denatured DNA is dissolved in a solution of $6.6 \times SSC$ at a concentration of 5–10 γ/ml. Nitrocellulose filters (Schleicher and Schuell, B-6,27 mm) are first soaked in $6.6 \times SSC$, and 10 ml of the same buffer is then passed through the filter, following which the appropriate amount of the DNA solution is filtered through. The amount of DNA actually retained is always monitored by measuring the O.D. of the DNA solution before and after the filtration. The loaded filters are allowed to dry at room temperature and then incubated in a vacuum dessicator at 80° for approximately 4 hr. When monitored with radioactive DNA, no detectable DNA (less than 1%) is lost during any of the subsequent steps involving washing, hybridization, or enzyme treatment. *Step 2— Hybridization:* The hybridization is carried out by immersing the loaded filter into a $2 \times SSC$ solution (4 ml) of the labeled RNA at the desired concentration and incubated in a stoppered vial at 65°. In some cases solvents of higher ($4 \times SSC$) ionic strength were used. The time required for completion of the hybridization must be determined by preliminary kinetic experiments to ensure that the saturation plateau is attained and kept. In the present instance, it was found that a 7–10 hr incubation period was adequate for the combinations tested. *Step 3—Removal of the unpaired RNA:* The filters are removed from the hybridizing mixture and washed with $2 \times SSC$ on *both* sides. They are then placed in a solution containing RNAase (free of DNAase) at a level of 20 γ/ml in $2 \times SSC$ and allowed to digest at 30° for 1 hr. After the digestion, the filters are again removed and washed with 60 ml of $2 \times SSC$ on *both* sides, dried, and counted.

It is of obvious importance to monitor the degree of contaminating "noise" and this was accomplished by including in the hybridizing mixture a heterologous RNA (*E. coli*) carrying a different identifying isotopic label (P^{32}). The amount of P^{32} found on the filter after step 3 provides a measure of the noise level. In the experiments to be described, this noise level is negligible.

(f) *Source of DNA:* Four different stocks of Drosophila were used in the present studies to prepare DNA containing various proportions of the "NO" region. One is the Standard[21] Urbana wild type. Individuals of this strain carry one "NO" region on each X and one on each Y chromosome, hence both males and females carry two. The DNA derived from this stock is designated either $\male(2)$ or $\female(2)$. Another is the G-21 of the Oak Ridge National Laboratory which has the following relevant genetic constitution: In(1) sc$^{4L, 8R}$, y sc^{4+8} cv v f/RA,yf/Y. The males of this line lack the "NO" region on their X but carry one on the Y. The corresponding DNA preparations are designated in the text as $\male(1)$. The third is G-31 from the Oak Ridge National Laboratory which has the following genetic constitution: In(1) sc$^{81L, 4R}$, sc^{81}vB/RA,yf/B^{8}Y. The males of this stock carry two "NO" regions on the X and another on the Y chromosome, making three in all. DNA preparations from the males of this stock are, therefore, designated as $\male(3)$. The fourth stock was designed to provide a DNA containing 4 "NO" regions per genome. To obtain it, males of G-31 were crossed with wild-type females. The females of the F1 were backcrossed with the males of stock G-31. From these, females were selected exhibiting, because of homozygosity, the genetic markers B,sc^{4+8}v. Here, advantage was taken of the presence of the inversion in the X-chromosome which, by suppressing crossing-over, permits use of the markers linked to "NO" to select the proper combinations with respect to "NO." The females chosen necessarily contain two X chromosomes, both of which carry two copies of the nucleolar organizer region. The stock was maintained with males of G-31 which, of course, also possess duplicates of "NO" on their X. DNA derived from the females of this stock is designated by $\female(4)$.

Results.—(a) *Numerical details of a saturation experiment:* To exemplify the absolute amounts of materials and radioactivity levels being dealt with, a typical experiment is detailed in Table 1. H^3-labeled ribosomal RNA at various input levels are incubated with a constant amount of $\male(1)$ DNA. The internal "noise" control is provided by including P^{32}-ribosomal RNA of *E. coli* in quantities which approximate the H^3-marked RNA from Drosophila. "Noise" correction is made by subtracting from the H^3 counts the proportion of P^{32} counts observed which survive the purification [section (d), *Methods*] process of step 3, account being taken of difference in specific activities. It is evident that the contamination is negligible compared to the counts hybridized.

(b) *Saturation curves with DNA containing various dosages of the nucleolar organizer*

TABLE 1

DETAILS OF A SATURATION EXPERIMENT

Drosophila DNA	Drosophila H³-RNA	E. coli P³²-RNA	Cpm-H³	Cpm-P³²	H³-noise	% of Genome
36.5	0.7	1	3284	10	83	0.113
37.3	1.4	2	4283	11	91	0.145
37.1	2.1	3	4838	15	124	0.164
37.3	2.1	3	4680	16	132	0.157
37.3	2.8	4	4609	20	166	0.154
37.3	3.5	5	4923	23	190	0.164

Incubations were carried out in 4 ml of 2 × SSC at 65°C. The indicated amounts of ♂(1) DNA were prefixed on the membrane filters as in *Methods*. The specific activity of the H³-RNA of Drosophila was 77,285 cpm/µg and that of P³²-RNA of *E. coli* was 9200 cpm/µg. Removal of irrelevant RNA by washing and RNAase treatment as described in *Methods*. H³-"noise" is calculated from the finally observed P³²-counts corrected for difference in specific activities. All recorded counts are corrected for background. Numbers in the first 3 columns refer to nucleic acid added in µg.

region: The four types of DNA carrying different proportions of "NO" were subjected to saturation curves and the results obtained are described in Figure 1. A majority of these experiments contained internal "noise" controls, but they are not recorded since they all yielded results identical to those shown in Table 1.

We may first focus our attention on the results with wild type (curve labeled ♀ (2), ♂(2) DNA) since they settle an issue which could have complicated numerical interpretation of the outcome. It will be noted from section (e) of *Methods* that males of some stocks and females of others were employed to achieve the desired dosage of the "NO" region. Consequently, we had to know whether the nucleolar organizer segments on the X chromosome and the Y chromosomes contribute equally to the observed proportion of complementarity between DNA and ribosomal RNA. For this purpose wild-type male and female DNA were tested in separate hybridization experiments. It is clear that the open circles (♀) and the half-shaded circles (♂) fall on the same curve and approach the same plateau of 0.27 per cent.

Since the XY and XX combinations contribute equivalently to the DNA which is complementary to ribosomal RNA, we can with confidence estimate expected plateaus without regard to sex. We now assume that the plateau attained by the wild type represents a dose of two "NO" segments, from which we can predict the plateaus which should be achieved by the others, *if all the DNA complements of the ribosomal RNA are confined to the "NO" segment.* These predictions are indicated by the solid horizontal lines of Figure 1. The corresponding numerical values are recorded on the right-hand ordinate.

Comparison of the observed and predicted plateau values in Figure 1 suggests that experimental support has been provided for the assertion that the DNA which is complementary to the ribosomal RNA is confined to the region of the nucleolar organizer.

Discussion.—(a) *Nature of the stocks:* The steps involved in the production of the stocks [see *Methods*, (f)] from the original inversion mutants are summarized diagrammatically in Figure 2. The deletion and duplications arise as complementary products of the cross between the two inversions. Thus, the segment appearing as a duplicate in one is equivalent to that deleted in the other. The deletion extends from the proximal break of inversion scute[4] to the proximal break of inversion scute,[8] encompassing the nucleolar organizer and the "bobbed" locus.

The experiments described indicate that all of the DNA complementary to

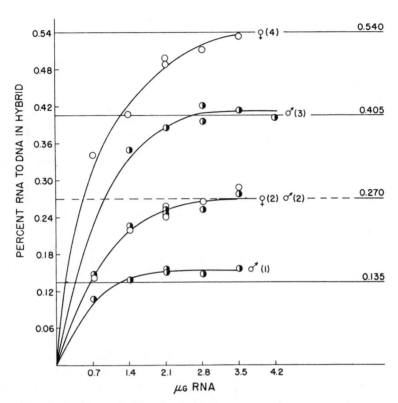

Fig. 1.—Saturation levels of DNA containing various dosages of nucleolar organizer region. Dosage of "NO" is indicated by the number in parentheses. The dotted horizontal line at 0.270 is assumed to be a correct estimate for a dosage of 2, and the solid horizontal lines represent predicted plateaus for dosages of 1, 3, and 4, respectively. Numerical values of the plateaus are given on the right. Replicate determinations are indicated by multiple points at the same input levels. All hybridizations and subsequent RNAase and washing treatments were carried out as described in *Methods*.

ribosomal RNA is confined to the relatively small region involved in the deletion employed. The cytological facts establish that the nucleolus is invariantly associated with the "NO" segment. In view of these observations, arguments on whether the relevant complementary DNA is in fact located precisely at the "NO" locus or at some neighboring site within the deleted segment are not likely to generate a particularly useful dialogue.

It would have been interesting to test a DNA completely lacking in nucleolar organizer. With Drosophila, death occurs rather early in the development of stocks homozygous for the deletion, making it difficult to obtain DNA uncontaminated with cytoplasmic DNA of maternal origin. The anucleolate mutant of *Xenopus laevis*[20] ultimately may furnish more suitable material, providing it is a deletion and not an operator mutation.

(b) *The plateau value:* It should be evident that the successful completion of the experiments described required the use of a hybrid detection method which

would be comparatively "noise"-free
and accurate enough to estimate 25
per cent differences at levels involving
0.1 per cent of the input DNA. The
method[31] used involves hybridization
with DNA previously fixed irreversibly
to nitrocellulose membranes. It combines the advantages of RNAase treatment[1] to eliminate noise, DNA im-
mobilization[32, 33] to avoid DNA-DNA
interaction, and the convenience of
membrane filters.[34] Since each hybridization curve of Figure 1 acts as a
quantitative control for the others, the
nature of the experiments provided a
useful challenge of the quantitative
adequacy of the method which was
successfully met. Further, the value
of 0.27 per cent for the wild type is in
agreement with an earlier[8] independent
estimate carried out by a somewhat

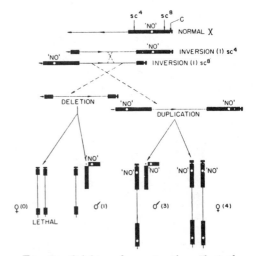

FIG. 2.—Origin and construction of stocks.
Symbols identify nucleolar organizer (NO), centromere (C), scute[4] (sc^4), and scute[8] (sc^8). A
cross between the two inversions yields the desired deletion and duplication from which the
various stocks can be developed as described in
Methods.

different procedure. It must be emphasized that all such plateau measurements
are more likely to be below than above the true value.

The multiplicity of ribosomal cistrons can be estimated from the saturation
plateau and the DNA content of a haploid genome which lies between 0.2 and 1 $\times$
10^{-12} gm.[35, 36] Conversion of the lower[35] value to equivalent molecular weights
yields 1.2×10^{11} daltons which must be divided by two for our purpose since available evidence[37-39] indicates that RNA is found complementary to only one of the
two DNA strands in any given region. Thus a wild-type haploid genome contains
1.6×10^8 daltons of DNA complementary to ribosomal RNA, which is equivalent
to about 100 stretches for each of the two ribosomal RNA components. If the
higher estimate of the genome is taken, this number becomes 500. It is interesting
to note that plateau values clustering around 0.3 per cent of the genome for the ribosomal RNA cistrons have been found for several bacteria,[3, 4] a higher plant,[5] and
now for an insect.

(c) *Implications of the findings:* The experiments reported here locate all the
DNA complementary to both 18S and 28S ribosomal components in the region of
the nucleolar organizer. They therefore decide the issue in favor of a clustered
rather than a disperse distribution. The data further specify a location of the relevant DNA templates which supports the notion that the nucleolus is the site of
ribosomal RNA synthesis. The fact that approximately 0.3 per cent of the DNA
is set aside for this purpose in genomes varying over several orders of magnitude
raises an interesting problem. It may, however, be worth pointing out that this
may turn out to be a simple numerical consequence of supply and demand. Ribosomal RNA constitutes the bulk of cellular RNA (85%) and in bacteria the corresponding cistrons must turn out as much RNA per generation as all the other cistrons
put together. If this is general, and no cistron can be made to work more than

300 times faster than the average, about 0.3 per cent of any genome would have to be set aside for ribosomal cistrons.   The availability of the Drosophila material makes possible a host of potentially informative experiments along lines similar to those described here.   Among these are the interrelation of 18S and 28S cistrons, the location of the transfer-RNA cistrons, and others.   Finally, we may note that it is a source of some satisfaction to be able to record here an example illustrating the successful union of classical genetic material and the more recent preoccupations with "molecular matching."

Summary.—Experiments were designed to see whether DNA complementary to ribosomal RNA was confined to the nucleolar organizer (NO) region.   DNA was prepared from four stocks of *Drosophila melanogaster* carrying 1, 2, 3, and 4 doses of the "NO" segment and hybridized to isotopically labeled ribosomal RNA. The results obtained support the following conclusions.   (1) The wild-type genome saturates at 0.27 per cent of the DNA indicating that it contains approximately 200 sites per diploid set for each of the two ribosomal components.   (2) The nucleolar organizer regions on the X and Y chromosomes contribute equally to the proportion of DNA complementary to ribosomal RNA.   (3) The proportions of the DNA found to be complementary to ribosomal RNA in the different stocks correspond to that predicted from the genetic constitution and the assumption that all the DNA complements of ribosomal RNA are confined to the nucleolar organizer locus.   (4) By identifying the "NO" segment as the site of the required DNA templates the data support the assertion that the nucleolus is the site of ribosomal RNA synthesis.

We would like to express our deep appreciation to Dr. D. L. Lindsley for the two inversion stocks and his invaluable counsel on the genetic operations of the investigations described.

* This investigation was supported by U.S. Public Health Service research grant CA-01094 from the National Cancer Institute, and by a grant from the National Science Foundation.

† On leave from the International Laboratory of Genetics and Biophysics, Naples, Italy.
[1] Hall, B. D., and S. Spiegelman, these Proceedings, 47, 137 (1961).
[2] Yankofsky, S. A., and S. Spiegelman, these Proceedings, 48, 1069 (1962).
[3] *Ibid.*, p. 1466.
[4] *Ibid.*, 49, 538 (1963).
[5] Chipchase, M. I. H., and M. L. Birnstiel, these Proceedings, 50, 1101 (1963).
[6] McConkey, E. H., and J. W. Hopkins, these Proceedings, 51, 1197 (1964).
[7] Perry, R. P., P. R. Srinivasan, and D. E. Kelley, *Science*, 145, 504 (1964).
[8] Vermeulan, C., and K. C. Atwood, in preparation.
[9] Yankofsky, S. A., and S. Spiegelman, in preparation.
[10] Casperson, T., *Cell Growth and Cell Function* (New York: Norton, 1950).
[11] Vincent, W. S., *Intern. Rev. Cytol.*, 4, 269 (1955).
[12] Busch, H. P., P. Byoret, and K. Smetana, *Cancer Res.*, 23, 313 (1963).
[13] Bernhard, W., and N. Graboulan, *Exptl. Cell Res.*, Suppl. 9, 25 (1963).
[14] Porter, K. R., in *Proceedings of the Fourth International Conference on Electron Microscopy* (Berlin: Springer-Verlag, 1960), vol. 2, p. 186.
[15] LaFontaine, J. G., *J. Biophys. Biochem. Cytol.*, 4, 777 (1958).
[16] Birnstiel, M. L., M. I. H. Chipchase, and B. B. Hyde, *Biochim. Biophys. Acta*, 76, 454 (1963).
[17] Edstrom, J. E., W. Grampp, and N. Schor, *J. Biophys. Biochem. Cytol.*, 11, 549 (1961).
[18] Perry, R. P., these Proceedings, 48, 2179 (1962).
[19] Perry, R. P., A. Hall, and M. Errera, *Biochim. Biophys. Acta*, 49, 47 (1961).
[20] Brown, D. D., and J. B. Gurdon, these Proceedings, 51, 139 (1964).
[21] Bridges, C. B., and K. Brehme, *Carnegie Inst. of Wash. Publ.*, J 62 (1944).

[22] Sidorov, B. P., *Proc. Fourth Congr. Zool.* (Russian), **20**, 251 (1930).

[23] Sturtevant, A. H., and G. W. Beadle, *Genetics*, **21**, 554 (1936).

[24] Muller, H. J., and A. A. Prokofyeva, *Dokl. Akad. Nauk SSSR*, **4**, 74 (1934).

[25] Agol, I. J., *J. Exptl. Biol.* (Russian), **5**, 84 (1929).

[26] Muller, H. J., D. Raffel, S. M. Gershenson, and A. A. Prokofyeva-Belgovskaya, *Genetics*, **22**, 87 (1937).

[27] Gershenson, S., *J. Genet.*, **28**, 297 (1934).

[28] Mead, C. G., *J. Biol. Chem.*, **239**, 550 (1964).

[29] Scherrer, K., H. Latham, and J. E. Darnell, these PROCEEDINGS, **49**, 240 (1963).

[30] Marmur, J., *J. Mol. Biol.*, **3**, 208 (1961).

[31] Gillespie, D., and S. Spiegelman, in press.

[32] Bautz, E. K. F., and B. D. Hall, these PROCEEDINGS, **48**, 400 (1962).

[33] Bolton, E. T., and B. J. McCarthy, these PROCEEDINGS, **48**, 1390 (1962).

[34] Nygaard, A. P., and B. D. Hall, *Biochem. Biophys. Res. Commun.*, **12**, 98 (1963).

[35] Swift, H., personal communications.

[36] Rudkin, G. T., "Genetics today," in *Proc. 11th Intern. Congr. Genet.*, in press; and personal communication.

[37] Hayashi, M., M. N. Hayashi, and S. Spiegelman, these PROCEEDINGS, **50**, 664 (1963).

[38] Tocchini-Valentini, G. P., M. Stodolsky, A. Aurisicchio, M. Sarnat, F. Graziosi, S. B. Weiss, and E. P. Geiduschek, these PROCEEDINGS, **50**, 935 (1963).

[39] Greenspan, C., and J. Marmur, *Science*, **142**, 387 (1964).

49

Reprinted from *Natl. Acad. Sci. (U.S.A.) Proc.* **63**(2):378–383 (1969)

FORMATION AND DETECTION OF RNA-DNA HYBRID MOLECULES IN CYTOLOGICAL PREPARATIONS*

By Joseph G. Gall and Mary Lou Pardue

KLINE BIOLOGY TOWER, YALE UNIVERSITY

Communicated by Norman H. Giles, March 27, 1969

Abstract.—A technique is described for forming molecular hybrids between RNA in solution and the DNA of intact cytological preparations. Cells in a conventional tissue squash are immobilized under a thin layer of agar. Next they are treated with alkali to denature the DNA and then incubated with tritium-labeled RNA. The hybrids are detected by autoradiography. The technique is illustrated by the hybridization of ribosomal RNA to the amplified ribosomal genes in oocytes of the toad *Xenopus*. A low level of gene amplification was also detected in premeiotic nuclei (oogonia).

Several techniques are currently used for annealing RNA molecules to their complementary DNA sequences. For certain purposes both the RNA and DNA can be in solution,[1, 2] but it is often more convenient to have the DNA immobilized in a solid or semisolid matrix,[3, 4] or attached to a nitrocellulose membrane filter.[5] The hybrids are generally detected by scintillation counting of radioactive RNA after treatment with ribonuclease to remove unhybridized RNA.

The hybridization of RNA to the DNA in a cytological preparation should exhibit a high degree of spatial localization, since each RNA species hybridizes only with sequences to which it is complementary. The general principles of a cytological hybridization technique are not difficult to lay down. The chromosomes or nucleus should be fixed in as lifelike a fashion as possible; basic proteins should be removed, since they are known to interfere with the hybridization procedure;[5] the DNA should be denatured in such a way that cytological integrity is not lost; the hybridization should be carried out with radioactive RNA of very high specific activity, since the number of hybridized molecules at a given locus will be small; and detection should be by tritium autoradiography to permit maximal cytological resolution.

This communication describes a cytological hybridization technique applicable to conventional squash preparations. It is illustrated by the hybridization of rRNA to the extrachromosomal rDNA in oocytes of the toad *Xenopus*. A preliminary report on the technique was presented in December 1968 at the International Symposium of Nuclear Physiology and Differentiation, Belo Horizonte, Brazil.[6]

Materials and Methods.—The cytological hybridization technique combines certain features of the agar column[4] and filter methods.[5] It should be generally applicable to any material that can be examined as a squash or smear. The following procedure was used in making the preparation shown in Figure 1.

(1) Ovaries from recently metamorphosed *Xenopus laevis* were fixed for a few minutes in ethanol-acetic acid (3:1).

(2) The tissue was transferred to a drop of 45% acetic acid on a microscope slide and

teased with jewelers' forceps. The larger bits of tissue were removed, a cover slip was added, and the cells were squashed. The slides had been previously subbed by dipping into a solution of 0.1% gelatin in 0.1% chrome alum and draining until dry.

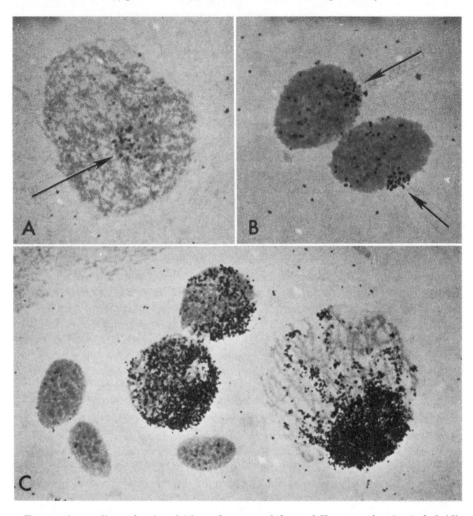

FIG. 1.—Autoradiographs of nuclei from the ovary of the toad *Xenopus*, after *in situ* hybridization with radioactive ribosomal RNA. The preparation was covered with agar, denatured in 0.07 *N* NaOH for 2 min, and hybridized with rRNA having a specific activity of 200,000 cpm/μg. Details are given in *Materials and Methods*. All nuclei from the same slide, exposed for 52 days, stained with Giemsa.

(*A*) Oogonial nucleus showing silver grains above the centrally placed nucleolus (*arrow*). This number of grains indicates that the nucleus contains some 20–40 copies of the nucleolus organizer. ×1500.

(*B*) Two leptotene nuclei with silver grains located over the eccentrically placed nucleolus (*arrows*). ×1500.

(*C*) Three unlabeled follicle nuclei and three labeled pachytene nuclei. The pachytene nuclei illustrate the progressive increase in extrachromosomal rDNA that occurs as the nucleus enlarges. The largest pachytene nucleus contains 25–30 pg of rDNA (approximately 3000 nucleolus organizers). The technique is not sensitive enough to demonstrate the small amount of rDNA in the two nucleolus organizers of the diploid follicle nuclei. ×1200.

(3) The slide was frozen on dry ice, and the cover slip was removed with a razor blade.[7] The slides were transferred to 95% ethanol for a few moments and then dried in air.

(4) The slides were dipped in 0.5% agar held molten at 60°C in a water bath. They were removed and drained vertically at room temperature. In this way a very thin but uniform agar layer covered the slide. The agar was allowed to gel but not to dry completely before the next step.

(5) The slide was placed for 2 min in 0.07 N NaOH at room temperature to denature the DNA. It was then transferred to 70 and 95% ethanol for a few minutes each and dried in air. Slides were often stored at this point.

(6) For the hybridization step about 200 μl of rRNA solution was placed directly onto the slide, and a large cover slip was added. The preparation was incubated at 66°C for 12 hr or longer in a moist chamber made from a Petri plate. We used 4-inch plastic plates into which were placed a few sheets of filter paper and enough 6X SSC to moisten thoroughly. The slide was supported above this on two rubber grommets. We used mixed 28S and 18S rRNA at a concentration of 1–2 μg/ml in 6X SSC. The rRNA was extracted by a detergent-phenol procedure from cultures of *Xenopus* cells[3] and was purified by sucrose density gradient centrifugation. It had a specific activity of 2×10^5 cpm/μg. The specific activity was determined by spotting known amounts of RNA-H[3] on nitrocellulose filters, drying, and counting in toluene-PPO-POPOP in a scintillation counter. The counter had an efficiency of about 40% for *unquenched* H[3] samples.

(7) After incubation the slides were washed in 6X SSC and placed in RNase for 1 hr at room temperature. Pancreatic RNase (1 mg/ml in 0.02 M Na acetate, pH 5) was boiled 5 min to remove protease activity and then made up to 20 μg/ml in 2X SSC.

(8) The slides were rinsed in 6X SSC and then in 70 and 95% ethanol before air-drying.

(9) The preparations were covered with Kodak NTB-2 liquid emulsion diluted 1:1 with distilled water. They were developed for 2 min in Kodak D-19, rinsed briefly in 2% acetic acid, and fixed 2 min in Kodak Fixer. After they were rinsed for 10 min in several changes of distilled water, they were stained 10 min with Giemsa, rinsed in distilled water, and air-dried. A drop of Permount medium and a cover slip were added.

Results.—The development of the cytological hybridization technique was facilitated by the use of oocytes of the toad *Xenopus laevis* as a test object. During pachytene of meiosis these cells carry out a differential synthesis of the genes coding for ribosomal RNA.[6, 9–12] Each pachytene nucleus, which contains 12 picograms (pg) of chromosomal DNA, produces about 30 pg of extrachromosomal rDNA. The extra DNA is cytologically detectable as a densely staining cap on one side of the nucleus. During diplotene the rDNA spreads over the inner surface of the nuclear envelope, where it produces the multiple nucleoli that characterize these cells.

The ovaries of recently metamorphosed toads contain many oocytes in the early meiotic stages. Squashes of these ovaries were denatured, hybridized with tritium-labeled rRNA, and subsequently autoradiographed. Heavy label was found in the oocyte nuclei, where it was limited to the extra DNA. Figure 1 (*B* and *C*) shows several stages in the formation of the nuclear cap from leptotene to late pachytene. In each case the label follows the distribution of the extra DNA, the chromosomes being without detectable radioactivity. The nuclei of follicle cells, connective tissue, and red blood cells are unlabeled (Fig. 1*C*), presumably because the technique is not sufficiently sensitive to demonstrate the small amount of rDNA in the normal genome. The unlabeled DNA in these nuclei provides a useful built-in control of the specificity of the hybridization reaction.

Attempts were made to hybridize preparations that had not been denatured with alkali. In autoradiographs exposed for one or two weeks, such control preparations showed no detectable radioactivity in any nuclei. However, most control slides showed weak labeling when exposed for periods of one to two months. In these cases the label displayed the same specific localization seen in alkali-denatured slides, namely, over the rDNA of oocytes. These results suggest that a small amount of DNA is denatured during the fixing and squashing steps.

Two additional tests of specificity have been made. Ovary squashes were treated with DNase (0.3 mg/ml in 0.01 M Tris buffer containing 10^{-3} M MgCl$_2$, pH 7.2, 37°C, 3 hr). Some were stained with the Feulgen reaction to assess the removal of DNA, and the remainder were covered with agar and hybridized as usual. The preparations showed no detectable Feulgen stain and they gave negative autoradiographs, an indication of the failure to bind rRNA. Some protein remains in such preparations, since cytological details are visible either by phase contrast microscopy or after staining with fast green at pH 2.

We have found that a large excess of heterologous rRNA has no effect on the hybridization reaction (Table 1). Hybridizations were carried out with 2 μg/ml of radioactive *Xenopus* rRNA in the presence of nonradioactive rRNA from *Xenopus* or *Escherichia coli*. The *E. coli* rRNA had no effect on the binding of radioactive *Xenopus* rRNA, even when present at 800 μg/ml. By contrast, nonradioactive *Xenopus* rRNA reduced the binding to low levels. A small fraction of the radioactivity (8%) was not competed by homologous *Xenopus* rRNA. We do not know why competition was incomplete, although it should be noted that the radioactive rRNA was derived from cultured kidney cells, whereas the nonradioactive rRNA came from mature ovaries.

In order to assess the sensitivity of the hybridization procedure, we exposed preparations for periods up to two months. Our aim was to detect the earliest stages of rDNA amplification, which was thought to begin in leptotene or early pachytene.[11, 12] In such preparations the mid- and late pachytene nuclei showed total film blackening above the extra DNA. We were surprised to find easily detectable label not only over all leptotene nuclei (Fig. 1*B*), but also over most oogonial nuclei (Fig. 1*A*). In the leptotene nuclei the label was generally

TABLE 1. *Hybridization of Xenopus rRNA-H³ in the presence of excess nonradioactive rRNA.*

Competing unlabeled rRNA	Silver grains per nucleus after 28 hr exposure, background subtracted (mean ± SEM)	Nuclei counted (no.)
None	185 ± 20	19 (4 preparations)
50–800 μg/ml *E. coli*	221 ± 10	43 (9 preparations)
50–800 μg/ml *Xenopus*	15 ± 2	50 (10 preparations)

In each case 2 μg/ml of *Xenopus* rRNA-H³ was present during the hybridization step. *E. coli* rRNA showed no competition even at 800 μg/ml. Unlabeled *Xenopus* rRNA reduced the binding of radioactive rRNA to about 8% of the control value. The small residual binding was unrelated to the quantity of competing rRNA at the levels used here (50–800 μg/ml). The labeled *Xenopus* rRNA was from cultured kidney cells, whereas the unlabeled rRNA came from mature ovaries.

localized at the periphery of the nucleus, whereas in the oogonia it was more often found in one or two clusters within the nucleus. The distribution of silver grains parallels the position of the nucleoli in both cases. These results indicate that a low level of rDNA amplification is present in premeiotic nuclei.

We have begun to examine variables affecting the level of hybridization. The NaOH concentration in the denaturing step has been varied from 0.01 to 0.10 N. Concentrations of 0.01–0.02 N have given either no hybridization or only a low level. We presume that this is due to inadequate denaturing during the two-minute treatment. Concentrations of 0.05, 0.07, and 0.10 N gave roughly comparable levels of hybridization. However, the morphological disruption in preparations treated with 0.10 N NaOH was often extensive.

We have had some success in replacing the agar with a collodion (nitrocellulose) layer during the denaturing step. The advantage of collodion is that it can be removed by dipping the slide in ethanol-ether (1:1) before hybridization. If the collodion method proves reliable, it should permit more accurate quantitation, since the nuclei are in direct contact with the autoradiographic emulsion.

Discussion.—The results with *Xenopus* oocytes show that RNA can be hybridized with the DNA of cytological preparations under conditions that preserve the morphological integrity of the nucleus. The following features of the reaction indicate that we are dealing with true hybrid molecules. (1) The DNA must be treated with a denaturing agent to obtain the full reaction. (2) Prior removal of DNA by DNase eliminates the reaction. (3) The complex of RNA with the nucleus is stable to RNase. (4) The reaction is competed by unlabeled *Xenopus* RNA but is unaffected by heterologous *E. coli* RNA. (5) The reaction with rRNA is limited to the nucleoli of oogonia and to the amplified rDNA of oocytes. Its absence from normal diploid nuclei can be explained by the small amount of rDNA in these cells.

In the filter technique of Gillespie and Spiegelman,[5] contaminating basic proteins bind RNA nonspecifically. At the outset, therefore, we were concerned that nuclear histones would interfere with any cytological hybridization procedure. Our first experiments involved pronase digestion after formaldehyde fixation; however, we found such material difficult to denature, even though it retained good morphology. In the method described here, most of the basic proteins are removed by the ethanol-acetic fixative and the 45 per cent acetic acid treatment. This we have demonstrated by acrylamide gel electrophoresis, thus confirming the earlier experiments of Dick and Johns.[13]

Quantitation of our results has been difficult, since we have had no adequate control over self-absorption within the specimen and shielding by the agar layer. The effect of both factors was easy to demonstrate. Shielding was seen when the preparation was covered with 2 per cent agar instead of 0.5 per cent. In this case no autoradiograph was obtained. Self-absorption within the specimen was suggested by a comparison of the silver grains over the large diplotene nuclei and the more compact late pachytenes. Both contain the same amount of rDNA[12] and hence presumably hybridize the same extent; but the number of grains was greater over the larger, more flattened diplotene nuclei.

Keeping these complications in mind, we can make a rough estimate of the sensitivity of the technique. In autoradiographs exposed for one month, the larger diplotene nuclei display approximately 3000 silver grains. These nuclei contain 25–30 pg of extrachromosomal DNA,[9, 12] of which about 18 per cent is complementary to 28S and 18S rRNA in filter experiments.[6] Thus, these preparations display 15 grains per day per picogram of hybridizable DNA. This can be translated into roughly 0.02–0.03 grain per day per nucleolus organizer.[6] At this level of sensitivity a single nucleolus organizer would be barely detectable after autoradiographic exposure for several months. For this reason the follicle nuclei in our preparations are negative (Fig. 1C).

The sensitivity of the hybridization technique can probably be increased by the use of RNA made *in vitro*. We are now preparing RNA enzymatically, using the *Xenopus* rDNA satellite as template. If we can increase the specific activity of the RNA by a factor of 10 or 100 over what is now available, we should be able to demonstrate the locus of rDNA in metaphase chromosomes of most higher organisms.

It should be relatively easy to demonstrate the rDNA in the polytene chromosomes of Diptera, which contain several hundred times as much DNA as a metaphase chromosome. Calculations are somewhat less reliable when dealing with other types of RNA. The genes for transfer RNA and for 5S ribosomal RNA can possibly be localized under the favorable conditions afforded by polytene chromosomes. Certain special problems, such as the cytological localization of the mouse satellite DNA[14, 15] and the integrated form of the SV40 viral DNA[16] should be approachable. In the latter case, highly radioactive complementary RNA has already been used effectively in filter hybridization experiments. It is difficult to predict what success may be had with messenger RNA species; hopefully it may be possible to characterize mixtures of messengers from different tissues or from different developmental stages. For such experiments the polytene chromosomes of Diptera offer the best chance of precise cytological localization.

The technical assistance of Mrs. Cherry Barney is gratefully acknowledged.

Abbreviations: rRNA, ribosomal RNA; rDNA, the DNA sequences coding for rRNA; SSC, 0.15 M NaCl, 0.015 M Na citrate, pH 7.0; toluene-PPO-POPOP, 4 gm of 2,5 diphenyloxazole and 50 mg of 1,4 bis [2-(4-methyl-5-phenyloxazolyl)]-benzene in 1 liter of toluene.

* Aided by USPHS grants GM 12427 and GM 397 from the National Institute of General Medical Sciences.

[1] Hall, B. D., and S. Spiegelman, these PROCEEDINGS, **47**, 137 (1961).

[2] Nygaard, A. P., and B. D. Hall, *J. Mol. Biol.*, **9**, 125 (1964).

[3] Bautz, E. K. F., and B. D. Hall, these PROCEEDINGS, **48**, 400 (1962).

[4] Bolton, E. T., and B. J. McCarthy, these PROCEEDINGS, **48**, 1390 (1962).

[5] Gillespie, D., and S. Spiegelman, *J. Mol. Biol.*, **12**, 829 (1965).

[6] Gall, J. G., *Genetics*, in press.

[7] Conger, A. D., and L. M. Fairchild, *Stain Technol.*, **28**, 281 (1953).

[8] The culture was kindly furnished by Dr. Keen Rafferty, The Johns Hopkins University School of Medicine, Baltimore.

[9] Brown, D. D., and I. B. Dawid, *Science*, **160**, 272 (1968).

[10] Evans, D., and M. Birnstiel, *Biochim. Biophys. Acta*, **166**, 274 (1968).

[11] Gall, J. G., these PROCEEDINGS, **60**, 553 (1968).

[12] Macgregor, H. C., *J. Cell Sci.*, **3**, 437 (1968).

[13] Dick, C., and E. W. Johns, *Biochem. J.*, **105**, 46 P (1967).

[14] Kit, S., *J. Mol. Biol.*, **3**, 711 (1961).

[15] Flamm, W. G., M. McCallum, and P. M. B. Walker, these PROCEEDINGS, **57**, 1729 (1967).

[16] Sambrook, J., H. Westphal, P. R. Srinivasan, and R. Dulbecco, these PROCEEDINGS, **60**, 1288 (1968).

AUTHOR CITATION INDEX

SUBJECT INDEX

About the Editors

RONALD L. PHILLIPS is Professor in the Department of Agronomy and Plant Genetics at the University of Minnesota. He has taught a graduate course in cytogenetics for seven years and is the leader of the Minnesota Agricultural Experiment Station project on cytogenetics. His research interests include studies of higher plant ribosomal RNA genetic systems, histone genetics, somatic cell culture, recombination, duplication-deficiency chromosomes, cytogenetic location of genes, interspecific hybridization, and genetic regulation of the lysine-threonine-methionine biosynthetic pathway of plants. Professor Phillips received his B.S. and M.S. from Purdue University, studying under Dr. W. F. Keim, Ph.D. from the University of Minnesota in 1966 majoring in genetics under Dr. C. R. Burnham's direction, and then was an NIH Fellow with Dr. A. M. Srb at Cornell University studying the cytogenetics of *Neurospora*. In 1967 he joined the University of Minnesota faculty and assumed Professor Burnham's position upon his retirement in 1972.

CHARLES R. BURNHAM, Emeritus Professor, 1972, University of Minnesota, born at Hebron, Wisconsin, began his undergraduate studies at the University of Minnesota in 1920, completed them at the University of Wisconsin in 1924, and received his Ph.D. there in 1929 under Dr. R. A. Brink. He then held a National Research Council Fellowship at Cornell University, the Bussey Institution at Harvard University, and the California Institute of Technology; then a Sterling Fellowship in Botany at Yale University. He then became a member of the Department of Agronomy and Genetics at West Virginia University where he taught cytogenetics and conducted cytogenetic and breeding research from 1934 to 1938. He then took the position of Associate Professor in the Department of Agronomy and Plant Genetics at the University of Minnesota where he conducted cytogenetic research and taught graduate courses in advanced genetics and cytogenetics. He was a Gosney Fellow at the California Institute of Technology while on sabbatical leave (1947–48) and a Visiting Professor at Purdue University for one semester in 1961–62.

His publications include the book *Discussions in Cytogenetics*, first published in 1962, with three subsequent reprintings, and papers that deal with chromosomal aberrations in plants and their utilization in studies of chromosome behavior and in plant breeding.

This volume brings together two editors with 42 years' combined experience in teaching cytogenetics.